Barley: Production, Improvement, and Uses

Barley: Production, Improvement, and Uses

Editor:

Valeria Severino, Ph.D.

Delve Publishing

www.delvepublishing.com

Barley: Production, Improvement, and Uses

Editor: Valeria Severino, Ph.D.

708 3rd Avenue, 6th Floor
New York, NY 10017
United States of America

www.delvepublishing.com

ISBN: 978-1-68095-776-1

Library of Congress Control Number: 2016951532

About the Editor

Valeria Severino, Ph.D.

Valeria Severino obtained her PhD in Biological Processes and Biomolecules from the Second University of Naples in 2009. Her scientific interests concern the structural and functional characterization of proteins and peptides by using biochemical approaches. Fields of expertise comprises food chemistry, cellular biology, natural products and related applications. She published more than 40 papers on international peer-reviewed journals.

Contents

List of Contributors

Zhi-Ming Zhong
Lhasa Plateau Ecosystem Research Station, Key Laboratory of Ecosystem Network Observation and Modeling, Institute of Geographic Sciences and Natural Resources Research, Chinese Academy of Sciences, Beijing, 100101 China

Zhen-Xi Shen
Lhasa Plateau Ecosystem Research Station, Key Laboratory of Ecosystem Network Observation and Modeling, Institute of Geographic Sciences and Natural Resources Research, Chinese Academy of Sciences, Beijing, 100101 China

Gang Fu
Lhasa Plateau Ecosystem Research Station, Key Laboratory of Ecosystem Network Observation and Modeling, Institute of Geographic Sciences and Natural Resources Research, Chinese Academy of Sciences, Beijing, 100101 China

Poets AM
Corresponding authors: 1991 Upper Buford Circle, 411 Borlaug Hall, Saint Paul, MN 55108.

Mohammadi M
Department of Agronomy, Purdue University, West Lafayette, IN 47906.

Seth K
Department of Agronomy, Purdue University, West Lafayette, IN 47906.

Wang H
Bayer CropScience, Morrisville, NC 27560.

Kono TJ
Department of Agronomy, Purdue University, West Lafayette, IN 47906.

Fang Z
Department of Agronomy, Purdue University, West Lafayette, IN 47906.

Muehlbauer GJ
Department of Agronomy, Purdue University, West Lafayette, IN 47906.

Smith KP
Department of Agronomy, Purdue University, West Lafayette, IN 47906.

Morrell PL
1991 Upper Buford Circle, 411 Borlaug Hall, Saint Paul, MN 55108.

Luca Marchiol
DI4A—Department of Agriculture, Food, Environment and Animal Sciences—University of Udine, via delle Scienze 206, I-33100 Udine, Italy

Alessandro Mattiello
DI4A—Department of Agriculture, Food, Environment and Animal Sciences—University of Udine, via delle Scienze 206, I-33100 Udine, Italy

Filip Pošćić
DI4A—Department of Agriculture, Food, Environment and Animal Sciences—University of Udine, via delle Scienze 206, I-33100 Udine, Italy

Guido Fellet
DI4A—Department of Agriculture, Food, Environment and Animal Sciences—University of Udine, via delle Scienze 206, I-33100 Udine, Italy

Costanza Zavalloni
DI4A—Department of Agriculture, Food, Environment and Animal Sciences—University of Udine, via delle Scienze 206, I-33100 Udine, Italy
Agriculture Studies Department, California State University Stanislaus, One University Circle, Turlock, CA 95382, USA

Elvio Carlino
IOM-CNR Laboratorio TASC, Area Science Park Basovizza, Bld MM, SS 14, Km 163.5, 34149 Trieste, Italy

Rita Musetti
DI4A—Department of Agriculture, Food, Environment and Animal Sciences—University of Udine, via delle Scienze 206, I-33100 Udine, Italy

Kazuhiro Sato
Institute of Plant Science and Resources, Okayama University, 2-20-1, Chuo, Kurashiki, Okayama 710-0046, Japan.

Miki Yamane
Institute of Plant Science and Resources, Okayama University, 2-20-1, Chuo, Kurashiki, Okayama 710-0046, Japan.

Nami Yamaji
Institute of Plant Science and Resources, Okayama University, 2-20-1, Chuo, Kurashiki, Okayama 710-0046, Japan.

Hiroyuki Kanamori
National Institute of Agrobiological Sciences, Tsukuba 305-8602, Japan.

Akemi Tagiri
National Institute of Agrobiological Sciences, Tsukuba 305-8602, Japan.

Julian G. Schwerdt
ARC Centre of Excellence in Plant Cell Walls, School of Agriculture, Food and Wine, University of Adelaide, Waite Campus, Glen Osmond, South Australia 5064, Australia.

Geoffrey B. Fincher
ARC Centre of Excellence in Plant Cell Walls, School of Agriculture, Food and Wine, University of Adelaide, Waite Campus, Glen Osmond, South Australia 5064, Australia.

Takashi Matsumoto
National Institute of Agrobiological Sciences, Tsukuba 305-8602, Japan.

Kazuyoshi Takeda
Institute of Plant Science and Resources, Okayama University, 2-20-1, Chuo, Kurashiki, Okayama 710-0046, Japan.

Takao Komatsuda
National Institute of Agrobiological Sciences, Tsukuba 305-8602, Japan.

Qisen Zhang
Australian Export Grains Innovation Centre, 3 Baron-Hay Court, South Perth, WA 6155, Australia

Xiaoqi Zhang
Western Barley Genetics Alliance, Murdoch University, 90 South Street, Murdoch, WA 6150, Australia

Songbo Wang
BGI-Shenzhen, Beishan Industrial Zone, Yantian District, Shenzhen, 518083, China

Cong Tan
Western Barley Genetics Alliance, Murdoch University, 90 South Street, Murdoch, WA 6150, Australia

Gaofeng Zhou
Western Barley Genetics Alliance, Murdoch University, 90 South Street, Murdoch, WA 6150, Australia

Chengdao Li
Western Barley Genetics Alliance, Murdoch University, 90 South Street, Murdoch, WA 6150, Australia
Department of Agriculture and Food Western Australia, 3 Baron-Hay Court, South Perth, WA 6155, Australia

Zhenguo Ma
Department of Biology, Memorial University of Newfoundland, St. John's, NL, Canada
Genomics and Biotechnology, London Research and Development Centre, Agriculture and Agri-Food Canada, London, ON, Canada
Department of Biology, University of Western Ontario, London, ON, Canada

Frédéric Marsolais
Genomics and Biotechnology, London Research and Development Centre, Agriculture and Agri-Food Canada, London, ON, Canada
Department of Biology, University of Western Ontario, London, ON, Canada

Natalia V. Bykova
Morden Research and Development Centre, Agriculture and Agri-Food Canada, Morden, MB, Canada

Abir U. Igamberdiev
Department of Biology, Memorial University of Newfoundland, St. John's, NL, Canada

Nicole Staroske
Leibniz Institute of Plant Genetics and Crop Plant Research, D-06466 Gatersleben, Germany

Udo Conrad
Leibniz Institute of Plant Genetics and Crop Plant Research, D-06466 Gatersleben, Germany

Jochen Kumlehn
Leibniz Institute of Plant Genetics and Crop Plant Research, D-06466 Gatersleben, Germany

Götz Hensel
Leibniz Institute of Plant Genetics and Crop Plant Research, D-06466 Gatersleben, Germany

Ruslana Radchuk
Leibniz Institute of Plant Genetics and Crop Plant Research, D-06466 Gatersleben, Germany

Alexander Erban
Max-Planck-Institute of Molecular Plant Physiology, D-14476 Potsdam-Golm, Germany

Joachim Kopka
Max-Planck-Institute of Molecular Plant Physiology, D-14476 Potsdam-Golm, Germany

Winfriede Weschke
Leibniz Institute of Plant Genetics and Crop Plant Research, D-06466 Gatersleben, Germany

Hans Weber
Leibniz Institute of Plant Genetics and Crop Plant Research, D-06466 Gatersleben, Germany

Gwendolin Wehner
Julius Kühn-Institut (JKI), Federal Research Centre for Cultivated Plants, Institute for Resistance Research and Stress Tolerance, Rudolf-Schick-Platz 3, 18190 Sanitz, Germany.
Interdisciplinary Center for Crop Plant Research (IZN), Hoher Weg 8, 06120 Halle, Germany.

Christiane Balko
Julius Kühn-Institut (JKI), Federal Research Centre for Cultivated Plants, Institute for Resistance Research and Stress Tolerance, Rudolf-Schick-Platz 3, 18190 Sanitz, Germany.

Klaus Humbeck
Interdisciplinary Center for Crop Plant Research (IZN), Hoher Weg 8, 06120 Halle, Germany.

Martin-Luther-University Halle-Wittenberg, Institute of Biology, Weinbergweg 10, 06120 Halle, Germany.

Eva Zyprian
Julius Kühn-Institut (JKI), Federal Research Centre for Cultivated Plants, Institute for Grapevine Breeding, Geilweilerhof, 76833 Siebeldingen, Germany.

Frank Ordon
Interdisciplinary Center for Crop Plant Research (IZN), Hoher Weg 8, 06120 Halle, Germany.
Julius Kühn-Institut (JKI), Federal Research Centre for Cultivated Plants, Institute for Resistance Research and Stress Tolerance, Erwin-Baur-Str. 27, 06484 Quedlinburg, Germany.

Pingchuan Deng
State Key Laboratory of Crop Stress Biology in Arid Areas, College of Agronomy and Yangling Branch of China Wheat Improvement Center, Northwest A&F University, Yangling, Shaanxi, China

Le Wang
State Key Laboratory of Crop Stress Biology in Arid Areas, College of Agronomy and Yangling Branch of China Wheat Improvement Center, Northwest A&F University, Yangling, Shaanxi, China

Licao Cui
State Key Laboratory of Crop Stress Biology in Arid Areas, College of Agronomy and Yangling Branch of China Wheat Improvement Center, Northwest A&F University, Yangling, Shaanxi, China

Kewei Feng
State Key Laboratory of Crop Stress Biology in Arid Areas, College of Agronomy and Yangling Branch of China Wheat Improvement Center, Northwest A&F University, Yangling, Shaanxi, China

Fuyan Liu
State Key Laboratory of Crop Stress Biology in Arid Areas, College of Agronomy and Yangling Branch of China Wheat Improvement Center, Northwest A&F University, Yangling, Shaanxi, China

Xianghong Du
State Key Laboratory of Crop Stress Biology in Arid Areas, College of Agronomy and Yangling Branch of China Wheat Improvement Center, Northwest A&F University, Yangling, Shaanxi, China

Wei Tong
State Key Laboratory of Crop Stress Biology in Arid Areas, College of Agronomy and Yangling Branch of China Wheat Improvement Center, Northwest A&F University, Yangling, Shaanxi, China

Xiaojun Nie
State Key Laboratory of Crop Stress Biology in Arid Areas, College of Agronomy and Yangling Branch of China Wheat Improvement Center, Northwest A&F University, Yangling, Shaanxi, China

Wanquan Ji
State Key Laboratory of Crop Stress Biology in Arid Areas, College of Agronomy and Yangling Branch of China Wheat Improvement Center, Northwest A&F University, Yangling, Shaanxi, China

Song Weining
State Key Laboratory of Crop Stress Biology in Arid Areas, College of Agronomy and Yangling Branch of China Wheat Improvement Center, Northwest A&F University, Yangling, Shaanxi, China

Xingquan Zeng
Tibet Academy of Agricultural and Animal Husbandry Sciences, Lhasa 85000

Lijun Bai
Chengdu Life Baseline Technology Co., LTD, Chengdu 610041, China.

Zexiu Wei
Tibet Academy of Agricultural and Animal Husbandry Sciences, Lhasa 85000

Hongjun Yuan
Tibet Academy of Agricultural and Animal Husbandry Sciences, Lhasa 85000

Barley Improvement and Yak Breeding Key Laboratory of Tibet Autonomous Region, Lhasa 850002, China

Yulin Wang
Tibet Academy of Agricultural and Animal Husbandry Sciences, Lhasa 85000
Barley Improvement and Yak Breeding Key Laboratory of Tibet Autonomous Region, Lhasa 850002, China

Qijun Xu
Tibet Academy of Agricultural and Animal Husbandry Sciences, Lhasa 85000
Barley Improvement and Yak Breeding Key Laboratory of Tibet Autonomous Region, Lhasa 850002, China

Yawei Tang
Tibet Academy of Agricultural and Animal Husbandry Sciences, Lhasa 85000
Barley Improvement and Yak Breeding Key Laboratory of Tibet Autonomous Region, Lhasa 850002, China

Tashi Nyima
Tibet Academy of Agricultural and Animal Husbandry Sciences, Lhasa 85000
Barley Improvement and Yak Breeding Key Laboratory of Tibet Autonomous Region, Lhasa 850002, China

Megan C. Shelden
Australian Centre for Plant Functional Genomics, School of Agriculture, Food and Wine, University of Adelaide, Glen Osmond SA 5064, Australia

Daniel A. Dias
Metabolomics Australia, The University of Melbourne, Parkville VIC 3010, Australia

Nirupama S. Jayasinghe
Metabolomics Australia, The University of Melbourne, Parkville VIC 3010, Australia

Antony Bacic
Metabolomics Australia, The University of Melbourne, Parkville VIC 3010, Australia
ARC Centre of Excellence in Plant Cell Walls, School of BioSciences, The University of Melbourne, Parkville VIC 3010, Australia

Ute Roessner
Metabolomics Australia, The University of Melbourne, Parkville VIC 3010, Australia
School of BioSciences, The University of Melbourne, Parkville VIC 3010, Australia

Jolanta Kwasniewska
Department of Plant Anatomy and Cytology, Faculty of Biology and Environmental Protection, University of Silesia in Katowice, Katowice, Poland

Joanna Jaskowiak
Department of Plant Anatomy and Cytology, Faculty of Biology and Environmental Protection, University of Silesia in Katowice, Katowice, Poland

Baojian Guo
Jiangsu Key Laboratory of Crop Genetics and Physiology/Co-Innovation Center for Modern Production Technology of Grain Crops, Key Laboratory of Plant Functional Genomics of the Ministry of Education, Barley Research Institution of Yangzhou University, Yangzhou University, Yangzhou, China

Haiye Luan
Jiangsu Key Laboratory of Crop Genetics and Physiology/Co-Innovation Center for Modern Production Technology of Grain Crops, Key Laboratory of Plant Functional Genomics of the Ministry of Education, Barley Research Institution of Yangzhou University, Yangzhou University, Yangzhou, China
JiangSu Coastal Area Institute of Agricultural Sciences, Yancheng, China

Shen Lin
Jiangsu Key Laboratory of Crop Genetics and Physiology/Co-Innovation Center for Modern Production Technology of Grain Crops, Key Laboratory of Plant Functional Genomics of the Ministry of Education, Barley Research Institution of Yangzhou University, Yangzhou University, Yangzhou, China

Chao Lv
Jiangsu Key Laboratory of Crop Genetics and Physiology/Co-Innovation Center for Modern Production Technology of Grain Crops, Key Laboratory of Plant Functional Genomics of the Ministry of Education, Barley Research Institution of Yangzhou University, Yangzhou University, Yangzhou, China

Xinzhong Zhang
Jiangsu Key Laboratory of Crop Genetics and Physiology/Co-Innovation Center for Modern Production Technology of Grain Crops, Key Laboratory of Plant Functional Genomics of the Ministry of Education, Barley Research Institution of Yangzhou University, Yangzhou University, Yangzhou, China

Rugen Xu
Jiangsu Key Laboratory of Crop Genetics and Physiology/Co-Innovation Center for Modern Production Technology of Grain Crops, Key Laboratory of Plant Functional Genomics of the Ministry of Education, Barley Research Institution of Yangzhou University, Yangzhou University, Yangzhou, China

Vasilios M. E. Andriotis
John Innes Centre, Norwich Research Park, Norwich, United Kingdom

Gerhard Saalbach
John Innes Centre, Norwich Research Park, Norwich, United Kingdom

Robbie Waugh
James Hutton Institute, Invergowrie, United Kingdom

Robert A. Field
John Innes Centre, Norwich Research Park, Norwich, United Kingdom

Alison M. Smith
John Innes Centre, Norwich Research Park, Norwich, United Kingdom

Jéssica Rosset Ferreira
Department of Crop Plants, Faculty of Agronomy, Federal University of Rio Grande do Sul, Porto Alegre, RS, Brazil

Jorge Fernando Pereira
Embrapa Trigo, Passo Fundo, RS, Brazil.

Caroline Turchetto
Embrapa Trigo, Passo Fundo, RS, Brazil.

Euclydes Minella
Embrapa Trigo, Passo Fundo, RS, Brazil.

Luciano Consoli
Embrapa Trigo, Passo Fundo, RS, Brazil.

Carla Andréa Delatorre
Department of Crop Plants, Faculty of Agronomy, Federal University of Rio Grande do Sul, Porto Alegre, RS, Brazil

Yanling Ma
Tasmanian Institute of Agriculture and School of Land and Food, University of Tasmania, Kings Meadows, TAS 7249, Australia

Chengdao Li
Western Barley Genetics Alliance, Murdoch University, 90 South Street, Murdoch, WA 6150, Australia

Peter R. Ryan
CSIRO Agriculture, Canberra, ACT 2601, Australia. Australia

Sergey Shabala
Tasmanian Institute of Agriculture and School of Land and Food, University of Tasmania, Kings Meadows, TAS 7249, Australia

Jianfeng You
CSIRO Agriculture, Canberra, ACT 2601, Australia

Jie Liu
Tasmanian Institute of Agriculture and School of Land and Food, University of Tasmania, Kings Meadows, TAS 7249, Australia
CSIRO Agriculture, Canberra, ACT 2601, Australia

Chunji Liu
CSIRO Agriculture, 306 Carmody Road, St Lucia, QLD 4067, Australia

Meixue Zhou
Tasmanian Institute of Agriculture and School of Land and Food, University of Tasmania, Kings Meadows, TAS 7249, Australia

Goetz Hensel
Leibniz Institute of Plant Genetics and Crop Plant Research (IPK), Plant Reproductive Biology, Corrensstr. 3, 06466, Stadt Seeland/OT Gatersleben, Germany

Doreen M. Floss
Leibniz Institute of Plant Genetics and Crop Plant Research (IPK), Phytoantibodies, Corrensstr. 3, 06466, Stadt Seeland/OT Gatersleben, Germany
Medical Faculty, Institute of Biochemistry and Molecular Biology II, Heinrich-Heine-University, Universitätsstrasse 1, 40225, Düsseldorf, Germany

Elsa Arcalis
Department for Applied Genetics and Cell Biology, University of Natural Resources and Life Sciences, Muthgasse 18, 1190, Vienna, Austria

Markus Sack
Institute for Molecular Biotechnology, Biology VII, RWTH Aachen, Worringerweg 1, 52074, Aachen, Germany

Stanislav Melnik
Department for Applied Genetics and Cell Biology, University of Natural Resources and Life Sciences, Muthgasse 18, 1190, Vienna, Austria

Friedrich Altmann
Department of Chemistry, University of Natural Resources and Life Sciences, Muthgasse 18, 1190, Vienna, Austria

Twan Rutten
Leibniz Institute of Plant Genetics and Crop Plant Research (IPK), Structural Cell Biology, Corrensstr. 3, 06466, Stadt Seeland/OT Gatersleben, Germany

Jochen Kumlehn
Leibniz Institute of Plant Genetics and Crop Plant Research (IPK), Plant Reproductive Biology, Corrensstr. 3, 06466, Stadt Seeland/OT Gatersleben, Germany

Eva Stoger
Department for Applied Genetics and Cell Biology, University of Natural Resources and Life Sciences, Muthgasse 18, 1190, Vienna, Austria

Udo Conrad
Leibniz Institute of Plant Genetics and Crop Plant Research (IPK), Phytoantibodies, Corrensstr. 3, 06466, Stadt Seeland/OT Gatersleben, Germany

Preface

Barley (Hordeum vulgare L.) is a cereal crop belonging to the grass family of Poaceae. It represents one of the world's oldest cultivated grains, dating back around 13000 years ago in Eurasia. Along with wheat, corn and rice, barley is one of the world's most important crops, according to the U.S. Grains Council. Barley has played an important role in plant genetics and breeding, plant physiology, agronomy, cereal chemistry, human and animal nutrition, plant pathology, and entomology. Barley, as an experimental organism, has contributed to the development of scientific knowledge, and science has contributed to the improvement of barley as a crop. Barley, as the fifth most - produced crop in the world today, involves massive amounts of resources and people working in production agriculture; commodity transportation. This cereal grows in temperate climates, thus being cultivated worldwide, especially in Europe, North Africa, Russia, Canada, and Australia. Barley is widely used for several purposes, including food, beverages, brewing, animal feed, medicinal plant and as a model organism in molecular research. Interests in barley production and use continue to grow, causing a need for an up-to-date review of works related to the recent state-of-the-art of barley's production field. In this book, latest papers about barley cultivation, including improvements in production, selection and uses, are collected. In particular, the first couple of works report the effect of specific conditions on barley production, such as warming, genetic drift and exposition to nanoparticles. Then, genetic characterization, resistance and biochemical studies from seeds' germination to maturation are proposed. Moreover, some genomics, transcriptomics and proteomics papers about barley improvement have been selected. Finally, a very innovative use for transgenic production of anti-HIV antibodies in the barley endosperm is presented. Providing comprehensive information on topics ranging from fundamental aspects to specific applications, this book offers a useful resource for scientists, plant breeders, teachers and advanced students in the fields of molecular and plant cell biology, plant biotechnology, and agronomy.

Editor
Valeria Severino, Ph.D.

INTRODUCTION

Barley is a common staple in human and animal diets. Part of the grass family, it grows in over 100 countries and is one of the most popular cereal crops, surpassed only by wheat, corn and rice. Although barley is fairly adaptable and can be grown in many regions, it is a tender grain and care must be taken in all stages of its growth and harvest.

Barley is recognized as one of the very first crops to be domesticated for human consumption. It remains one of the major cereal crops grown in the world: barley is grown on every continent on which crops are grown. It is well adapted to diverse environmental conditions and thus it is produced across a broader geographic distribution than most other cereals. Relative to other cereal crops, barley ranks fourth in total grain production. The grain of barley enters the human food chain via distinctly different routes. First, barley is used as an animal feed and therefore makes an essential contribution to the human diet indirectly through meat production. Second, barley serves as a substrate for the production of alcoholic beverages, in particular beer. Third, a minor amount of barley is used to produce a diverse range of foodstuffs eaten by humans.

Barley is grown on nearly sixty million hectares of land worldwide, resulting in the production of approximately 140 million metric tons of grain. The top ten barley producing countries include Russia, Germany, Canada, Ukraine, France, Australia, the United Kingdom, Turkey, United States, and Denmark. Barley producers select the appropriate varieties to be grown and crop management schemes to produce grain well suited to a particular end use. Specific producer considerations vary widely, but include the choice of variety to plant, the timing of planting and grain harvesting, and agricultural inputs such as fertilizers, herbicides, fungicides, and insecticides. In the Northern Hemisphere, in regions where winters are too severe to allow winter barleys to survive overwintering, spring barleys are planted, usually in April or May. In warmer regions where winter barleys can overwinter, planting is done in September or November.

Given that the producer obtains only a small profit margin for barley produced, additional agricultural inputs are minimized. For instance, nitrogen fertilization is managed to maximize yield without compromising end-use quality. Barley is also subject to damage by a range of diseases, including powdery mildew, stem and leaf rust, smuts, leaf blotches, viral diseases, and head blights. Thus these diseases are managed by a combination of strategies including the use of disease-resistant varieties, application of fungicides, tillage practices, timing of crop planting, and crop rotation.

When the barley crop is mature, harvesting is accomplished either by direct combining or by first swathing, which entails cutting the culms and allowing the grain to dry in the field, and then harvesting. Technology for harvesting ranges from a simple sickle in developing countries to sophisticated mechanical combines that cut the culms above the soil, move the harvested plants between rollers to dislodge seeds, and pass this over mesh screens, allowing the grain to fall into a collection bin and the chaff to blow back out to the ground. Once the grain is harvested, it may either be stored in bins on the farm, or delivered to a local grain elevator where it is purchased from the producer. The price paid for barley grain depends on its intended end use. After purchasing the grain, the elevator cleans, dries, and stores it, and ultimately resells the grain to the various businesses that use barley (feed, food, and malt industries).

Biology of the Barley Plant

The scientific name of barley is Hordeum vulgare L. Barley is a flowering plant belonging to the family Poaceae (the grasses). In addition to barley, the grass family includes the crops most important to human existence, including rice, wheat, and maize, and other species such as sorghum, oats, rye, millet, and sugarcane. In European contexts, barley grain is often referred to as "corn." The genus Hordeum includes approximately thirty species that are indigenous to at least four continents. Barley is the only domesticated species to have emerged from the genus Hordeum, in contrast to other crop genera such as Triticum (the wheats) and Phaseolus (the dry beans), which each contain several domesticated species. The other members of the genus Hordeum exist as wild plants.

Growth and development: The life cycle of the barley plant first begins with seed germination underground. The first visible sign of germination is root emergence, followed by the emergence of the cylindrically shaped coleoptile, which is the first structure to appear above ground. Interestingly, most above-ground tissues of the barley plant initially develop from the crown, a structure located below ground. The first leaf grows upward within the cylinder of the coleoptile, and emerges above the soil. This and other leaves do not expand along their entire length; rather the outer section (the blade) does so, while the base (the sheath) remains formed in a hollow cylinder. New leaves emerge in succession up through the sheaths of the older leaves.

The barley plant is not restricted to the development of one main stem, as observed in maize. Rather, like most other small-grain cereals such as wheat and rice, barley produces several additional secondary stems termed tillers that emerge up from the crown beginning a few weeks after the emergence of the main stem. The number of tillers produced varies depending upon the barley genotype and the environment. For instance, under highly fertile conditions, plants will produce more tillers than if nutrient starved.

Reproduction: Barley varieties are classified as spring or winter types, depending on whether they need a cold treatment, ranging from two to several weeks before making the transition to the reproductive phase of growth. When barley switches to its reproductive phase, the true stems of some of the tillers, called culms, elongate upward. The flowering structure, known variously as the spike, ear, or head, is borne upward, ultimately emerging from the "boot," which is the sheath of the uppermost leaf on the culm (the flag leaf). The height of a barley plant when the spike has emerged varies greatly, but averages approximately eighty centimeters. The spike consists of a large number of individual flowers called florets, which are present in individual spikelets that are attached to a central stemlike structure called the rachis. Spikelets are attached in groups of three on opposite sides of the rachis. The total number of spikelets on a spike varies, but averages approximately sixty. The awns are a notable feature of the spike. They are hairlike extensions that emerge upward from the lemma, one of two thin sheets of cells that surround the floret (the other is the palea). Awn length varies considerably among barley varieties.

Barley is also classified based upon the fertility of the florets on the spike. In six-rowed barleys, all of the florets are fertile, leading to six vertical rows of seeds on the spike. In contrast, in two-rowed types only the central floret of the three at each node is fertile, and thus just two rows of seeds develop on opposite sides of the rachis. The fertile florets consist of both male and female reproductive structures, and fertilization occurs as the spikes are emerging from the boot. Barley is thus predominantly self-pollinated. Between twenty-five and sixty seeds per spike are produced, and for spring barley the seed matures three to five months after planting.

Characteristics of the barley grain: Barley seeds are approximately eight millimeters in length and weigh approximately fifty milligrams when mature, though there is a considerable range in these values between varieties. In most cases, the harvested barley grain includes the seed, a small structure called the rachilla, and both the palea and lemma, all of which adhere firmly to the seed. Barley grain in which these structures remain attached is referred to as covered barley, with the palea and lemma, collectively termed the hull. However, barleys in which the palea and lemma do not adhere to the seed are also well known. These hulless barleys share harvesting features similar to wheat, because when harvested the seed is cleanly separated from all other components of the spike.

The barley seed consists of the embryo, a series of outer layers of cells called the pericarp, and the endosperm. The endosperm contains different nutrients that the embryo draws upon as it grows into a plant. The principal compound found in the endosperm is starch, which represents about two-thirds of the mass of the seed. This starch serves as an energy source for the seedling. Another significant carbohydrate, the -glucans, are components of the endosperm cell walls. The second largest component of the barley endosperm is protein. The amount of protein present is generally inversely proportional to the amount of starch. This protein provides a source of amino acids that can be used for protein synthesis by the seedling. The amount of protein present in a seed is positively correlated with the amount of nitrogen fertilizer applied when the parent plant is being grown. Additionally, the barley grain contains a large variety of other compounds present in minor amounts, including mineral nutrients and different organic compounds, including various vitamins.

Varieties of Barley

Barley is a hearty plant, able to withstand many different growing conditions. However, it is least tolerant of hot, humid conditions, which makes it unsuitable for the subtropical regions. There are two basic types, based on the number of rows of grain seen when the heads of the stalks are viewed from above:

1. Two-row barley, because there are two rows of barley kernels on the head of the stalk. Each stalk produces between 15 and 30 kernels.

2. Six-row barley, because the head of the stalk contains six rows of barley kernels, each stalk produces between 25 and 60 kernels.

Cultivation Pattern

Barley is a temperate as well as a tropical crop and widely adaptable to various climatic conditions. It is cultivated as a pure crop and also with wheat, gram, lentil, pea, rape, mustard and linseed. It is considered as one of the most drought- and salt-tolerant crops. But for effective yields, it needs appropriate irrigation and rich fertile soil. Extremely dry weather or wet weather can damage the crop even if the land is properly ploughed. A slightly dry weather is a bit favorable for the crop at the time of harvesting. The crop takes about four months to mature.

Crop Harvesting Pattern

After physiological maturity, 10 or 15 days are required to harvest barley with combine in temperate dry lands. If this duration is exceeded, crop will get too dry and then cause shattering at harvest. Harvesting time should be decided when barley stem becomes dry enough to be broken by hand easily in semi-arid and arid areas. In humid regions seed moisture and hardness should be checked before deciding harvest by using teeth or using moisture meter. In some areas rainfalls may force to postpone the harvest, but harvest before rainfall should be preferred, as seed after drying following rainfall may be discolored. The crop is harvested manually either by pulling it out and or by cutting with the help of sickles. The grain is obtained through a process called 'threshing' (beating it up with sticks) and the barley plants are fed to the animals. In India, the sowing period of barley is around October and November, which makes it a rabi crop. The harvesting period is around March and April.

Processing of Barley

Transportation of Barley

The means of transportation of barley grains to market or to the granaries depend upon farm size, physical and geographical conditions, availability of transportation facilities. Small size farmers prefer to sell their crop in village. Generally the crops are transported to markets by tractors or trucks.

Threshing

Threshing with "doven: Wooden threshing sled with flint blades," which is driven by animals is not a common practice and particularly employed by small farmers in mountainous areas or in undeveloped areas of developing countries.

Drying

As the other grains, barley must be dried before putting in granaries. For this purpose harvested crop is left in open and sunny place for drying near the granary in village conditions. It is aerated by inverting the heap with shovel and covered with a material during the night.

Cleaning

Cleaning barley grains is an important process for malt and feeding industries. During the cleaning process of malting barley product in Turkey, separated materials such as stones, earth, weed seeds and broken grains, and amounted to 10 % of total grains.

Storage

Generally barley is stored for a short period under shelter or in depots in order not to be affected by rainfall or other adverse climatic events. Grains are stored in depots made by brick, cement, and wood or earth wells specially drilled. Storage types include wooden depots, adobe depots, cement depots, vertical wooden depots, vertical earth well depots etc.

Overall Losses

Durable foodstuffs with low moisture content form the basis of most human diets precisely because these commodities can be stored for extended periods and are continuously available, provided that there is no insect infestation or spoilage. However, losses occur at every stage of grain handling, storage and processing. These losses may be either quantitative or qualitative. The magnitude of losses is highly variable and in certain cases they may even reach 100%. Qualitative losses are more difficult to evaluate then quantitative ones. Qualitative losses for example, may consist of changes in the physical appearance, nutritional degradation, loss of germination, presence of fragments and insect infestation, contamination by mould or development of mycotoxins.

Uses of Barley

Barley finds a wide range of usage.

Some of them are:

Animal feed: Barley, being a good source of protein, is used as a feed for the livestock. For use as an animal feed, it is rolled, grounded, flaked or pelleted. These products are given in the form of grain, silage or straw to dairy and beef cattle. The by-products of malted barley are also used in the form of animal feed.

Human consumption: A wide variety of barley products are known to be suitable for human consumption. These include porridge, muesli and cookies made of barley flakes, cereals made of barley bran, and muffins, cookies, breads and pasta made of barley flour.

Malted barley: Barley is malted for use in alcoholic beverages like beer and wine. Apart from this, the malted barley is also used as flavor, sweetener, malt extract, malt flour, etc.

Other uses: Apart from being used as an eatable, barley is used in fields like industries and agriculture. The barley straw is used to make the bed for the livestock, while bales of barley are used in making paper, fibre board, etc. The starch present in barley is used in making paper, paper starch based detergents, bio degradable plastics, etc.

Health Benefits of Barley

Barley has numerous health benefits. It is known for its ability to help alleviate diabetes, high cholesterol and obesity. It is also effective against diseases like atherosclerosis, diabetes, insulin resistance etc. Given below are some of the benefits.

- Barley is low in fat and is known to reduce craving some kinds for food and therefore very useful to those who want to lose weight. It acts as an appetite suppressant making people feel like they have eaten more than they really have.
- Barley is high is soluble fiber and can reduce blood cholesterol and blood glucose levels. It helps in reducing the cholesterol level in the body as the dietary fiber in barley binds to bile acids and helps dispose them through the feces. Consuming Barley is said to soothe and calm the bowels. Dietary fiber in Barley is helpful in making your digestive system run smoothly and maintaining a healthy colon. The friendly bacteria present in the large intestine ferment the insoluble fiber content of the barley to form butyric acid which functions as the main fuel for intestinal cells. The insoluble fiber in barley is also said to help women fight against the formation of gallstones.
- The Copper present in barley is known to be helpful in reducing the symptoms of rheumatoid arthritis. Copper helps in maintaining the flexibility in blood vessels, bones and joints.
- Barley is rich in niacin, which is highly effective in reducing the risk of cardiovascular diseases and blood clots.
- Barley is also good for persons suffering from diabetes and can take the place of normal food where rice is prohibited. Barley may be an even better breakfast choice than oats for persons with Type 2 diabetes. Consuming Barley is known to substantially reduce the risk of diabetes.
- Barley can prove to be extremely helpful in curing childhood asthma.
- It is Easy to digest and can be given to invalids.

- Barley is diuretic, brings clear urine and is useful in fevers. It can be used as a nutrient in cases of mucus colitis or simple diarrhea with fever.
- Barley contains Selenium which is helpful in preserving the elasticity of the skin and protects it from radical damage. It also has Manganese which works with the B complex vitamins and gives an overall feeling of well being.
- Barley grass is used in traditional Chinese medicine. It rich in vitamins A, C, B1, B2, folic acid, B12, calcium, iron, potassium and chlorophyll. It has been prescribed to fight diseases of the spleen or poor digestion. It has also been utilized to treat conditions such as depression or emotional imbalance. Barley grass is a powerful antioxidant that is believed to help the body kill cancer cells and overcome a variety of ailments, including acne and ulcers.

Manufacturing Process

The capacities of the Brewery have been planned as 10,000 Kilolitres. Before Malt Extract Production begins, the barley is tested for extract content, extract color, protein content, water content, diastatic powder and grain size. Barley is tipped and conveyed to Silos. Before entering the mills the grain passes through a cleaner and destoner. Grain is crushed in a roller mill to produce grist. Grist is mixed with water and heated. For the best extraction, the mash is held for fixed times, known as rests at a series of specified temperatures. The malt own enzymes help to convert insoluble starches and proteins to soluble sugars, which dissolve in water to form worth. After extraction, the worth is separated from husks and residues known as "spent grains". Separation is carried out in the lauter by a natural process using the husks as a filter. The worth, consisting of 5-20% solids, is evaporated under vacuum to form liquid malt extract, containing around 80% solids. Liquid malt extract is stored in silos before processing or for sale to industrial customers. Beer is a fermented beverage. The sugar needed for the fermentation process comes from the transformation of the starch contained into the grain through the effect of specific enzymes contained into malt.

CHAPTER 1 RESPONSE OF SOIL RESPIRATION TO EXPERIMENTAL WARMING IN A HIGHLAND BARLEY

Zhi-Ming Zhong, Zhen-Xi Shen, and Gang Fu

Keywords: Infrared radiator, Soil moisture, Temperature sensitivity, Tibetan Plateau, Warming magnitude

BACKGROUND

The global surface temperature is predicted to increase by 1.0–3.7 °C by the end of this century and the Tibetan Plateau, "the Third Pole of the Earth", is one of the most sensitive regions to climatic warming (Fu et al. 2015; IPCC 2013). To better understand and predict the effect of such warming on alpine ecosystems, many warming experiments are performed; however, no studies are conducted in agricultural ecosystems on the Tibetan Plateau (Zhang et al. 2015). The croplands are experiencing obvious warming on the Tibetan Plateau (Shen et al. 2014) and about 80 % of the population lives in the cropland areas of the Tibet (Yang et al. 1996). These suggest that it remains unclear about how alpine agricultural ecosystems respond to future climatic warming on the Tibetan Plateau.

The Tibetan Plateau is one of the domestication centers of cultivated barley and the planting area of highland barley accounts for approximately 43 % of the grain crop area on the Tibetan Plateau (Dai et al. 2012; Zhao et al. 2015). Highland barley, the only crop which is grown at high altitudes, is one of the dominant crops on the Tibetan Plateau and is an important staple food of the Tibetan people (Wang et al. 2013). Highland barley can be made into a variety of conventional foods, such as fried noodles (i.e. zanba), wine, and health food due to its high beta-glucan (Zhang et al. 2002). The highland barley ecosystem is one sensitive system to climatic warming considering the significant relationship between the potential productivity of highland barley and air temperature (Zhao et al. 2015). However, no studies demonstrate the warming effects on highland barley ecosystem under controlled warming on the Tibetan Plateau. Therefore, it remains unclear about how highland barley system responds to future climatic warming on the Tibetan Plateau.

Soil respiration (R_s) is one primary component of the carbon cycling in terrestrial ecosystems (Rustad et al.2001). Soil temperature is one of the most important abiotic factors affecting R_s (Raich and Schlesinger1992). Soil respiration generally increases exponentially with increasing soil temperature and the Q_{10} is often used to evaluate the temperature sensitivity of R_s (Lin et al. 2011; Raich and Schlesinger 1992). The response of R_s and its temperature sensitivity to climatic warming are different among different terrestrial ecosystems with regard to vegetation types and climatic conditions. For example, the effects of experimental warming onR_s are negative in a semi-arid grassland in Inner Mongolia, China (Liu et al. 2009), positive in a cropland at the southern Germany (Reth et al. 2009) or neutral in an alpine meadow in the Northern Tibet (Shen et al.2015). The Q_{10} of R_s shows a negative correlation with experimental warming in a temperate agricultural ecosystem in Germany (Poll et al. 2013), positive in a subtropical cropland in China (Liu et al. 2012) or neutral in an alpine meadow in the Northern Tibet (Shen et al. 2015). Soil moisture (SM) is another one of the most important abiotic factors influencing R_s, especially in arid and semi-arid ecosystems (Liu et al. 2009; Shen et al. 2015). SM can not only directly affect R_s, but also influnence the Q_{10} of R_s (Lin et al. 2011; Shen et al. 2015). SM can also dampen the effect of soil temperature on R_s (Liu et al. 2009; Shen et al. 2015).

No studies discuss the response of R_s to climatic warming under controlled warming condition in agricultural ecosystems on the Tibetan Plateau. Therefore, it is still rudimentary about

Citation: Zhong Z-M, Shen Z-X, Fu G. Response of soil respiration to experimental warming in a highland barley. SpringerPlus. 2016;5:137. doi:10.1186/s40064-016-1761-0.

how climatic warming will affectR_s in the alpine croplands on the Tibetan Plateau. In this study, a field warming experiment using infrared heaters was conducted in a highland barley system of the Tibet. The main objective of this study was to analyze the response of R_s to warming.

METHODS

Study Area

The study area (91°21′E, 29°41′N, 3688 m above sea level) is located at the Lhasa Agro-ecosystem Research Station, Tibet Autonomous Region in China. Mean annual air temperature is 7.9 °C and mean annual precipitation is around 425 mm, with more than 90 % concentrated in the period from June to September (He et al. 2011). The annual air temperature was 8.5 °C and annual precipitation was 642.4 mm in 2014. That is, it is a warmer and wetter year (2014).

Experimental Design

The experimental soils have been used for crop planting since 1970s. Infrared heaters were used to increase temperature during the whole study period from May 26 to September 14 in 2014. There were three warming treatments with three replicates: the control (CK), low (1000 W) and high (2000 W) warming treatments with a total of nine 2 m × 2 m experimental plots. A 165 cm × 15 cm infrared heater (Kalglo Electronics Inc., Bethlehem, PA, USA) was suspended approximately 1.7 m above the ground in the center of each 2 m × 2 m plot. There was approximately 6–7 m distance between plots.

The highland barley was sown in May 26, 2014 and harvested in September 14, 2014. There was approximately 0.25 m between seeding rows and the seeding was 18.75 g m^{-2}. There were no highland barley outside the 2 m × 2 m plot and there were no other vegetation types within each 2 m × 2 m plot. There were approximately 1150 plants per 2 m × 2 m plot.

Soil Temperature and soil Moisture Monitor

Soil temperature (T_s) and SM sensors were set a depth of 0.05 m in the center of each plot. For each plot, the two sensors were connected to a data logger (HOBO weather station, Onset Computer, Bourne, MA, USA). The measurements were taken every minute, and the data was processed to provide an average every 5 min.

Soil respiration Measurements

A CO_2 flux system (LI-8100, LI-COR Biosciences, Lincoln, NE, USA) with a 20 cm diameter opaque survey chamber was used to measure R_s (Fu et al. 2014). A polyvinyl chloride (PVC) collar (diameter, 20 cm; height, 5 cm) was inserted about 2–3 cm into the soil in the center of each plot (the soil temperature and soil moisture sensors were just beneath the PVC collar) in May 2014. The PVC collar was left the same place during the whole study period. The soil temperatures and moistures were consistent with those beneath the chambers. We started to measure R_s in May 2014. The opaque survey chamber was manually mounted on the PVC collar in each plot for the measurement of R_s. Daily cycles of R_s measurements were generated from 8:00 to 8:00 on June 5–6 (approximately 3–4 days after sprouting), July 26–27 (approximately flag leaf stage), August 6–7 (approximately blooming stage), August 26–27 (approximately waxy ripeness stage) and September 6–7 (from yellow ripeness stage to full ripeness stage). The measuring interval was 2 and 3 h during the daytime (8:00–20:00) and nighttime (20:00–8:00), respectively.

Statistical Analysis

A repeated-measures ANOVA with experimental warming as the between subject factor and measuring date and time as the within subject factors was conducted for R_s. Duncan multiple comparisons were performed among the three warming treatments.

Exponential regression analyses were conducted between R_s and T_s, whereas linear regression analyses were conducted between R_s and SM for each treatment. For each treatment, a stepwise multiple regression analysis was used to analyze the relationships between R_s and T_s and SM, before which natural-logarithm transformations were made for R_s and SM.

We analyzed the sensitivity of R_s to soil temperature for each treatment using all measurement data according to

$$R_s = ae^{bT_s}, \quad 1$$

where a is the intercept of R_s when T_s is 0 °C (i.e. the R_s value when T_s is 0 °C), and b reflects the temperature sensitivity of R_s (Shen et al. 2015). The b values were used to calculate the respiration quotient (Q_{10})

$$Q_{10} = \frac{R_{t+10}}{R_t} = e^{10b} \quad 2$$

where t is a given reference soil temperature, R_{t+10} and R_t are the R_s values when soil temperature is t + 10 and t °C, respectively.

To decrease the disturbance of SM on Q_{10} of R_s, we also chose data to compare the Q_{10} of R_s among the three treatments by the two following rules: (1) there were no correlations between R_s and T_s when SM <0.17 m^3 m^{-3} for the high warming treatment. The R_s was obviously suppressed when SM <0.17 m^3 $m^{-3}$for the low warming treatment. Therefore, only measuring data when SM >0.17 m^3 m^{-3} were used; and (2) measuring date and time was consistent among the three treatments. That is, 26 groups of measured R_s, $T_s$and SM were used for each treatment.

All the statistical analyses were performed using the SPSS software (version 16.0; SPSS Inc., Chicago, IL, USA).

RESULTS

Compared to the control, the low and high warming treatments significantly increased daily average T_s by 1.52 and 1.98 °C, respectively, over the whole study period (Fig. 1).The high warming treatment significantly decreased daily average SM by 16.1 % (−0.03 m^3 m^{-3}) over the whole study period (Fig. 1).

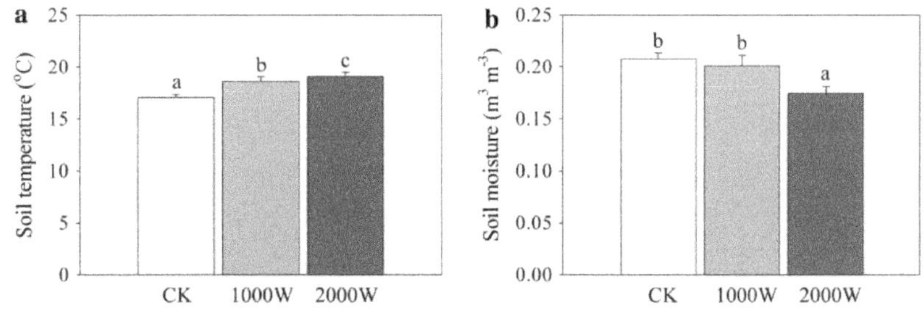

Figure 1: Effects of experimental warming on **a** soil temperature and **b** soil moisture over the whole study period in a highland barley of the Tibet. CK: the control, 1000 W: the low warming treatment, and 2000 W: the high warming treatment.

There were significantly temporal variations of R_s (Table 1; Fig. 2). The average R_s rates were 4.31, 5.41 and 4.85 μmol CO_2 m^{-2} s^{-1} for the control, low and high warming treatments over the five measuring dates, respectively. There were no significant differences of R_s among the three warming treatments (Table 1).

Table 1: Repeated measures ANOVA for the main and interactive effects of experimental warming (W), measuring date (D) and time (T) on soil respiration (R_s, μmol CO_2 m^{-2} s^{-1}) in a highland barley of the Tibet (n = 3)

Model	*df*	*F*	*p*
W	2, 6	0.99	0.43
D	4, 24	20.82	<0.001
T	10, 60	42.38	<0.001
W × D	8, 24	1.36	0.26
W × T	20, 60	0.57	0.72
D × T	40, 240	3.76	<0.05
W × D × T	80, 240	0.76	0.63

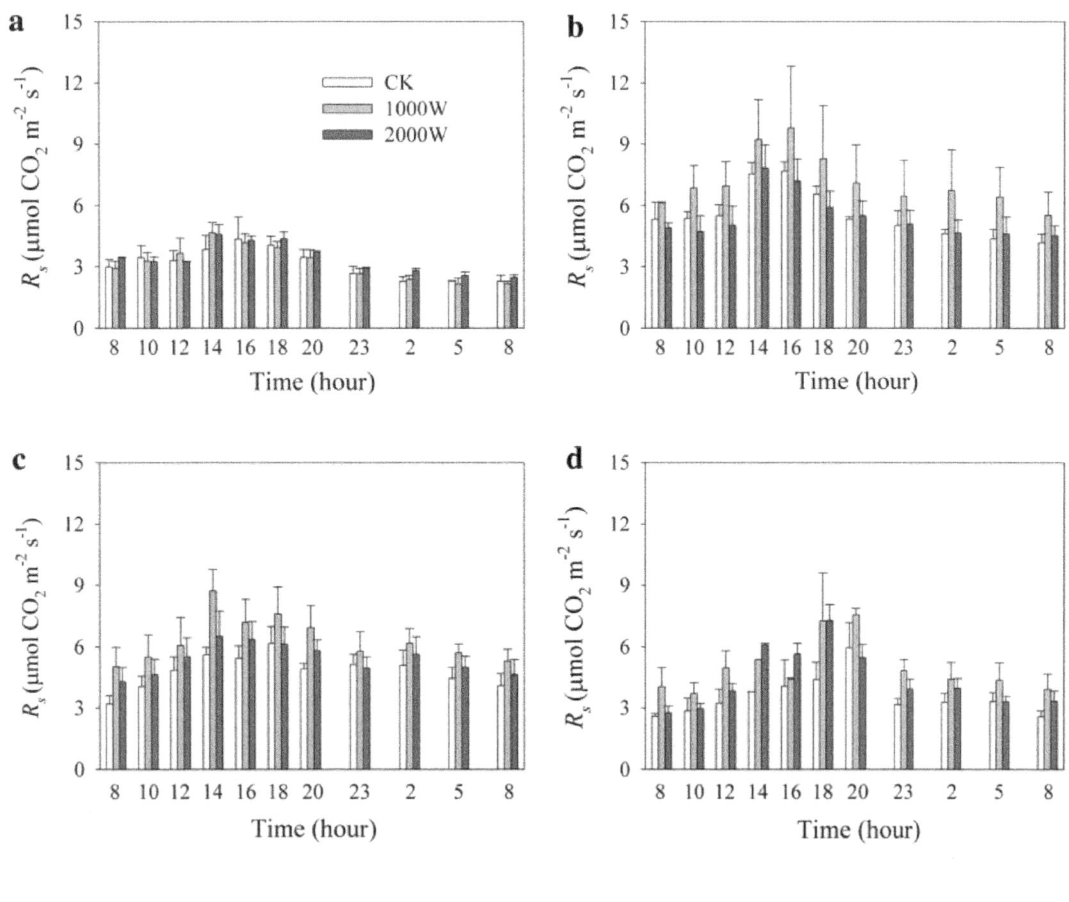

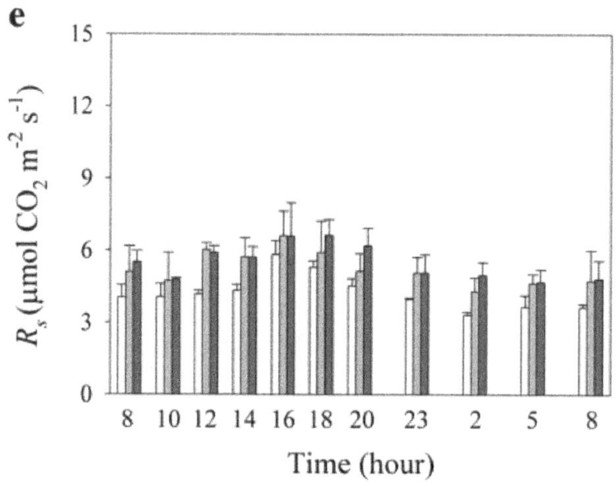

Figure 2: Effects of experimental warming on soil respiration (R_s) in a highland barley of the Tibet (mean ± SE, n = 3). **a** June 5–6, **b** July 26–27, **c**August 6–7, **d** August 26–27, and **e** September ...

Soil respiration significantly increased with increasing T_s (Fig. 3a–f). There was no significant effect of experimental warming on the apparent Q_{10} values when all the measuring data were used (control: 1.70; low warming treament: 1.65; high warming treament: 1.41). In contrast, there was significant effect of experimental warming on the Q_{10} of R_s when only the 26 groups data were used (df = 2, 75; F = 5.84;p < 0.01). In detail, the low and high warming treatments significantly increased the Q_{10} by approximately 0.67 and 1.22, while there was a negligible difference between the low and high warming treatments. The experimental warming-induced change of R_s significantly increased with that of SM (Fig. 4).

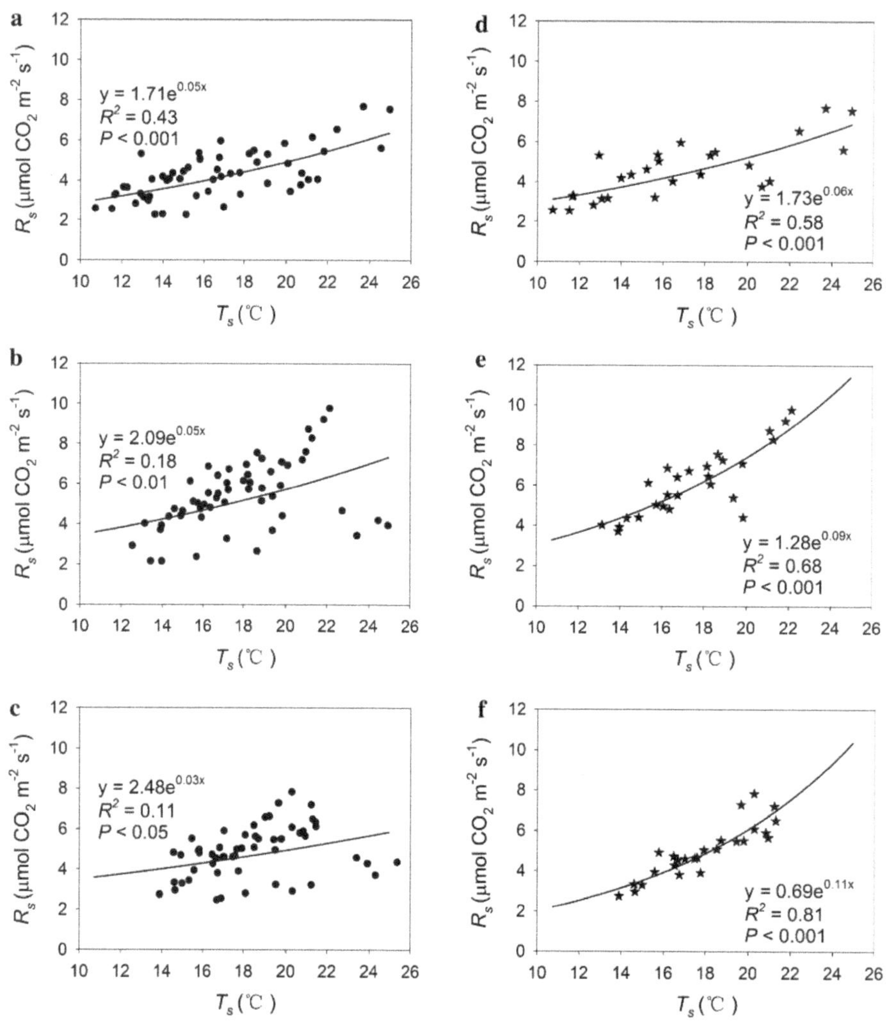

Figure 3: Relationship between soil respiration (R_s) and soil temperature (T_s) across all measuring data for **a** the control, **b** low warming treatment and **c** high warming treatment; and data included when soil moisture was >0.17 m^3 m^{-3} ...

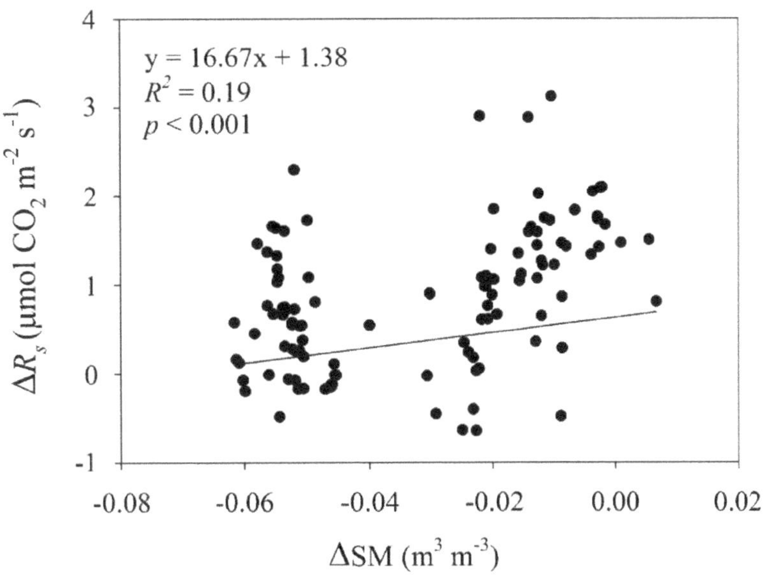

Figure 4: Relationship between the change of soil respiration (ΔR_s) and the change of soil moisture (ΔSM) caused by the low and high warming treatment.

Stepwise regression analyses indicated that the variation of R_s was explained by T_s for the control plots, but explained by SM for the high warming treatment (Table 2). Soil temperature and SM together explained the variation of R_s for the low warming treatment (Table 2). However, the partial correlation between R_s and SM was larger than that between R_s and T_s for the low warming treatment (Table 2).

Table 2: Stepwise multiple regression analyses between soil respiration (R_s) and soil temperature (T_s) and soil moisture (SM), showing changes in the regression coefficient, significance probability (p), coefficient of determination (R^2) and partial correlation

Treatment	Coefficient	R^2	Partial correlation	p
CK				
Constant	0.54			<0.001
T_s	0.05	0.43	0.65	<0.001
1000 W				
Constant	5.27			<0.001
T_s	0.06	0.24	0.75	<0.001
SM	2.88	0.57	0.88	<0.001
2000 W				
Constant	5.43			<0.001
SM	2.20	0.39	0.62	<0.001

DISCUSSION

For the first time to our knowledge, this study quantified the warming effects on R_s and its temperature sensitivity in agricultural ecosystems of the Tibetan Plateau. Our finding that experimental warming did not significantly change R_s in this highland barley system was in line with our one previous study performed in alpine meadows of the Northern Tibet (Shen et al. 2015). However, there were significant positive responses of R_s to experimental warming in the alpine meadow of the Haibei station (Lin et al. 2011) and the Songpan County (Shi et al. 2012), an alpine steppe of the Northern Tibet (Lu et al. 2013) and forest ecosystems (Xu et al. 2010) on the Tibetan Plateau. These findings implied that the alpine ecosystems on the Tibetan Plateau did not always show a positive response of R_s to climatic warming. These results also indicated that the effect of climatic warming on R_s varied with not only ecosystems but also regions on the Tibetan Plateau. Compared to alpine grasslands and forests, the soil CO_2 production of alpine agricultural ecosystem may have a lower response to climatic warming on the Tibetan Plateau. Our finding was also not in line with some previous studies which demonstrated that experimental warming significantly increased R_s in temperate and subtropical croplands (Liu et al. 2012; Reth et al. 2009). This phenomenon implied that alpine agricultural soil may not always have higher temperature sensitivity than temperate and subtropical agricultural soils.

In this study, the no significant response of R_s to experimental warming was most likely due to the experimental warming-induced soil drying (Table 2; Fig. 3). A meta-analysis showed that drying had a significant negative effect on R_s (Wu et al. 2011). Experimental warming-induced soil drying suppressed the effect of increased temperature on R_s in croplands (Poll et al. 2013; Wall et al. 2013), alpine and temperate grasslands (Liu et al. 2009; Shen et al. 2015). Soil drying can suppress soil microbial activity and microbial respiration (Fu et al. 2012; Liu et al. 2009), and plant photosynthesis and primary production (Fu et al. 2013; Xu et al. 2013), all of which are positively correlated with R_s (Fu et al. 2014; Iqbal et al. 2010). The finding that the effect of experimental warming on R_s did not correlate with warming magnitudes may be attributed to the finding that a higher experimental warming resulted in a greater soil drying in our study system.

Different from many previous studies which exhibited experimental warming significantly decreased temperature sensitivity of soil respiration (Luo et al. 2001; Poll et al. 2013; Suseela and Dukes 2013; Zhou et al. 2012), experimental warming did not significantly affect the temperature sensitivity of R_s across all the measuring data in this highland barley system (Fig. 3a–c). In contrast, when the disturbance of SM was dampened (i.e. when SM >0.17 $m^3\ m^{-3}$), experimental warming significantly increased the temperature sensitivity of R_s in this highland barley system (Fig. 3d–f). Soil respiration did not significantly correlate with soil temperature when SM was smaller than 0.17 $m^3\ m^{-3}$ for the high warming treatment. Similarly, the temperature sensitivity of R_s declined with drought in a New England old-field ecosystem (Suseela and Dukes 2013) and increased with increasing SM in an alpine meadow in the Northern Tibet (Shen et al.2015). Therefore, soil water availability affected the temperature sensitivity of soil respiration and the negligible response of temperature sensitivity derived from all the measuring data may be due to soil drying caused by experimental warming.

CONCLUSIONS

Experimental warming-induced soil drying masked the effect of increased soil temperature on soil respiration in the highland barely system. Water availability will be a limited factor for the carbon emission of alpine agricultural soil under climatic warming on the Tibetan Plateau. The alpine soil may not show a positive feedback to climatic warming, which contrasted with the finding as most of previous studies have suggested on the Tibetan Plateau.

AUTHORS' CONTRIBUTIONS

Z-MZ and GF designed the study; Z-XS provided valuable suggestions about the data analysis. GF wrote the first draft of the manuscript, aided by Z-XS and Z-MZ, while they contributed substantially to revisions. All authors read and approved the final manuscript.

ACKNOWLEDGEMENTS

We thank the editor and reviewers for their insightful and valuable comments, which greatly improved the quality of this manuscript. This work was funded by the National Natural Science Foundation of China (Nos. 31370458, 41171084), the Chinese Academy of Science Western Light Talents Program (Response of livestock carrying capability to climatic change and grazing in the alpine meadow of Northern Tibetan Plateau), the Natural Science Foundation of Tibet Autonomous Region (Response of species richness and above-ground biomass to warming in the alpine meadow of Tibet), the National Science and Technology Plan Project of China (No. 2011BAC09B03) and the Science and Technology Plan Projects of Tibet Autonomous Region (Forage Grass Industry).

REFERENCES

1. Dai F, Nevo E, Wu DZ, Comadran J, Zhou MX, Qiu L, Chen ZH, Beiles A, Chen GX, Zhang GP. Tibet is one of the centers of domestication of cultivated barley. Proc Natl Acad Sci. 2012;109:16969–16973. doi: 10.1073/pnas.1215265109.
2. Fu G, Shen Z, Zhang X, Zhou Y. Response of soil microbial biomass to short-term experimental warming in alpine meadow on the Tibetan Plateau. Appl Soil Ecol. 2012;61:158–160. doi: 10.1016/j.apsoil.2012.05.002.
3. Fu G, Zhang X, Zhang Y, Shi P, Li Y, Zhou Y, Yang P, Shen Z. Experimental warming does not enhance gross primary production and above-ground biomass in the alpine meadow of Tibet. J Appl Remote Sens. 2013
4. Fu G, Zhang X-Z, Zhou Y-T, Yu C-Q, Shen Z-X. Partitioning sources of ecosystem and soil respiration in an alpine meadow of Tibet Plateau using regression method. Pol J Ecol. 2014;62:17–24. doi: 10.3161/104.062.0103.
5. Fu G, Shen ZX, Sun W, Zhong ZM, Zhang XZ, Zhou YT. A meta-analysis of the effects of experimental warming on plant physiology and growth on the Tibetan Plateau. J Plant Growth Regul. 2015;34:57–65. doi: 10.1007/s00344-014-9442-0.
6. He YT, Sun W, Zhang ZX, Shi PL, Qu YC, Zhong ZM, Hu J, Zhao HZ. Effect of organic/inorganic compound fertilizer on the yield of crop and fodder double-cropping system in the Tibetan Plateau. Chin J Eco-Agric.2011;19:568–573. doi: 10.3724/SP.J.1011.2011.00568.
7. IPCC . Summary for policymakers. In: Stocker TF, Qin D, Plattner G-K, Tignor M, Allen SK, Boschung J, Nauels A, Xia Y, Bex V, Midgley PM, editors.Climate change 2013: the physical science basis. Contribution of working group I to the fifth assessment report of the intergovernmental panel on climate change. Cambridge: Cambridge University Press; 2013.
8. Iqbal J, Hu RG, Feng ML, Lin S, Malghani S, Ali IM. Microbial biomass, and dissolved organic carbon and nitrogen strongly affect soil respiration in different land uses: a case study at Three Gorges Reservoir Area, South China. Agric Ecosyst Environ. 2010;137:294–307. doi: 10.1016/j.agee.2010.02.015.
9. Lin XW, Zhang ZH, Wang SP, Hu YG, Xu GP, Luo CY, Chang XF, Duan JC, Lin QY, Xu B, Wang YF, Zhao XQ, Xie ZB. Response of ecosystem respiration to warming and grazing during the growing seasons in the alpine meadow on the Tibetan plateau. Agric For Meteorol. 2011;151:792–802. doi: 10.1016/j.agrformet.2011.01.009.
10. Liu WX, Zhang Z, Wan SQ. Predominant role of water in regulating soil and microbial respiration and their responses to climate change in a semiarid grassland. Glob

Chang Biol. 2009;15:184–195. doi: 10.1111/j.1365-2486.2008.01728.x.

11. Liu Y, Chen ST, Hu ZH, Ren JQ, Shen XH. Effects of simulated warming on soil respiration in a cropland under winter wheat-soybean rotation. Environ Sci. 2012;33:4205–4211.

12. Lu XY, Fan JH, Yan Y, Wang XD. Responses of soil CO_2 fluxes to short-term experimental warming in alpine steppe ecosystem, Northern Tibet. PLoS One. 2013;8:e59054. doi: 10.1371/journal.pone.0059054.

13. Luo YQ, Wan SQ, Hui DF, Wallace LL. Acclimatization of soil respiration to warming in a tall grass prairie. Nature. 2001;413:622–625. doi: 10.1038/35098065.

14. Poll C, Marhan S, Back F, Niklaus PA, Kandeler E. Field-scale manipulation of soil temperature and precipitation change soil CO_2 flux in a temperate agricultural ecosystem. Agric Ecosyst Environ. 2013;165:88–97. doi: 10.1016/j.agee.2012.12.012.

15. Raich JW, Schlesinger WH. The global carbon dioxide flux in soil respiration and its relationship to vegetation and climate. Tellus Ser B Chem Phys Meteorol. 1992;44:81–99. doi: 10.1034/j.1600-0889.1992.t01-1-00001.x.

16. Reth S, Graf W, Reichstein M, Munch JC (2009) Sustained stimulation of soil respiration after 10 years of experimental warming. Environ Res Lett 4. doi:10.1088/1748-9326/4/2/024005

17. Rustad LE, Campbell JL, Marion GM, Norby RJ, Mitchell MJ, Hartley AE, Cornelissen JHC, Gurevitch J, Gcte N. A meta-analysis of the response of soil respiration, net nitrogen mineralization, and aboveground plant growth to experimental ecosystem warming. Oecologia. 2001;126:543–562. doi: 10.1007/s004420000544.

18. Shen ZX, Fu G, Yu CQ, Sun W, Zhang XZ. Relationship between the growing season maximum enhanced vegetation index and climatic factors on the Tibetan Plateau. Remote Sens. 2014;6:6765–6789. doi: 10.3390/rs6086765.

19. Shen ZX, Li YL, Fu G. Response of soil respiration to short-term experimental warming and precipitation pulses over the growing season in an alpine meadow on the Northern Tibet. Appl Soil Ecol. 2015;90:35–40. doi: 10.1016/j.apsoil.2015.01.015.

20. Shi FS, Chen H, Chen HF, Wu Y, Wu N. The combined effects of warming and drying suppress CO_2 and N_2O emission rates in an alpine meadow of the eastern Tibetan Plateau. Ecol Res. 2012;27:725–733. doi: 10.1007/s11284-012-0950-8.

21. Suseela V, Dukes JS. The responses of soil and rhizosphere respiration to simulated climatic changes vary by season. Ecology. 2013;94:403–413. doi: 10.1890/12-0150.1.

22. Wall GW, McLain JET, Kimball BA, White JW, Ottman MJ, Garcia RL. Infrared warming affects intrarow soil carbon dioxide efflux during vegetative growth of spring wheat. Agron J. 2013;105:607–618. doi: 10.2134/agronj2012.0356.

23. Wang J, Li HR, Li YH, Yu JP, Yang LS, Feng FJ, Chen Z. Speciation, distribution, and bioavailability of soil selenium in the Tibetan Plateau kashin-beck disease area-a case study in Songpan County, Sichuan Province, China. Biol Trace Elem Res. 2013;156:367–375. doi: 10.1007/s12011-013-9822-5.

24. Wu ZT, Dijkstra P, Koch GW, Penuelas J, Hungate BA. Responses of terrestrial ecosystems to temperature and precipitation change: a meta-analysis of experimental manipulation. Glob Chang Biol. 2011;17:927–942. doi: 10.1111/j.1365-2486.2010.02302.x.

25. Xu ZF, Wan CA, Xiong P, Tang Z, Hu R, Cao G, Liu Q. Initial responses of soil CO_2 efflux and C, N pools to experimental warming in two contrasting forest ecosystems, Eastern Tibetan Plateau, China. Plant Soil.2010;336:183–195. doi: 10.1007/s11104-010-0461-8.

26. Xu WF, Yuan WP, Dong WJ, Xia JZ, Liu D, Chen Y. A meta-analysis of the response of soil moisture to experimental warming. Environ Res Lett. 2013

27. Yang GH, Du ES, Xu ZY, Wang CF. Productivity of land resources and population carrying capacity in Xizang. Lhasa: Tibetan People's Publishing House; 1996. pp. 44–72.

28. Zhang GP, Wang JM, Chen JX. Analysis of beta-glucan content in barley cultivars from different locations of China. Food Chem. 2002;79:251–254. doi: 10.1016/S0308-8146(02)00127-9.

29. Zhang XZ, Shen ZX, Fu G. A meta-analysis of the effects of experimental warming on soil carbon and nitrogen dynamics on the Tibetan Plateau. Appl Soil Ecol. 2015;87:32–38. doi: 10.1016/j.apsoil.2014.11.012.

30. Zhao XY, Wang WJ, Wan WY, Li H. Influence of climate change on potential productivity of naked barley in the Tibetan Plateau in the past 50 years.Chin J Eco-Agric. 2015;23:1329–1338.

31. Zhou JZ, Xue K, Xie JP, Deng Y, Wu LY, Cheng XH, Fei SF, Deng SP, He ZL, Van Nostrand JD, Luo YQ. Microbial mediation of carbon-cycle feedbacks to climate warming. Nat Clim Chang. 2012;2:106–110. doi: 10.1038/nclimate1331.

CHAPTER 2

THE EFFECTS OF BOTH RECENT AND LONG-TERM SELECTION AND GENETIC DRIFT ARE READILY EVIDENT IN BARLEY BREEDING POPULATIONS

Poets AM, Mohammadi M, Seth K, Wang H, Kono TJ, Fang Z, Muehlbauer GJ, Smith KP, Morrell PL

Keywords: Introgression, breeding, evolution of crops, genetics, identity by state

INTRODUCTION

The evolution of breeding populations encompasses processes that reasonably mimic the evolution of natural populations, but in accelerated time (Ross-Ibarra*et al.* 2007). Comparative population genetic approaches can be used for the identification of genes underlying adaptive variation and to understand the effects of demographic patterns on diversity without specific phenotypic information (Nielsen *et al.* 2007; Ross-Ibarra *et al.* 2007).

Breeding populations are subject to demographic effects, such as founder events associated with their initiation (Martin *et al.* 1991), and potentially bottlenecks, and gene flow from other breeding populations or exotic sources. Although the primary goal of breeding programs is the selection of progeny with improved agronomic adaptation, these demographic effects also contribute to differences in allele frequency among breeding populations. Thus the differences among populations are a combination of genetic drift and selection for both agronomic traits and for improved local adaptation. This selection also involves some degree of concomitant linked selection. Understanding the selection and demographic history, including the effects of genetic drift and migration in breeding populations, can accelerate crop improvement (Ross-Ibarra *et al.*2007), for example through the identification of loci involved in domestication and improvement (*e.g.*, Cavanagh *et al.* 2013; van Heerwaarden *et al.* 2012;Hufford *et al.* 2012), identification of introgression between domesticates and wild relatives (*e.g.*, Hufford *et al.* 2013), and the determination of specific donor individuals contributing to disease resistance variants (*e.g.*, Fang *et al.*2013).

The history of population differentiation due to selection can be investigated by the identification of loci with large differences in allele frequency (Cavalli-Sforza 1966). Selection at different timescales can be detected by complementary techniques. For older and longer-term selective events we can use fixation indices or *F*-statistics (Nielsen *et al.* 2007) to measure allele frequency differences between populations. Variants with large differences in allele frequency among subpopulations, as measured by F_{ST}, may have been subject to selection (Lewontin and Krakauer 1973). A high degree of haplotype sharing between individuals can be indicative of more recent selection (Pritchard *et al.* 2010; Horton *et al.* 2012). Haplotype sharing based analyses can identify specific haplotypes (and their underlying sequence variants) subject to selection based on their relative frequencies in a population (Hudson *et al.*1994; Innan *et al.* 2005).

Migration between populations can contribute to adaptive variation (Slatkin 1987). Identity by state (IBS) analysis is a sensitive approach for the identification of specific genomic regions involved in migration between distinct populations. IBS analyses have been used in studies of barley and maize to identify specific genomic regions involved in adaptation to new environments (Hufford *et al.* 2013) and disease resistance genes derived from exotic germplasm (Fang *et al.* 2013).

In the present study we use SNP genotype data from the Barley Coordinated Agricultural Project to investigate the breeding history of North American barley breeding populations. Barley (*Hordeum vulgare* subsp. *vulgare*) was introduced to North America by European colonists as early as 1602 as a crop es-

Citation: Poets AM, Mohammadi M, Seth K, et al. The Effects of Both Recent and Long-Term Selection and Genetic Drift Are Readily Evident in Barley Breeding Populations. Doi:10.1534/g3.115.024349.

sential for beer production (Weaver 1950). Early successes in barley production in North America involved introduction of cultivars adapted to similar environmental conditions. In the eastern growing region of the United States barley was introduced from northern Europe, including English malting varieties; and in the western growing region from Mediterranean feed cultivars (Weaver 1943). Notable exceptions from outside northern Europe included "Manchuria" from northeastern China, which was well adapted to the Upper Midwest growing environment, "Stavropol" from Russia, which was the most important cultivar in the Lower Midwest (especially Kansas), and "Trebi" from the southern shores of the Black Sea in Turkey, which was found to be particularly productive in western regions under irrigated conditions (Weaver 1943; Wiebe and Reid 1961). Despite the use of germplasm from multiple Old World sources, the set of founder cultivars was relatively narrow, which likely contributed to reduced diversity in modern North American cultivars (Martin *et al.* 1991).

Barley production is frequently divided into spring and winter growth habit. Spring barley dominates in North America, while winter barley is grown in more southerly latitudes and moderate coastal climates (Weaver 1950). Spring and winter barley have been bred separately by breeding programs located in distinct geographic regions. Further trait requirements by end-use markets (*i.e.*, malting, feed, and food) and the establishment of breeding programs for two- and six-rowed spike-type contribute to highly structured barley populations (Cuesta-Marcos *et al.* 2010; Hamblin *et al.* 2010; Wang *et al.* 2012; Zhou *et al.*2012).

Our analysis focuses on four major questions. First, which of the factors, including breeding populations, growth habit, and spike-type, contribute most directly to genetic differentiation among our samples? Second, to what extent are loci identified as major contributors to phenotypic variance in Old World barley germplasm contributing to allele frequency differences in North American breeding populations? Third, can we identify evidence of recent or long-term selection acting on breeding populations and what loci are involved? Fourth, how have patterns of shared ancestry and migration contributed to diversity and relatedness in current breeding populations?

Allele frequency-based analyses of the North American breeding programs indicate a large contribution of genetic drift and linked selection to genetic variation. Patterns of allele frequencies observed in loci known to be involved in adaptive traits in the Old World are recapitulated in North American breeding programs. Additionally we were able to identify populations and specific genomic segments involved in more recent gene flow.

MATERIALS AND METHODS

Plant Materials

Genotypic data from a total of 3971 barley accessions including cultivars, advanced lines, and genetic stocks from barley breeding programs that participated in the Barley Coordinated Agricultural Project (CAP) were downloaded from The Triticeae Toolbox (T3) (http://triticeaetoolbox.org/). Barley samples are representatives of 4 yr of germplasm enhancement (2006–2009) from 10 breeding programs. These breeding programs include most barley growing regions and market end-uses in the United States. For more information about the programs see Hamblin *et al.* (2010) and Wang *et al.*(2012).

The breeding of two- and six-rowed barley within a single program often involves different objectives; thus we consider these types as independent populations resulting in 16 breeding populations (see Table 1). Hereafter, the following notation was used when referring to each breeding population: "breeding program's name abbreviation" followed by 2 or 6 for two- and six-rowed spikes, respectively, as described in Table 1. The Busch Agricultural Resources international program referred to here as BAI2, was separated from the North American two-rowed lines, referred to here as BA2. There were 10 six-rowed accessions from Oregon mislabeled in T3 as two-rowed (P. Hayes, personal communication); we used the corrected spike-type (see Supporting Information, for more details). The current sample is divided hierarchically at the highest level into spring and winter growth habits and then within each growth habit by breeding population (Figure 1).

Table 1: Descriptive statistics by breeding populations

Breeding Program	Abbreviation	Spike-Type	No. Markers	Sample Size	Average Pairwise Diversity	Mean F_{IS}	Private SNP
Spring barley			2392	2952	0.331		72
University of Idaho in Aberdeen	AB2	2	2103	239	0.266	0.983	3
	AB6	6	1791	142	0.257	0.984	1
Busch Agricultural Resources, Inc.	BA2	2	2029	172	0.203	0.997	0
	BA6	6	1770	147	0.166	0.997	0
Busch Agricultural Resources, Inc.(International)	BAI2	2	1508	60	0.295	0.997	0
University of Minnesota	MN6	6	1650	386	0.148	0.989	0
Montana State University	MT2	2	2041	317	0.264	0.989	1
North Dakota State University Two-row	N2	2	2177	353	0.241	0.959	5
North Dakota State University Six-row	N6	6	1722	380	0.197	0.98	2
Washington State University	WA2	2	2032	351	0.248	0.971	1
	WA6	6	1665	32	0.256	0.976	0
Winter barley			2336	661	0.291		16
Oregon State University	OR2	2	2052	73	0.293	0.960	0
	OR6	6	2272	268	0.274	0.968	11
Utah State University	UT2	2	1650	30	0.360	0.975	1
	UT6	6	2283	343	0.303	0.970	6
Virginia Polytechnic Institute and State University	VT6	6	2025	320	0.246	0.985	2
Average			1972.1	401.4	0.258	0.980	6.7

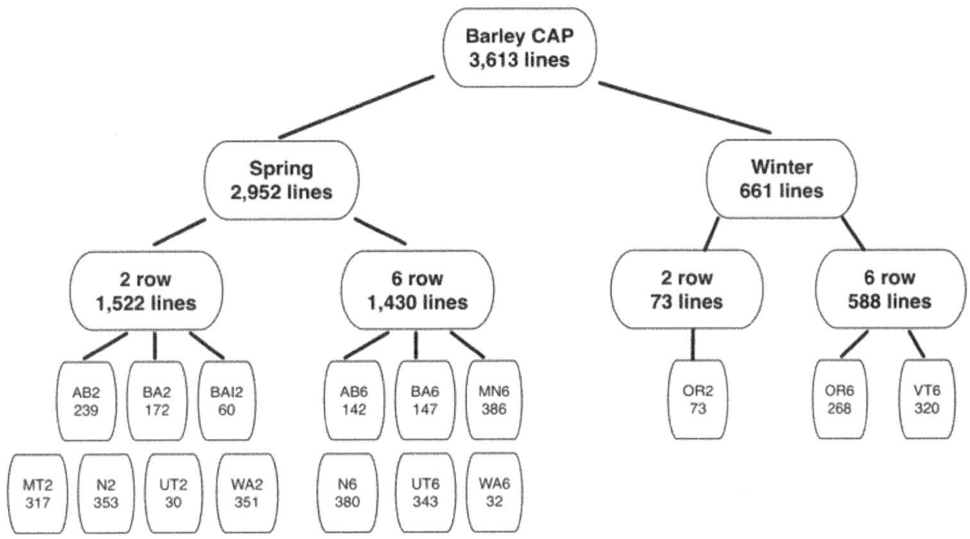

Figure 1: Breeding programs. University of Idaho in Aberdeen (AB), Busch Agricultural Resources, Inc. (BA), Busch Agricultural Resources, Inc. International lines (BAI), Oregon State University (OR), Utah State University (UT), Washington State University (WA), ...

Genotyping

All 3971 accessions were genotyped with 2882 Barley Oligo Pooled Assay Single Nucleotide Polymorphisms, referred to as BOPA SNPs (Close *et al.*2009) using Illumina GoldenGate Technology (Illumina, San Diego, CA). BOPA SNPs were identified primarily from resequencing of expressed sequenced tags (ESTs) (Close *et al.* 2009). SNP order along each linkage group is based on the consensus genetic map of Muñoz-Amatriaín *et al.* (2011). Sampled accessions were self-fertilized to at least the F_4 generation before genotyping (Hamblin *et al.* 2010; Wang *et al.* 2012), resulting in average expected heterozygosity of 6.25%. Genotyping information was downloaded from (http://triticeaetoolbox.org/) using only the OPA 2011 markers. The options to download the dataset were set to minimum MAF ≥0% and to remove markers and accessions with >99.9% missing data.

SNP annotations and metadata information, including gene names and whether SNPs occur in genic or nongenic regions, were obtained using SNPMeta (Kono*et al.* 2014).

Quality Control

The dataset was filtered for monomorphic SNPs, and SNPs or accessions with more than 25% missing data. Additionally, we removed accessions with incomplete sample information (*i.e.*, spike-type or growth habit) and accessions where single lines represented a population. We also removed accessions with heterozygosity that exceeded the expected 6.25%, a conservative threshold intended to filter out accessions that might have genotyping errors. Finally, we removed genetic stocks or near-isogenic lines such that one accession from each near-isogenic line set was retained in the dataset.

Summary Statistics

Descriptive statistics were calculated for each of the 16 breeding populations (Table 1). The degree of inbreeding was estimated by the inbreeding coefficientF_{IS} $(1 - H_o/H_e)$ using a custom Python script. The similarity between samples was estimated by the percent pairwise diversity calculated using the *compute*program from the libsequence library (Thornton 2003). Monomorphic SNPs in each population were excluded and heterozygous and ambiguous calls were treated as missing data. The SharedPoly program from libsequence (Thornton 2003) was used to count the number of private SNPs in each breeding population.

Derived site frequency spectrum (SFS)

Derived site frequency spectra (*i.e.*, the distribution of the derived allele frequencies) were calculated and plotted for each of the 16 populations and for the complete dataset. To infer ancestral state, SNP states from *Hordeum bulbosum* as reported for BOPA SNPs were used (Fang *et al.* 2014). This involved alignment of RNAseq data from one diploid accession of *H. bulbosum*(Cb2920/4) to the Morex draft assembly (Mayer *et al.* 2012), and calling the *H. bulbosum* nucleotide at the BOPA SNP positions. Ambiguous nucleotide calls,*trans*-specific polymorphisms, and sites at which *H. bulbosum* segregates for different nucleotides than barley (*H. vulgare* subsp. *vulgare)* were treated as missing data.

Joint Derived SFS

To compare the change in frequency of derived alleles between partitions of the data we used the joint derived SFS. Following the same procedure as for the derived SFS for each breeding population, we calculated the derived SFS within winter or spring accessions and within two- or six-rowed accessions using a custom R script. The derived SFS was compared between growth habit and between spike-types. The joint derived SFSs were plotted using the R package grDevices (R Core Team 2014).

Population Structure

The degree of differentiation among individuals from all breeding programs was estimated by Principal Component Analysis (PCA). The analysis was performed using the SmartPCA program from the EIGENSOFT package

(Patterson *et al.* 2006). SmartPCA permits PCA analysis with SNPs that include missing data.

Maximum Likelihood tree of Relatedness and Migration

The population relatedness and patterns of gene flow between breeding populations was inferred using a maximum likelihood approach implemented in TreeMix (Pickrell and Pritchard 2012). To polarize the divergence among populations we used genotyping data for 803 landrace accessions from the Old World (Poets *et al.* 2015) (see https://github.com/AnaPoets/BarleyLandraces). The 2021 SNPs shared between datasets were used to build a population tree using landraces as the outgroup. We ran 25 replicates of the tree, bootstrapping with 75 SNP windows. We used the replicate with the lowest standard error for the residuals as the base tree topology and inferred the likelihood of having between one and five migration events among breeding populations. The plot of the residuals was used to evaluate which tree best fit the data. Candidates for admixture can be identified by those population pairs with residuals above zero standard error, which represent populations that are more closely related to each other in the data than in the best-fit tree (Pickrell and Pritchard 2012).

Changes in Allele Frequency

To identify putative targets of long-term selection involved in the spike-type and growth habit differentiation among breeding programs we used the Weir and Cockerham (1984) measure of F_{ST} referred to as θ, as implemented in the R package hierfstat (Goudet 2005). F_{ST} was calculated for the following partitions of the data: (1) spring *vs.* winter, (2) two- *vs.* six-rowed, and (3) among breeding populations. Owing to the high level of inbreeding in the dataset, a haploid model for F_{ST} estimation was used as this better represented the sample size in our study. Heterozygous SNPs were treated as missing data. An empirical genome-wide threshold for the top 2.5% of F_{ST} values was used to identify SNPs with large differences in frequency relative to the genome-wide average. To identify the degree of differentiation that each breeding population has with respect to other populations we report F_{ST} for all pairwise comparisons. To determine the degree to which our F_{ST} results are robust to either the specific set of SNPs or individuals used here, we used bootstrapping to remove either 20% of SNPs or samples per iteration. We ran 100 iterations in each case.

To characterize average allele frequency divergence for breeding populations we used the F_{ST}-like measure of Nicholson *et al.* (2002), reported here as "*c*." This estimate differs from other F_{ST}-like measures in that it attempts to estimate the degree of divergence of a population from ancestral allele frequencies. For this analysis, the dataset was divided into two groups: spring six-rowed and spring two-rowed. We do not report *c* values for winter barleys owing to limited sampling. For this analysis, monomorphic SNPs were removed from each group. The *c* value was calculated using the popdiv program from the popgen package in R (Marchini 2013). We used a burn-in period of 1000 iterations followed by a run length of 10,000 iterations with the scale parameter of a Dirichlet distribution used to update global allele frequencies $m = 10$.

Analysis of Resequencing data for Known genes Contributing to Phenotypic Differentiation

Resequencing data for 10 accessions for vernalization sensitivity loci (*Vrn-H3,Sgh3*) (von Zitzewitz *et al.* 2005) and 96 accessions for the *Vrs1* gene controlling spike-type differentiation (Komatsuda *et al.* 2007) were obtained from NCBI Popsets [UID #157652625 (Karsai *et al.* 2008) and 219664771 (unpublished data)]. Contextual sequences for SNPs known to occur in these genes were downloaded from T3. To determine the position of SNPs within these genes and their correlation with growth habit and spike-type differentiation, SNP contextual sequence for individual SNPs were aligned to each resequencing dataset in Geneious v.7.1.9 (Kearse *et al.* 2012).

Identity by State

We used an IBS analysis to identify shared genomic segments among individuals between breeding populations potentially indicative of recent introgression. The analysis used PLINK v.1.90 (Chang *et al.* 2015) with window sizes of 50 and 100 SNPs, allowing for up to 10% mismatch. The frequency of shared segments between two populations was estimated for each SNP window. This analysis made use of phased genotyping data, with phase inferred using fastPHASE v1.2 (Scheet and Stephens 2006). Missing genotypic state was treated as missing (*i.e.*, genotypic state inferred during phasing was ignored). The phased data were only used for IBS and pairwise haplotype sharing analyses.

Pairwise Haplotype Sharing

To explore recent events of selection within populations, we used the pairwise haplotype sharing (PHS) approach (Toomajian *et al.* 2006). A shared haplotype is defined as a genomic segment that extends out from a focal SNP, and is shared among individuals in a population. PHS is a form of IBS analysis that compares the extent of shared haplotypes among individuals normalized by genome-wide sharing. A PHS score depends on the length of the shared haplotype and its frequency in the population. Extended shared haplotypes are potentially suggestive of recent or ongoing selection, owing to limited potential for recombination to break down genomic regions subject to recent selection (Horton *et al.* 2012). PHS was calculated within each breeding population using a customized Perl script (Cavanagh *et al.* 2013). An empirical threshold of the upper 2.5% of PHS values and a minimum focal SNP frequency of 10% within each population were used to identify outliers in the distribution of PHS.

Four-Population test for GENE Flow Detection

The robustness of patterns of migration inferred using TreeMix (see *Results* for details) was assessed using the four-population test (f_4-test) (Keinan *et al.* 2007;Reich *et al.* 2009). The f_4-test is designed to distinguish introgression from incomplete lineage sorting. The test evaluates trees of relatedness among populations and measures genetic drift along lineages quantitatively based on the variance in allele frequencies. Significant deviations from zero in three possible tree topologies indicate that the tree evaluated does not fit the data, suggesting the presence of gene flow.

We used the topology inferred in TreeMix as our hypothesized relationship between populations. We inferred that a tree of relatedness with three migration events provides a better fit to the data than topologies that include either a lesser or greater number of migration events (see *Results* for more details). Following the [(A, B),(C,D)] notation for a tree topology with four populations, we assessed migration among the following sets of populations: [(N2, X),(Y_1, Y_2)], [(UT6, UT2), (Y_1, Y_2)] and [(OR2, X), (OR6,VT6)], where X was replaced iteratively for any two-rowed barley and Y_1 and Y_2 were replaced by a pair of six-rowed barley populations. The populations UT6, OR2, and N2 were chosen because of an indication of gene flow based on TreeMix results. Pairs of populations from the same branch of the TreeMix topology were considered on the inference that these populations are more similar to each other due to shared ancestral polymorphisms. The second pair of populations was selected from the branch containing the population putatively involved in the migration event (populations connected by the arrows in Figure 6). Since the population putatively involved in migration was not clearly defined, we ran the analysis iteratively between each possible pair of populations on the branch. We used the fourpop option in the TreeMix software (Pickrell and Pritchard 2012) to estimate the f_4*-value* for each configuration. Significance of f_4*-values* was determined at $P < 0.05$. We infer that migration has occurred when the three possible trees had a significant nonzero f_4*-value*.

RESULTS

The original dataset included 3971 accessions genotyped with 2882 SNPs representative of 10 breeding programs across the United States. We removed 340 SNPs and 241 accessions with ≥25% missing data, 22 accessions with heterozygosity >6.25% across SNPs, 13 accessions with missing growth habit (spring *vs.* winter) or spike-type (two- *vs.* six-row spikes) information, 79 near-isogenic lines, and three accessions with single accessions representing a population. After quality control our dataset consisted of 3613 barley accessions and 2542 SNPs.

Characterization of North American Breeding Populations

Analysis of the structure of the North American populations using PCA revealed that the primary population structure (PC1 = 19.6% variance) is explained by differences in spike-type, which corresponds to an average F_{ST} of 0.23. PC2 indicates that 9.1% of the variance among lines is explained by differentiation in growth habit, with an average F_{ST} of 0.17. These results are congruent with earlier analyses on a subset of these populations (Cuesta-Marcos *et al.* 2010; Hamblin *et al.* 2010; Wang *et al.* 2012) and on a comparable set of samples analyzed by Zhou *et al.* (2012).

The 16 breeding populations (*i.e.*, breeding programs separated by spike-type) were represented by an average sample size of 225 lines with a minimum of 30 lines (from UT2), and a maximum sample size of 386 for MN6 (Table 1 and Figure 1). On average, only two SNPs were private to each of the breeding populations, with a maximum of 11 private SNPs in OR6. Winter populations had 16 private SNPs with respect to spring populations, which in turn had 72 private SNPs. The average inbreeding coefficient (F_{IS}) across populations was 0.98 as expected after four generations of self-fertilization. Percent pairwise diversity ranged from 0.15 to 0.36 and averaged 0.25 across breeding populations.

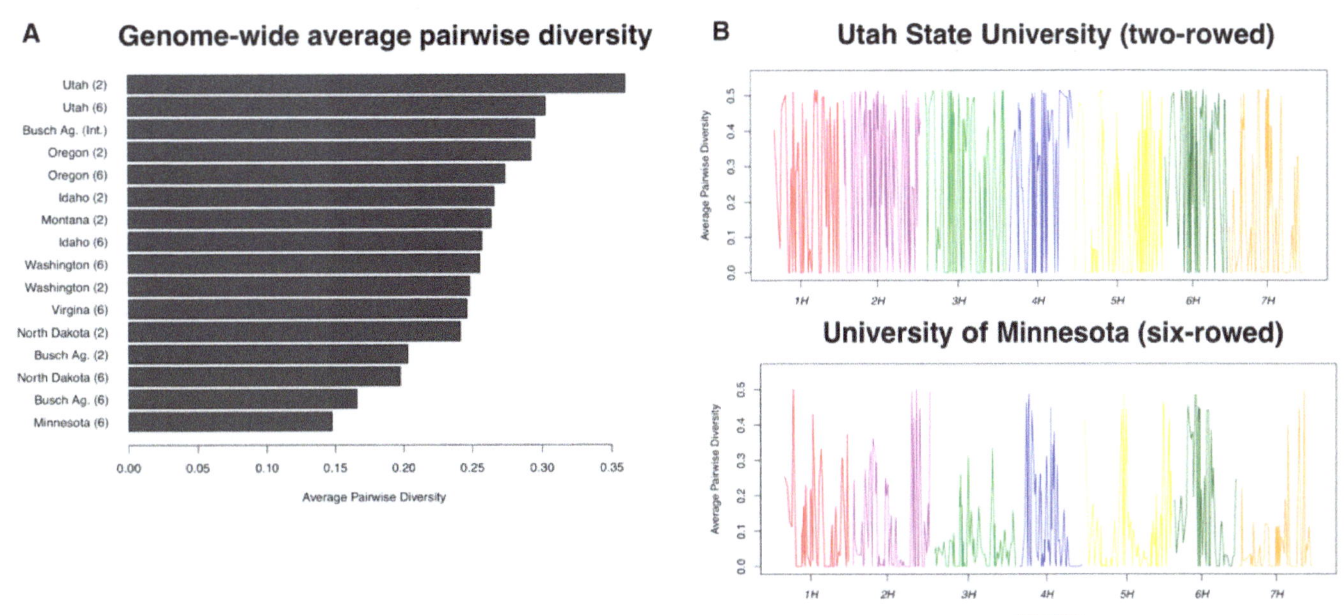

Figure 2: Genetic diversity in breeding populations. (A) Average pairwise diversity in each breeding population. (B) Genome-wide pairwise diversity for the most and least diverse populations according to average diversity, Utah State University (two-rowed) and .

The joint unfolded SFS showed that there are slightly more rare variants segregating in spring than winter programs and slightly more rare variants segregating in six-rowed than in two-rowed programs, reflecting minimal impact of ascertainment bias in each of these partitions. In individual breeding populations there is an excess of rare and high-frequency variants relative to neutral expectations based on Kimura's model of a population at equilibrium (Kimura

1983). OR2, OR6, UT6, and UT2 have a higher proportion of midfrequency variants than other populations. When the breeding populations are considered jointly, the derived SFS displays an elevated number of midfrequency variants, consistent with retention of variants segregating at an average minor allele frequency of 24% in the discovery panel (Close *et al.* 2009).

Genome-wide scan for evidence of selection in North American breeding programs

The distribution of F_{ST} statistics for comparisons of growth habit, spike-type, and breeding populations showed that among the three classifications, the greatest differentiation in allele frequencies is found among breeding populations with an average F_{ST} of 0.37, *vs.* 0.23 for spike-type and 0.17 for growth habit. For each of the three partitions, we identified 61 SNPs in the upper ≥0.975 of F_{ST} values. Of these, 34 SNPs were outliers for both the spike-type and breeding population comparisons, seven were common between breeding populations and growth habit comparisons, and there were no common outlier SNPs between growth habit and spike-type comparisons.

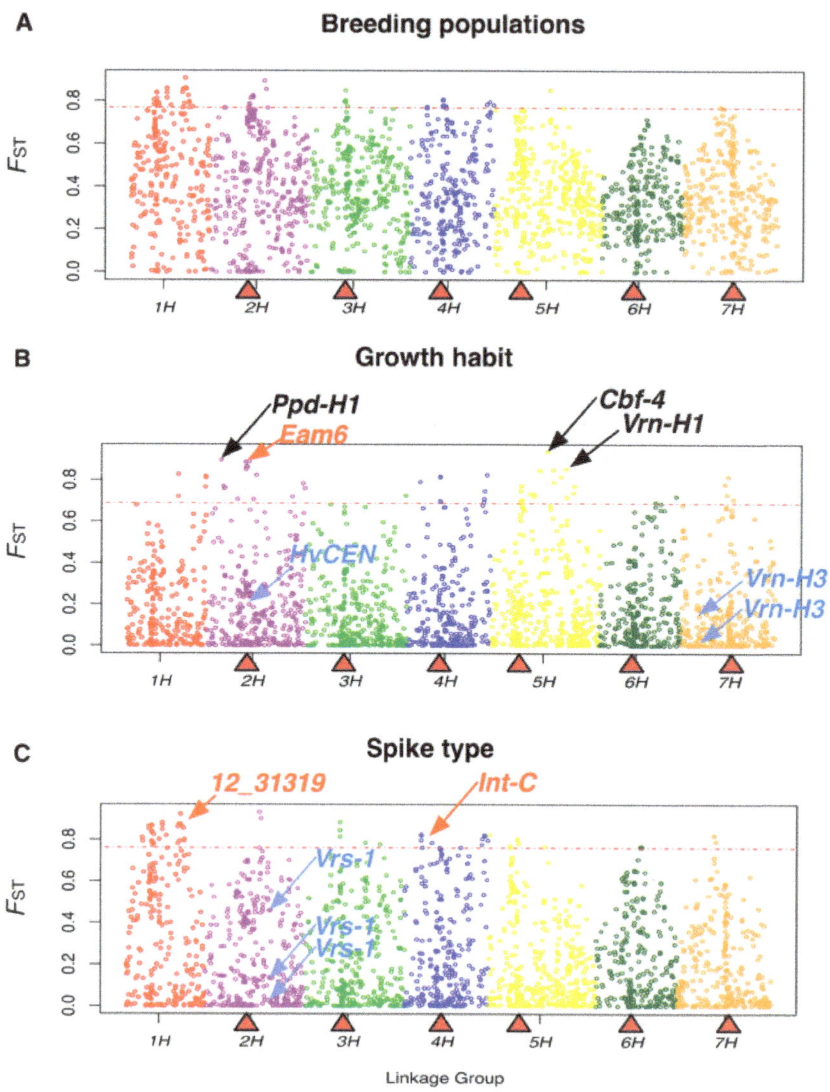

Figure 3: Allele frequency differentiation between comparisons of partitions of the data. (A) Among breeding populations; (B) between spring and winter types; and (C) between six- and two-rowed spike-types. Triangles indicate the genetic position of centromeres.

Bootstrapping, by removing 20% of individuals in each partition of our sample or 20% of SNPs, was used to determine whether F_{ST} outlier detection was dependent on the particular individual's sampled or set of markers used. Across samples we recovered an average of 95% of SNPs outliers in any given comparison. Outliers include SNPs near or in genes mentioned below. Across SNPs we recovered an average of 87% of the genetic regions where SNPs were identified as outliers in the full dataset. This average excludes linkage group 1H where the large number of outliers precludes identification of outliers plausibly attributable to selection

Known genes Contributing to Phenotypic Differentiation

We identified a SNP outlier (12_30883, linkage group 5H) in the F_{ST}comparison for growth habit located in the vernalization sensitivity locus (*Vrn-H1*, *Sgh2*) (von Zitzewitz *et al.* 2005) known to be involved in the growth habit differentiation. SNPs within two additional well-characterized genes including photoperiod response-1H (*Ppd-H1*, *Eam1*) (Turner *et al.* 2005; Jones *et al.*2008) and c-repeat binding factor 4 (*Cbf4*) (Haake *et al.* 2002) were also found to be F_{ST} outliers (Figure

3B). Both of these genes have reported functions related to growth habit differentiation. *Ppd-H1* alters the flowering time response under long-day conditions, thus making it possible to avoid unfavorable seasonal conditions (Lister *et al.* 2009) whereas *Cbf4* contributes to cold acclimation (Skinner *et al.* 2005).

While vernalization sensitivity and photoperiod response are important determinants that contribute to growth habit through an environmental response, there are also loci that contribute to flowering time independent of environmental cues. This class of genes in barley is referred to as Earliness *per se* (EPS) loci; one characterized example is the early maturity 6 locus (*Eam6,eps2S, mat-c*) (Boyd *et al.* 2003; Comadran *et al.* 2012). A linkage mapping study placed early maturity 6 locus (*Eam6*) (Laurie *et al.* 1995) near the centromere of linkage group 2H (cM 67.8), a region where we identify six SNPs with elevated F_{ST} (average F_{ST}= 0.87). The two SNPs (11_20438 and 11_20366) most strongly associated with flowering time in a recent GWAS study (Comadran *et al.* 2011) that resulted in the cloning of the barley *Centroradialis* gene (HvCEN) responsible for flowering time variation (Comadran *et al.* 2012) were not identified as outliers in our F_{ST}comparison of growth habit. However, both SNPs 11_20438 and 11_20366 have an above average F_{ST} of 0.52 and 0.25, respectively.

In an association mapping (AM) study Cuesta-Marcos *et al.* (2010) identified two significant SNPs in linkage groups 1H and 2H (12_31319 and 11_10213, respectively) associated with spike-type differentiation. In our analysis, 12_31319 has an outlier F_{ST} value of 0.86, while 11_10213 has an above average F_{ST} = 0.47 but is not defined as an outlier (Figure 3C). Additionally, there are two SNPs (11_20422 and 11_20606) in linkage group 4H that have been identified near the *Intermedium-C* gene (*Int-c*) (Ramsay *et al.* 2011), which is a modifier of lateral spikelet development in barley. After quality control, only 11_20422 remained in our dataset. This SNP has an F_{ST} value of 0.81 and is thus considered an outlier in the comparisons between spike-types (Figure 3C).

Other SNPs occurring in well-characterized genes did not appear as outliers despite the previously reported contribution to function. There were three SNPs in our dataset (12_30893, 12_30894, and 12_30895 on linkage group 7H) occurring within the vernalization sensitive locus 3 (*Vrn-H3)* (Yan *et al.* 2006) that were not outliers in the F_{ST} comparison between spring and winter types (Figure 3). This gene is an ortholog of the *Arabidopsis thaliana* flowering locus T (*FT*) that promotes flowering time under long days (Turck *et al.* 2008). In barley, nine linked polymorphisms in the first intron have been predicted to be responsible for the variation in flowering time at this locus (Yan *et al.* 2006). Our alignment of the three outlier SNPs to resequencing data of this gene from 10 barley accessions (Karsai *et al.* 2008) identified three of the SNPs segregating in the first intron of *Vrn-H3* . In a study by Yan *et al.*(2006) nucleotide states "A" and "G" in 12_30894 and 12_30895, respectively, were associated with spring barley types (two accessions), while "T" and "C" were associated with winter types (eight accessions). The resequencing data in part support this association having two out of four spring barleys with the inferred haplotype while five out of six winter barleys carried the correct inferred haplotype. However, in our much larger dataset of spring and winter accessions these SNPs are segregating at an average allele frequency of 50% in both spring and winter growth habits, showing no association between these SNPs and spring *vs.* winter growth habit (maximum F_{ST} = 0.04).

The *Vrs1* gene, a well-characterized contributor to spike-type (Komatsuda *et al.*2007) was not identified in the F_{ST} comparison between two- and six-rowed accessions. Our SNP panel included five SNPs (12_30896, 12_30897, 12_30899, 12_30900, and 12_30901, linkage group 2H) in *Vrs1*. However, none of the SNPs reached the empirical cutoff for F_{ST} in this comparison, with a maximum observed F_{ST} of 0.45. The "G" nucleotide state at SNP 12_30900 results in an amino acid substitution associated with the six-rowed phenotype (*vrs1.a3* allele); however, the alternative nucleotide state "C" can be found in either six- or two-rowed accessions (Komatsuda *et al.* 2007; Youssef *et al.*2012). In a panel of 96 European accessions of cultivated barley (Popset ID 219664771) the "G" state always resulted in a six-rowed phenotype. In our sample of North American breeding programs 2% of individuals that carry this variant state were reported as two-rowed barleys, which is similar to previous results that found this SNP significant for spike-type differentiation segregating in two-rowed accessions with an allele frequency of 1% (Cuesta-Marcos *et al.* 2010). The "C" state segregated at 50% frequency in each of the row-type partitions.

Haplotype Sharing and Evidence for Recent Selection

The PHS analysis permits the identification of genomic regions that are putatively involved in more recent selection. PHS analysis within individual breeding populations identified a total of 775 SNPs in the upper ≥0.975 of the PHS distribution. In a small number of cases, focal SNPs in the PHS analysis were identified as outliers in more than one breeding population. Sharing of PHS outliers is greatest for OR6 and UT6 (22), BA6 and WA6 (17), and AB2 and OR2 (14). The number of focal SNPs that were PHS outliers shared among populations were an average of three SNPs shared within two-rowed populations and four SNPs within six-rowed populations. An average of two focal SNPs were shared between two- and six-rowed populations. Out of the 775 SNPs, 77 were in genes with known function. The haplotypes for these significant PHS values varied in length from 19.7 cM to 139.6 cM with mean 55.9 cM across breeding populations. The frequency of the SNP state with significant PHS ranged from 10% (the minimum value we consider) to 61% with mean 24%.

Within BAI2, BA6, MN6, UT2, UT6, WA6, and OR2 we observed long runs of haplotype sharing (average length 112.45 cM) at an average frequency of 24%, significantly exceeding genome-wide similarities. These regions were putatively subject to recent selection. Among outliers for PHS, BA2 showed significant PHS surrounding three SNPs (12_30893, 12_30894, 12_30895, in linkage group 7H) with all three SNPs occurring within *Vrn-H3*. AB2, MT2, OR2, and OR6 had an outlier PHS value for SNP 12_30901 (linkage group 2H) in the *Vrs1* gene. The SNPs in *Vrn-H3* and *Vrs1* were at an average frequency of 22.25% in each of these populations, with an average length of shared haplotype of 66.98 cM. There are three SNPs (12_20368, 12_20593, and 12_21049 in linkage group 2H) with significant PHS value in OR2 (frequency 11% and haplotype length 138.9 cM). SNPMeta annotations identified SNP 12_20593 within the nicotianamine synthase 2 (*nashor2*) gene in barley, which is a key enzyme for iron homeostasis (Herbik*et al.* 1999), and SNP 12_20368 within a gene with sequence similarity to galactinol synthase 2 gene in wheat (*TaGolS* 2), which in turn is orthologous to the characterized gene *TaGolS* 2 in *A. thaliana* (Taji *et al.* 2002). In *A. thaliana*,*TaGolS* 2 has been identified as playing an important role in drought-stress tolerance (Taji *et al.* 2002).

Divergence from Ancestral Allele Frequencies

Quantification of allele frequency divergence based on *c* (Nicholson *et al.* 2002) showed that six-rowed breeding populations have experienced more divergence than two-rowed populations, with mean *c* of 23.1% and 17.1%, respectively. Among the two-rowed populations, Utah two-rowed has diverged the most while Idaho two-rowed resembles ancestral allele frequencies, suggesting reduced effects of drift or linked selection in the latter population. Among the six-rowed populations, Utah six-rowed was the closest to ancestral allele frequencies while Minnesota six-rowed has experienced the most divergence. Utah two-rowed diverged more from ancestral allele frequencies than its six-rowed counterpart (contrasting the pattern observed in other spike-type pairs of populations). This is expected because the two-rowed individuals came from crosses between two-rowed and six-rowed varieties (data not shown).

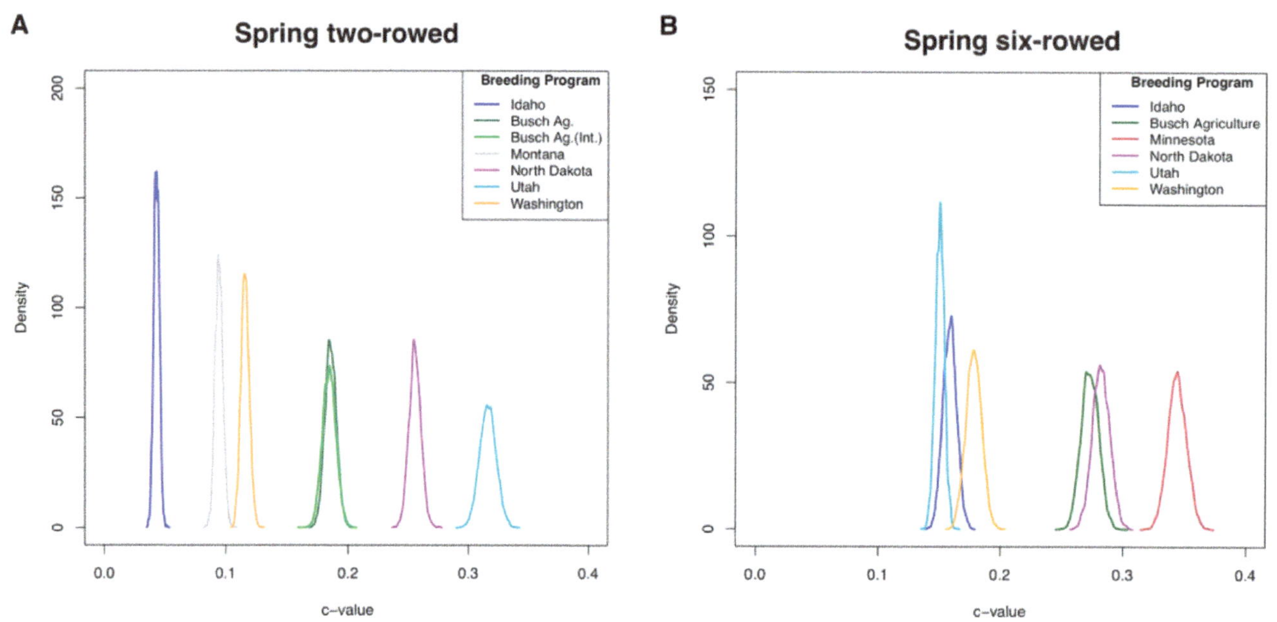

Figure 4: Marginal posterior density plots obtained for the *c* estimator for spring barley populations. Density distribution after 10,000 iterations. Plot shows the amount of drift estimated from ancestral allele frequencies (0,0 coordinates). (A) Two-rowed barley .

Gene flow Between Breeding Populations

To identify genomic segments subject to recent introgression, we tested for regions with a high degree of IBS between breeding populations. Window sizes of 50 (0.63–44.96 cM) and 100 (13.0–62.25 cM) SNPs were used, and we permitted up to 10% mismatch among haplotypes. As expected, populations within growth habit and spike-type had a higher degree of haplotype sharing than between these partitions. The degree of haplotype sharing was lower within spike-type than within growth habit. There was a high frequency of shared haplotypes among two-row populations at 50 SNP windows, but this was reduced when 100 SNP windows were considered. At 100 SNP windows BAI2 and N2 presented the lowest degree of shared haplotypes with other two-rowed populations, while BA2 shared the most haplotypes. For both 50 and 100 SNP windows, high-frequency shared haplotypes were common for all six-rowed populations – see for example Figure 5A, except for the Utah populations). This contrasted with VT6, which showed a very low degree of haplotype sharing, particularly with the various spring breeding populations. Haplotype sharing for VT6 occurred primarily with the OR2 and OR6 populations, *i.e.*, with the only other populations with similar growth habit. However, OR2 and OR6 showed higher levels of allele frequency similarity with other breeding populations than VT6. These results are consistent with differentiation in allele frequency observed in F_{ST}comparisons between breeding populations. For 50 SNP windows, there were a high number of IBS segments shared between spike-types. However, when windows of 100 SNPs were considered, the majority of the IBS haplotypes were at low frequency (<20%) in one of the two populations compared. An exception to this was the comparison between OR2 and OR6, where shared haplotypes were quite common. Sharing occurred at every 100 SNP window, with shared haplotypes at frequencies as high as 90% in the two populations with an average frequency within populations of 0.46 and 0.58 (for OR2 and OR6, respectively). IBS with 50 SNP windows for N2 and N6 spanned 94% of all windows, with shared haplotypes occurring at average frequencies of 22% and 50%, respectively. Increasing the window size to 100 SNPs identified many fewer shared haplotypes between N2 and N6, leaving 70% of the genome shared at an average frequency of 8% in N2 and 35% in N6.

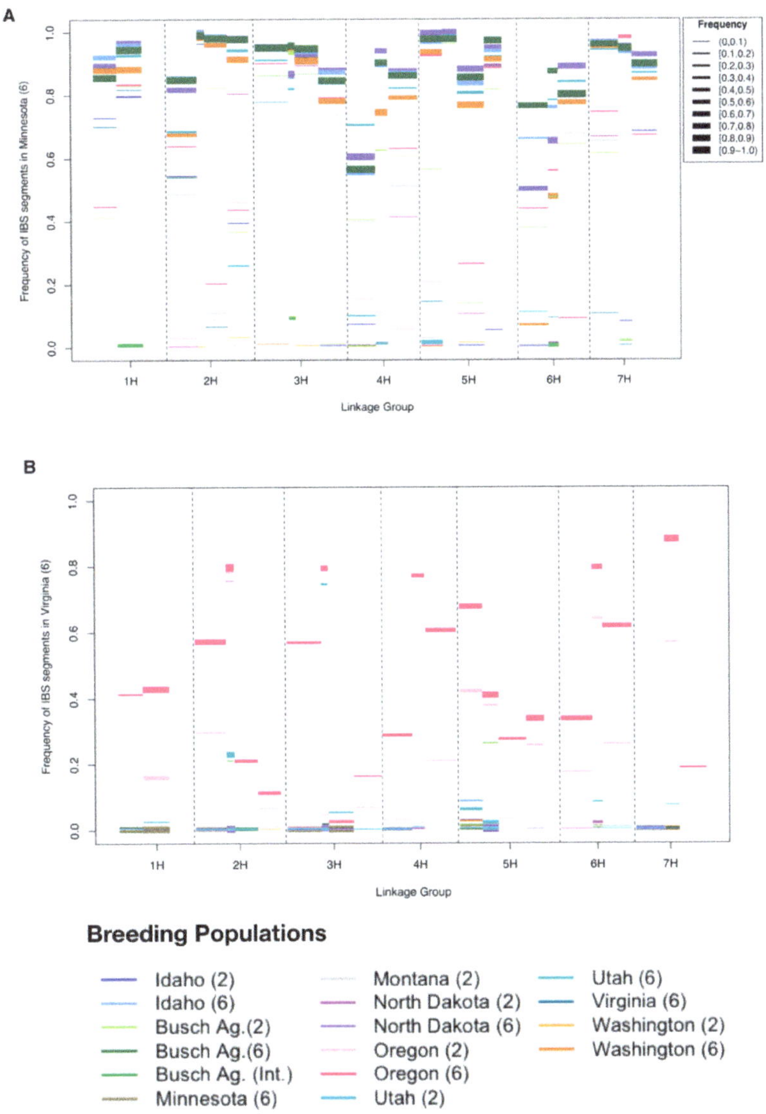

Figure 5: Frequency of IBS haplotypes in breeding populations from (A) University of Minnesota and (B) Virginia Polytechnic Institute and State University. Using 100 SNP windows. Linkage groups are in the X-axis. The Y-axis is the frequency of the haplotype in .

Maximum Likelihood tree of Relatedness and Migration

We determined that the tree topology that best describes the 16 populations separates the populations first by spike-type followed by growth habit, consistent with the F_{ST} and PCA results, with the exception of UT2 and UT6 that are not separated by spike-type. In the Tree Mix analysis, the two Utah and winter populations were more similar to ancestral allele frequencies, while MN6 was the most diverged.

Adding three migration events resulted in the lowest residuals and standard errors relative to trees with no migration, or one, two, four, or five migrations. With three migrations, we infer exchange between spring six-rowed populations and N2; VT6/OR6, and OR2, and spring six-rowed populations and UT6 (Figure 6).

Based on the f_4-test, six different topologies yielded significant results suggesting gene flow between UT6 and spring six-rowed populations, with z-scores significantly different from zero (significance at $P < 0.05$). The f_4-test for introgression between N2 and any spring six-rowed population resulted in values significantly different from zero, consistent with gene flow. The f_4-test did not support introgression between OR2 and VT6 and/or OR6 since none of the possible tree topologies involving these populations were significant.

DISCUSSION

Comparative analyses of allele frequency differentiation and extended haplotype sharing in a sample of 3613 barley accessions representing 16 barley breeding populations result in five primary conclusions: (i) barley breeding programs in North America have strongly differentiated allele frequencies among breeding populations; across programs, inflorescence type followed by growth habit account for the greatest proportion of variance in allele frequency; (ii) a number of loci previously identified as major contributors to growth habit adaptation in barley (primarily in Europe) are readily identifiable as outliers in allele frequency in our F_{ST} analyses of North American breeding populations; (iii) we identify putative signals of recent and long-term selection similar in magnitude to those at previously isolated genes of known function; (iv) the average per SNP allele frequency divergence at the population level appears consistent with the dominant action of genetic drift and linked selection; and (v) we identify populations that are recently subject to exchange of genetic material. There are low levels of genetic exchange across spike-type boundaries.

Identification of loci Putatively Subject to Recent and long-term Selection

Allele frequency differentiation sufficient to be detected in an F_{ST} outlier analysis is generally the result of long-term directional selection (Beaumont and Balding 2004). The SNPs identified as outliers for spring *vs.* winter growth habit include previously isolated genes of large effect, for example *Vrn-H1* and*Ppd-H1*, congruent with a previous F_{ST} comparison in European barleys (Comadran *et al.* 2012). Allele frequency differentiation between spike-type, which explains a larger portion of the allele frequency divergence in our sample, identified two SNPs in the genomic region of genes contributing to spike-type; however, SNPs within the well-characterized *Vrs1* locus were not identified as F_{ST} outliers (see below in the *Limitations of allele frequency comparisons* section). Many additional F_{ST} outliers are newly identified as putative targets of selection.

Despite a relatively recent introduction to North America, multiple genes found to contribute to agronomic traits in Eurasia appear to recapitulate allele frequencies in current, highly structured breeding programs. This appears to be particularly true when the phenotype is conferred by a single allele as opposed to an allelic series, as occurs at *Vrs1* where the six-rowed phenotype is conferred by multiple alleles (see Komatsuda *et al.* 2007). Perhaps more surprising is that many of the same allele frequency outliers are identified in European and North American samples despite very different demographic and breeding histories.

PHS analysis in the North American breeding programs detected a series of loci that are putatively targets of recent selection within barley breeding populations, comparable to findings in wheat (Cavanagh *et al.* 2013). The length of the haplotypes shared (average 81.38 cM) and their frequencies (average 25%) are consistent within recent generations, limiting the potential for recombination to break down the haplotypes. This analysis indicates that selection and linked selection are altering patterns of haplotype diversity across the genome. However, there are some chromosomes that have little or no evidence of selection in recent generations.

Limitations of Allele Frequency Comparisons

Despite having five SNPs positioned in the *Vrs1* gene (controlling the fertility of lateral spikelet) (Komatsuda and Mano 2002), including one SNP (12_30900) variant cosegregating with the six-rowed phenotype, the comparison of allele frequencies differentiation between spike-type fails to identify outliers in *Vrs1*. One important contributing factor may be that the six-rowed phenotype can result from multiple disruptions of the *Vrs1* gene (Komatsuda *et al.* 2007). The phenomenon of multiple mutational paths to the same phenotype has been identified as "functionally equivalent mutations" (Ralph and Coop 2010) and can result in multiple variants at modest frequency controlling a phenotype. The maintenance of alleles of large effect at modest frequency in the population reduces the power to detect phenotypic associations, in part because any given variant explains a small portion of phenotypic variation (Thornton *et al.* 2013).

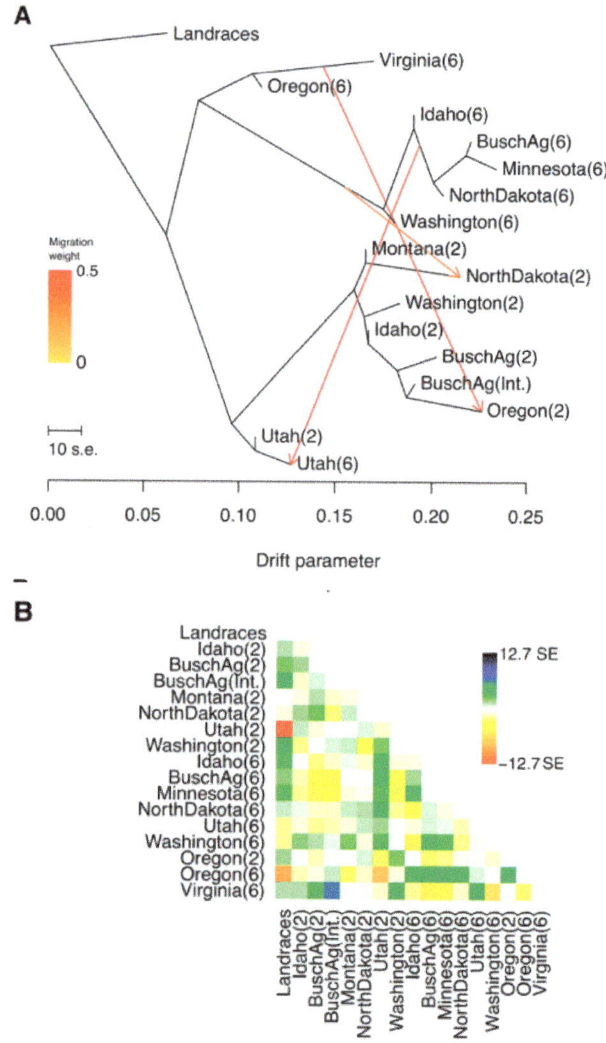

Figure 6: Tree of relatedness among North American breeding populations. (A) Plotted is the structure of the graph inferred by TreeMix for the 16 North American breeding populations, allowing three migration events. The arrows (migration events) are colored according ...

The BOPA SNP platform used here is largely derived from variants identified from cDNA libraries, from cultivated accessions (Close *et al.* 2009), and thus is highly enriched for SNPs in genic rich regions (Kono *et al.* 2014). The BOPA SNPs are 89% synonymous or noncoding and 11% are nonsynonymous (data not shown). However, SNPs genotyped within a locus can have limited correlation with a phenotype, depending on the haplotype on which they occur (see Nordborg and Tavaré (2002) for discussion of this issue). This may be the case for the previously cloned *Vrn-H3* where we see limited correlation between SNPs in this gene and allele frequency difference in the growth habit comparison. It should also be noted that the resequencing panel in which *Vrn-H3* alleles were found to be associated with the phenotype included only five spring and eight winter individuals (Yan *et al.* 2006). It is perhaps not surprising that sampling error (*i.e.*, small sample sizes) accounts for inconsistencies in the phenotype–genotype associations reported in previous studies and our study using a panel of >3000 accessions.

The effects of Linked Selection and Drift in Breeding Populations

The derived SFS in individual breeding populations identifies an excess of rare and high-frequency derived variants with respect to neutral model of a population at equilibrium. During selection, alleles linked to the target variant can be carried to high frequencies. Therefore, an excess of rare and high-frequency derived variants can be associated with the effects of linked selection (Fay and Wu 2000) as has been observed in a resequencing study of*Oryza sativa* (Caicedo *et al.* 2007). The PHS results presented here suggest considerable potential to detect the effect of linked selection, particularly when the selection was recent, impacting large genomic regions. One of the largest genomic regions detected as an outlier in the PHS analysis is found in OR2; this haplotype involves 138.9 cM on linkage group 2H, observed at 11% frequency. Individuals carrying this haplotype were derived from a two- by six-rowed cross (Merlin × Strider) and were selected to recover the two-rowed phenotype (P. Hayes, personal communication). This resulted in a long shared haplotype centered on the *Vrs1* gene contributing to spike-type (identified by SNP 12_30901). Two of the SNPs that permitted the identification of the outlier haplotype occur within an iron homeostasis (*nashor 2*) and drought-stress tolerance genes (*TaGolS 2*).

Our comparisons of allele frequency differences among individual barley breeding populations suggest that genetic drift (and likely linked selection) plays a major role in differentiation among these populations. One major factor that can reduce effective population size, and thus increase the role of drift in breeding populations, is the frequent reuse of popular cultivars as parents in breeding programs. Historically, this included founders such as Betzes, Hanna, Lion, Manchuria, and Trebi (Martin *et al.*1991) but more recently lines such as Bonanza, Cheri, and Chevron occur frequently in pedigrees (Condón *et al.* 2008).

The cumulative effects of selection and genetic drift over several generations of breeding may result in reduced response to selection (Hanrahan *et al.* 1973), with larger effects of drift when the population is small and/or highly inbred (Robertson 1960). The magnitude of drift suggests that the majority of barley breeding programs have small effective population sizes. Gerke *et al.* (2015) noted that genetic drift played a major role in changes in allele frequency over the history of maize breeding in North America. Although, in the larger partitions of barley populations, signals of directional selection are clearly evident at loci of large effect, drift may be extremely important to the loss of variation for variants of small effect (or for phenotypes determined by multiple genes), especially when the selective pressure is sporadic (*e.g.*, drought tolerance).

Gene Flow Among North American Breeding Populations

Gene flow between populations can be detected by the presence of large chromosomal regions in (admixture) linkage disequilibrium (Chakraborty and Smouse 1988; Chakraborty and Weiss 1988; Briscoe *et al.* 1994) or extended genomic regions of shared ancestry (Gusev *et al.* 2009, 2012). Shorter shared haplotypes indicate shared ancestry a larger number of generations before present (more generations of recombination). Considering the high frequency of IBS segments (and associated low F_{ST}) across the genome between populations of similar inflorescence type, we speculate that these segments could reflect a history of shared ancestry among these populations, possibly predating the separation of breeding programs (Martin *et al.* 1991).

Gene flow that involves adaptive variation can result in differential retention of genomic segments in the recipient population, and can also be evident as shared genomic regions with reduced diversity (Hufford *et al.* 2013). For example, MN6 demonstrates a high degree of IBS with other six-rowed populations (Figure 6A), and these genomic regions have lower average pairwise diversity than other regions when the six-rowed populations are analyzed together.

With regard to IBS, VT6 is the most clearly differentiated from other North American populations, showing similarity only to the other winter populations, OR2 and OR6 (for 50 SNP windows, Figure S10). Pairwise F_{ST} with other breeding populations and genetic differentiation as measured by PCA suggest that this isolation has been maintained over many generations. Large IBS segments (100 SNP windows) shared between VT6 and OR2 and OR6 (Figure 5B) indicate that winter populations have recently exchanged genetic material. The demographic history inferred in the TreeMix analysis supports this relationship (Figure 6), particularly when invoking three migration events, the topology best supported by the data. However, the formal examination of migration using the f_4-test failed to support gene flow among two- and six-rowed winter populations, thus suggesting that haplotype similarities across spike-types are due to ancestral history rather than recent introgression. In addition to the postulated shared ancestry, genetic similarity is expected between these populations under the premise that both Virginia and Oregon breeding populations have been subject to relative recent introgression from leaf rust resistance lines from the International Maize and Wheat Improvement Center (CIMMYT) (C. Griffey and P. Hayes, personal communication). Isolated populations, like VT6, can carry adaptive variants absent in other populations (Kristiansson *et al.* 2008). Therefore, VT6 could be used for the identification of new variants for yield or disease resistance, as well as a source of genetic variation to increase diversity in other breeding populations.

The large number of IBS segments shared between the Utah populations and most of the six-rowed populations yielded a significant signal of introgression based on the f_4-test. It is possible that the signal detected here is due to the ongoing genetic exchange between UT2 and UT6 (as documented in pedigrees) since 2001 (D. Hole, personal communication) and the genetic history shared between six-rowed barleys. This may also account for the elevated pairwise diversity in the Utah populations. Additionally, there is evidence of introgression between N2 and two six-rowed populations (WA6 and AB6), but not with N6. Introgression between the six-rowed populations and N2 are uncommon (J. Franckowiak, personal communication), thus this result needs more exploration.

Using the tree of relatedness as our hypothetical relationship, a necessary component of the f_4-test, forces populations of the same spike-type to be assumed as related by shared history. Therefore, gene flow within spike-types is not investigated. Although crosses across spike-type are not common in barley breeding programs due to concerns of recombining desirable alleles with less favorable ones, introgression across spike-type could increase diversity (Martin*et al.* 1991). We detect two instances of introgression across spike-type that result in higher levels of genetic diversity.

Caveats of the Analysis

It is important to note that the results presented here are dependent on the samples submitted for genotyping (https://triticeaetoolbox.org) by each breeding program. Given that one of the goals of the Barley Coordinated Agricultural Project (http://www.barleycap.org) was to evaluate the diversity in the North American barley breeding program, each program was encouraged to submit representative lines. We assume that the data represent the diversity within each breeding program and that the sampling scheme across breeding programs is comparable. Deviations from these assumptions could influence the results in four primary ways: (i) underestimation of the diversity within breeding populations; (ii) overestimation of the role of drift in breeding populations due to reduced representation of the parental lines used in the breeding programs; (iii) over- or underestimation of allele frequency differentiation between partitions of the data; and (iv) excess of shared haplotypes due to accessions highly related by pedigree (*e.g.*, sibs and half-sibs). With these caveats in mind, we made the best use of the dataset to uncover the underlying genetic response to breeder's efforts. We encourage readers of this paper and barley breeders to evaluate the results with an understanding of the limitations of the sampling.

Implications

Comparative population genetic approaches have the potential to uncover breeding history at a level of detail not previously possible. Our applications of allele frequency differentiation analyses in barley breeding populations are able to identify candidate genes or linked markers controlling major traits. Though the nature of the target of selection identified by individual SNPs is not always readily apparent, the identification of F_{ST} outliers reported here provides a reasonable first step toward the discovery of genes underlying agronomic adaptation or potential markers linked to these genes.

F_{ST} analyses also reveal the effects of linked selection and drift on breeding populations. Linkage has the potential to reduce the efficacy of selection by forcing selection to act on the net effect of linked variants (Hill and Robertson 1966; Robertson 1970). Furthermore, beneficial variants could be in linkage with deleterious mutations (Felsenstein 1974). Gains from selection in breeding programs will depend on the disassociation of unfavorable linkage (Morrell *et al.* 2012; Rodgers-Melnick *et al.* 2015). This can be achieved by increasing the effective amount of recombination. Hill and Robertson (1966) suggested that a generation of relaxed selection with a larger number of contributing parents between each generation of selection could increase the amount of recombination in the population, increasing the potential to disassociate these variants.

ACKNOWLEDGMENTS

The authors thank Michael Kantar, Ron Okagaki, and Paul Hoffman for helpful comments on an earlier version of the manuscript. We thank Kate Crosby, Justin Gerke, and Timothy Beissinger for thoughtful reviews that improved the final text. We thank Yaniv Brandvain for helpful suggestions on data analysis. We thank Dr. David Hole, Dr. Patrick Hayes, and Dr. Jerome Franckowiak for sharing information about their breeding programs. This work was performed using computing resources at the University of Minnesota Supercomputing Institute. We acknowledge funding from the US Department of Agriculture National Institute for Food and Agriculture USDA NIFA 2011- 68002-30029 (to P.L.M.), Barley Coordinated Agricultural Project: Leveraging Genomics, Genetics, and Breeding for Gene Discovery and Barley Improvement USDA-CSREES-NRI grant 2006-55606-16722 and USDA-NIFA grant 2009-85606-0570. We also acknowledge funding from University of Minnesota Doctoral Dissertation Fellowship (to A.M.P., T.J.Y.K., and Z.F.).

Footnotes

Supporting information is available online atwww.g3journal.org/lookup/suppl/doi:10.1534/g3.115.024349/-/DC1

Communicating editor: J. Ross-Ibarra

REFERENCES

1. Beaumont M. A., Balding D. J., 2004. Identifying adaptive genetic divergence among populations from genome scans. Mol. Ecol. 13(4): 969–980.
2. Boyd W. J. R., Li C. D., Grime C. R., Cakir M., Potipibool S., et al. , 2003. Conventional and molecular genetic analysis of factors contributing to variation in the timing of heading among spring barley (*Hordeum vulgare L.*) genotypes grown over a mild winter growing season. Crop Pasture Sci.54(12): 1277–1301.
3. Briscoe D., Stephens J. C., O'brien S. J., 1994. Linkage disequilibrium in admixed populations: applications in gene mapping. J. Hered. 85(1): 59–63.
4. Caicedo A. L., Williamson S. H., Hernandez R. D., Boyko A., Fledel-Alon A., et al. , 2007. Genome-wide patterns of nucleotide polymorphism in domesticated rice. PLoS Genet. 3(9): 1745–1756.
5. Cavalli-Sforza L. L., 1966. Population structure and human evolution. Proc. R. Soc. Lond. B Biol. Sci. 164(995): 362–379.
6. Cavanagh C. R., Chao S., Wang S., Huang B. E., Stephen S., et al. , 2013. Genome-wide comparative diversity uncovers multiple targets of selection for improvement in hexaploid wheat landraces and cultivars. Proc. Natl. Acad. Sci. USA 110(20): 8057–8062.
7. Chakraborty R., Smouse P. E., 1988. Recombination of haplotypes leads to biased estimates of admixture proportions in human populations. Proc. Natl. Acad. Sci. USA 85(9): 3071–3074.
8. Chakraborty R., Weiss K. M., 1988. Admixture as a tool for finding linked genes and detecting that difference from allelic association between loci.Proc. Natl. Acad. Sci. USA 85(23): 9119–9123.
9. Chang C. C., Chow C. C., Tellier L. C., Vattikuti S., Purcell S. M., et al. , 2015. Second-generation

PLINK: rising to the challenge of larger and richer datasets. Gigascience 4: 7 DOI: .10.1186/s13742-015-0047-8

10. Close T. J., Bhat P. R., Lonardi S., Wu Y., Rostoks N., et al. , 2009. Development and implementation of high-throughput SNP genotyping in barley. BMC Genomics 10: 582 DOI: 10.1186/1471-2164-10-582.

11. Comadran J., Russell J. R., Booth A., Pswarayi A., Ceccarelli S., et al. , 2011. Mixed model association scans of multi-environmental trial data reveal major loci controlling yield and yield related traits in *Hordeum vulgare* in Mediterranean environments. Theor. Appl. Genet. 122(7): 1363–1373.

12. Comadran J., Kilian B., Russell J., Ramsay L., Stein N., et al. , 2012. Natural variation in a homolog of *Antirrhinum* CENTRORADIALIS contributed to spring growth habit and environmental adaptation in cultivated barley. Nat. Genet. 44(12): 1388–1392.

13. Condón F., Gustus C., Rasmusson D. C., Smith K. P., 2008. Effect of advanced cycle breeding on genetic diversity in barley breeding germplasm.Crop Sci. 48(3): 1027–1036.

14. Cuesta-Marcos A., Szucs P., Close T. J., Filichkin T., Muehlbauer G. J., et al. , 2010. Genome-wide SNPs and re-sequencing of growth habit and inflorescence genes in barley: implications for association mapping in germplasm arrays varying in size and structure. BMC Genomics 11: 707 DOI: 10.1186/1471-2164-11-707.

15. Fang Z., Eule-Nashoba A., Powers C., Kono T. Y., Takuno S., et al. , 2013. Comparative analyses identify the contributions of exotic donors to disease resistance in a barley experimental population. G3 (Bethesda) 3(11): 1945–1953.

16. Fang Z., Gonzales A. M., Clegg M. T., Smith K. P., Muehlbauer G. J., et al. , 2014. Two genomic regions contribute disproportionately to geographic differentiation in wild barley. G3 (Bethesda) 4(7): 1193–1203.

17. Fay J. C., Wu C. I., 2000. Hitchhiking under positive Darwinian selection.Genetics 155(3): 1405–1413.

18. Felsenstein J., 1974. The evolutionary advantage of recombination. Genetics78(2): 737–756.

19. Gerke J., Edwards J., Ke G., Ross-Ibarra J., Mcmullen M. D., 2015. The genomic impacts of drift and selection for hybrid performance in maize.Genetics 201(3): 1201–1211.

20. Goudet J., 2005. Hierfstat, a package for R to compute and test hierarchical F statistics. Mol. Ecol. Notes 5(1): 184–186.

21. Gusev A., Lowe J. K., Stoffel M., Daly M. J., Altshuler D., et al. , 2009. Whole population, genome-wide mapping of hidden relatedness. Genome Res. 19(2): 318–326.

22. Gusev A., Palamara P. F., Aponte G., Zhuang Z., Darvasi A., et al. , 2012. The architecture of long-range haplotypes shared within and across populations. Mol. Biol. Evol. 29(2): 473–486.

23. Haake V., Cook D., Riechmann J. L., Pineda O., Thomashow M. F., et al. , 2002. Transcription factor *CBF4* is a regulator of drought adaptation in Arabidopsis. Plant Physiol. 130(2): 639–648.

24. Hamblin M. T., Close T. J., Bhat P. R., Chao S., Kling J. G., et al. , 2010. Population structure and linkage disequilibrium in US barley germplasm: implications for association mapping. Crop Sci. 50(2): 556–566.

25. Hanrahan J. P., Eisen E. J., Lagates J. E., 1973. Effects of population size and selection intensity of short-term response to selection for postweaning gain in mice. Genetics 73(3): 513–530.

26. Herbik A., Koch G., Mock H. P., Dushkov D., Czihal A., et al. , 1999. Isolation, characterization and cDNA cloning of nicotianamine synthase from barley. A key enzyme for iron homeostasis in plants. Eur. J. Biochem.265(1): 231–239.

27. Hill W. G., Robertson A., 1966. The effect of linkage on limits to artificial selection. Genet. Res. 8(3): 269–294.

28. Horton M. W., Hancock A. M., Huang Y. S., Toomajian C., Atwell S., et al. , 2012. Genome-wide patterns of genetic variation in worldwide Arabidopsis thaliana accessions from the RegMap panel. Nat. Genet. 44(2): 212–216.

29. Hudson R. R., Bailey K., Skarecky D., Kwiatowski J., Ayala F. J., 1994. Evidence for positive selection in the superoxide dismutase (*Sod*) region of *Drosophila melanogaster*. Genetics 136(4): 1329–1340.

30. Hufford M. B., Xu X., Van Heerwaarden J., Pyhäjärvi T., Chia J. M., et al. , 2012. Comparative population genomics of maize domestication and improvement. Nat. Genet. 44(7): 808–811.

31. Hufford M. B., Lubinksy P., Pyhajarvi T., Devengenzo M. T., Ellstrand N. C., et al. , 2013. The genomic signature of crop-wild introgression in maize.PLoS Genet. 9(5): e1003477.

32. Innan H., Zhang K., Marjoram P., Tavaré S., Rosenberg N. A., 2005. Statistical tests of the coalescent model based on the haplotype frequency distribution and the number of segregating sites. Genetics 169(3): 1763–1777.

33. Jones H., Leigh F. J., Mackay I., Bower M. A., Smith L. M. J., et al. , 2008. Population-based resequencing reveals that the flowering time adaptation of cultivated barley originated east of the Fertile Crescent. Mol. Biol. Evol.25(10): 2211–2219.

34. Karsai I., Szucs P., Koszegi B., Hayes P. M., Casas A., et al. , 2008. Effects of photo and thermo cycles on flowering time in barley: a genetical phenomics approach. J. Exp. Bot. 59(10): 2707–2715.

35. Kearse M., Moir R., Wilson A., Stones-Havas S., Cheung M., et al. , 2012. Geneious Basic: an integrated and extendable desktop software platform for the organization and analysis of sequence data. Bioinformatics 28(12): 1647–1649.

36. Keinan A., Mullikin J. C., Patterson N., Reich D., 2007. Measurement of the human allele frequency spectrum demonstrates greater genetic drift in East

Asians than in Europeans. Nat. Genet. 39(10): 1251–1255.

37. Kimura M., 1983. The Neutral Theory of Molecular Evolution. Cambridge University Press, Cambridge, United Kingdom.

38. Komatsuda T., Mano Y., 2002. Molecular mapping of the intermedium spike-c (*int-c*) and non-brittle rachis 1 (*btr1*) loci in barley (*Hordeum vulgare L.*).Theor. Appl. Genet. 105(1): 85–90.

39. Komatsuda T., Pourkheirandish M., He C., Azhaguvel P., Kanamori H., et al. , 2007. Six-rowed barley originated from a mutation in a homeodomain-leucine zipper I-class homeobox gene. Proc. Natl. Acad. Sci. USA 104(4): 1424–1429.

40. Kono T. J. Y., Seth K., Poland J. A., 2014. SNPMeta: SNP annotation and SNP Metadata collection without a reference genome. Mol. Ecol. Resour.14(2): 419–425.

41. Kristiansson K., Naukkarinen J., Peltonen L., 2008. Isolated populations and complex disease gene identification. Genome Biol. 9(8): 109.

42. Laurie D. A., Pratchett N., Snape J. W., Bezant J. H., 1995. RFLP mapping of five major genes and eight quantitative trait loci controlling flowering time in a winter× spring barley (*Hordeum vulgare L.*) cross. Genome 38(3): 575–585.

43. Lewontin R. C., Krakauer J., 1973. Distribution of gene frequency as a test of the theory of the selective neutrality of polymorphisms. Genetics 74(1): 175–195.

44. Lister D. L., Thaw S., Bower M. A., Jones H., Charles M. P., et al. , 2009. Latitudinal variation in a photoperiod response gene in European barley: insight into the dynamics of agricultural spread from 'historic' specimens. J. Archaeol. Sci. 36(4): 1092–1098.

45. Marchini J., 2013. *Statistical and Population Genetics*. CRAN; Available at:http://www.stats.ox.ac.uk/~marchini/software.html.

46. Martin J. M., Blake T. K., Hockett E. A., 1991. Diversity among North American spring barley cultivars based on coefficients of parentage. Crop Sci. 31(5): 1131–1137.

47. Mayer K. F., Waugh R., Langridge P., Close T. J., Wise R. P., et al. , 2012. A physical, genetic and functional sequence assembly of the barley genome. Nature 491(7456): 711–716.

48. Morrell P. L., Buckler E. S., Ross-Ibarra J., 2012. Crop genomics: advances and applications. Nat. Rev. Genet. 13(2): 85–96.

49. Muñoz-Amatriaín M., Moscou M. J., Bhat P. R., Svensson J. T., Bartoš J., et al. , 2011. An improved consensus linkage map of barley based on flow-sorted chromosomes and single nucleotide polymorphism markers. The Plant Genome Journal 4(3): 238.

50. Nicholson G., Smith A. V., Jónsson F., Gústafsson Ó., Stefánsson K., et al. , 2002. Assessing population differentiation and isolation from single-nucleotide polymorphism data. J. R. Stat. Soc. Series B Stat. Methodol.64(4): 695–715.

51. Nielsen R., Hellmann I., Hubisz M., Bustamante C., Clark A. G., 2007. Recent and ongoing selection in the human genome. Nat. Rev. Genet.8(11): 857–868.

52. Nordborg M., Tavaré S., 2002. Linkage disequilibrium: what history has to tell us. Trends Genet. 18(2): 83–90.

53. Patterson N., Price A. L., Reich D., 2006. Population structure and eigenanalysis. PLoS Genet. 2(12): e190.

54. Pickrell J. K., Pritchard J. K., 2012. Inference of population splits and mixtures from genome-wide allele frequency data. PLoS Genet. 8(11): e1002967.

55. Poets A. M., Fang Z., Clegg M. T., Morrell P., 2015. Barley landraces are characterized by geographically heterogeneous genomic origins. Genome Biol. 16(1): 173.

56. Pritchard J. K., Pickrell J. K., Coop G., 2010. The genetics of human adaptation: hard sweeps, soft sweeps, and polygenic adaptation. Curr. Biol.20(4): R208–R215.

57. R Core Team, editor. , 2014. R: A language and environment for statistical computing. R Foundation for Statistical Computing, Vienna, Austria: Available at: http://www.R-project.org/.

58. Ralph P., Coop G., 2010. Parallel adaptation: one or many waves of advance of an advantageous allele? Genetics 186(2): 647–668.

59. Ramsay L., Comadran J., Druka A., Marshall D. F., Thomas W. T., et al. , 2011. *INTERMEDIUM-C*, a modifier of lateral spikelet fertility in barley, is an ortholog of the maize domestication gene *TEOSINTE BRANCHED 1*. Nat. Genet. 43(2): 169–172.

60. Reich D., Thangaraj K., Patterson N., Price A. L., Singh L., 2009. Reconstructing Indian population history. Nature 461(7263): 489–494.

61. Robertson A., 1960. A theory of limits in artificial selection. Proc. R. Soc. Lond. B Biol. Sci. 153(951): 234–249.

62. Robertson, A., 1970 A theory of limits in artificial selection with many linked loci, pp. 246–288 in *Mathematical Topics in Population Genetics*, edited by K.- Kojima. Springer, Berlin-New York.

63. Rodgers-Melnick E., Bradbury P. J., Elshire R. J., Glaubitz J. C., Acharya C. B., et al. , 2015. Recombination in diverse maize is stable, predictable, and associated with genetic load. Proc. Natl. Acad. Sci. USA 112(12): 3823–3828.

64. Ross-Ibarra J., Morrell P. L., Gaut B. S., 2007. Plant domestication, a unique opportunity to identify the genetic basis of adaptation. Proc. Natl. Acad. Sci. USA 104: 8641–8648.

65. Scheet P., Stephens M., 2006. A fast and flexible statistical model for large-scale population genotype data: applications to inferring missing genotypes and haplotypic phase. Am. J. Hum. Genet. 78(4): 629–644.

66. Skinner J. S., Von Zitzewitz J., Szucs P., Marquez-Cedillo L., Filichkin T., et al. , 2005. Structural, functional, and phylogenetic characterization of a large CBF gene family in barley. Plant Mol. Biol. 59(4): 533–551.

67. Slatkin M., 1987. Gene flow and the geographic structure of natural populations. Science 236(4803): 787–792.

68. Taji T., Ohsumi C., Iuchi S., Seki M., Kasuga M., et al. , 2002. Important roles of drought- and cold-inducible genes for galactinol synthase in stress tolerance in Arabidopsis thaliana. Plant J. 29(4): 417–426.
69. Thornton K., 2003. libsequence: a C++ class library for evolutionary genetic analysis. Bioinformatics 19(17): 2325–2327.
70. Thornton K. R., Foran A. J., Long A. D., 2013. Properties and modeling of GWAS when complex disease risk is due to non-complementing, deleterious mutations in genes of large effect. PLoS Genet. 9(2): e1003258.
71. Toomajian C., Hu T. T., Aranzana M. J., Lister C., Tang C., et al. , 2006. A nonparametric test reveals selection for rapid flowering in the *Arabidopsis*genome. PLoS Biol. 4(5): e137.
72. Turck F., Fornara F., Coupland G., 2008. Regulation and identity of florigen: FLOWERING LOCUS T moves center stage. Annu. Rev. Plant Biol. 59: 573–594.
73. Turner A., Beales J., Faure S., Dunford R. P., Laurie D. A., 2005. The pseudo-response regulator *Ppd-H1* provides adaptation to photoperiod in barley. Science 310(5750): 1031–1034.
74. van Heerwaarden J., Hufford M. B., Ross-Ibarra J., 2012. Historical genomics of North American maize. Proc. Natl. Acad. Sci. USA 109(31): 12420–12425.
75. von Zitzewitz J., Szucs P., Dubcovsky J., Yan L., Francia E., et al. , 2005. Molecular and structural characterization of barley vernalization genes.Plant Mol. Biol. 59(3): 449–467.
76. Wang H., Smith K. P., Combs E., Blake T., Horsley R. D., et al. , 2012. Effect of population size and unbalanced data sets on QTL detection using genome-wide association mapping in barley breeding germplasm. Theor. Appl. Genet. 124(1): 111–124.
77. Weaver J. C., 1943. Barley in the United States: a historical sketch. Geogr. Rev. 33(1): 56–73.
78. Weaver J. C., 1950. American Barley Production: A Study in Agricultural Geography. Burgess Pub. Co., Minneapolis, MN.
79. Weir B. S., Cockerham C. C., 1984. Estimating F-statistics for the analysis of population structure. Evolution 38(6): 1358–1370.
80. Wiebe, G. A., and D. A. Reid, 1961 Classification of barley varieties growing in the United States and Canada in 1958. US Department of Agriculture 210.
81. Yan L., Fu D., Li C., Blechl A., Tranquilli G., et al. , 2006. The wheat and barley vernalization gene *VRN3* is an orthologue of *FT*. Proc. Natl. Acad. Sci. USA 103(51): 19581–19586.
82. Youssef H. M., Koppolu R., Schnurbusch T., 2012. Re-sequencing of vrs1 and int-c loci shows that labile barleys (*Hordeum vulgare convar. labile*) have a six-rowed genetic background. Genet. Resour. Crop Evol. 59(7): 1319–1328.
83. Zhou H., Muehlbauer G., Steffenson B. N., 2012. Population structure and linkage disequilibrium in elite barley breeding germplasm from the United States. J. Zhejiang Univ. Sci. B 13(6): 438–451.

CHAPTER 3

CHANGES IN PHYSIOLOGICAL AND AGRONOMICAL PARAMETERS OF BARLEY (HORDEUM VULGARE) EXPOSED TO CERIUM AND TITANIUM DIOXIDE NANOPARTICLES

Luca Marchiol, Alessandro Mattiello1, Filip Pošćić,
Guido Fellet, Costanza Zavalloni, Elvio Carlino, and Rita Musetti

Keywords: Cerium oxide nanoparticles, titanium oxide nanoparticles, barley, plant growth, food chain

INTRODUCTION

The useful properties of engineered nanoscale materials (ENMs) have resulted in the rapid development of nanotechnologies and large-scale production of nanoparticles or nanoparticles-containing products [1]. The increasing use of ENMs may result in the rise in the flux of ENMs discharged into the environment. Water bodies and soil are assumed to be the primary environmental recipients of nanomaterials [2].

Recent estimates included cerium oxide nanoparticles ($nCeO_2$) and titanium oxide nanoparticles ($nTiO_2$) among the 10 most commonly produced ENMs that are used worldwide. In the cosmetic industry, solar cells, paints, cements and coatings about 10,000 t of $nTiO_2$ are used per year [3]. A number of applications of $nTiO_2$ are in use in the food industry and agriculture, serving as nano-sensors and nano-agents for new delivery systems of plant protection products and fertilizers [4,5]. Also, $nCeO_2$ have a broad range of industrial application as additives in glass and ceramics, fuel-cell materials and the automotive industry [6]. On the other hand, $nCeO_2$ and $nTiO_2$ are both included in the list of ENMs for immediate priority testing by the Organization for Economic Cooperation and Development [7]. Vascular plants should be of particular concern as they interact closely with the environment and are conduits for bioaccumulation through the food chain [8,9,10,11].

Even though this subject is of primary importance, to date relatively few studies have been carried out on the responses of crops exposed to metal nanoparticles. Most of the currently available papers have reported data collected from experiments performed in hydroponic conditions [6,7,8,9,10,11,12,13]. Such approaches are not able to account for the complexity of the soil-plant system [14].

Early studies concerning the relationships between plants and $nCeO_2$ were mostly focused on the initial developmental stages of plants. López-Moreno *et al.* [12] observed seed germination and root elongation in cucumber, tomato, alfalfa and corn exposed to 0–4000 mg·L^{-1} $nCeO_2$ with contradictory results. Ma *et al.* [15] verified that the root growth of cabbage, cucumber, radish, rape, tomato, and wheat were not affected by 2000 mg L^{-1} $nCeO_2$. At the same concentration, the germination of soybean was undisturbed but indications of genotoxicity were reported by López-Moreno *et al.* [16]. The same authors demonstrated significant species-specific differential levels of plant uptake and translocation of Ce in plants exposed to 4000 mg L^{-1} of $nCeO_2$. Differences in root microstructures (e.g., pore size in root hairs) and physical and chemical interactions between $nCeO_2$ and root exudates in the rhizosphere could explain the differences [17]. The movements of $nCeO_2$ within the root tissues of kidney bean to the aerial tissues were verified by combining ICP-OES and μ-XANES analysis [18].

Citation: Marchiol, L.; Mattiello, A.; Pošćić, F.; Fellet, G.; Zavalloni, C.; Carlino, E.; Musetti, R. Changes in Physiological and Agronomical Parameters of Barley (Hordeum vulgare) Exposed to Cerium and Titanium Dioxide Nanoparticles. Int. J. Environ. Res. Public Health 2016, 13, 332. Doi:10.3390/ijerph13030332.

More recently it was reported $n\mathrm{CeO_2}$ induced compositional modifications in the root xylem in seedlings of rice, wheat and barley [19]. Finally, a life cycle study on barley grown in soil amended with 125–500 $\mu g \cdot g^{-1} n\mathrm{CeO_2}$, reported both beneficial and harmful effects of nanoceria [20].

The information on the effects of Ti nano-forms are controversial because several papers have demonstrated positive [21,22,23,24,25] or negative [26,27,28] effects of $n\mathrm{TiO_2}$ on plants. Recently, Frazier *et al.* [29] reported that in plantlets exposed to $n\mathrm{TiO_2}$ (range 1000–25,000 ppm) for three weeks the leaf count, root length and plant biomass significantly increased as Ti concentrations were raised. In contrast, Pakrashi *et al.* [30] observed a dose-dependent decrease in the mitotic index and an increase in the number of chromosomal aberrations in root tips of *Allium cepa* exposed to 12.5–100 mg $n\mathrm{TiO_2} \cdot mL^{-1}$.

Modelling studies have predicted that ENMs released to the environment are likely to be mostly found in water, sediments, and soils [31,32]. Considering the increasing speed of nanotechnology development it is plausible to assume that different ENMS might be present simultaneously in the environmental compartments (water, sediments, soil and biota). Therefore, living organisms could be exposed to a co-occurrence of EMNs. However, this issue is still poorly explored in literature and we are still lacking systematic and reliable information [33]. Moreover, most studies hitherto have only evaluated crop plants to the germination stages, and have not examined the complete developmental cycle.

Currently, nanotechnology is considered as an important tool in agriculture with the potential to provide new strategies to improve crop production for human consumption and animal feeding and promoting a reduction in the use of pesticides [34]. Since several scientific evidences suggest that nanomaterials may induce harmful environmental effects, it is crucial to investigate on the impact of nanomaterials on crops. In our study barley was considered as model crop since it is among the world's most important cereal crops [35]. In the present study, barley plants were grown for the whole crop cycle in a soil enriched with different levels of $n\mathrm{CeO_2}$ and $n\mathrm{TiO_2}$ and their combination in a fully factorial design. To the authors' knowledge, this is the first study that reports data on plants exposed simultaneously to different metal nanoparticles. We hypothesized that (i) the exposure of barley to $n\mathrm{CeO_2}$ and $n\mathrm{TiO_2}$ (also combined each other) would influence plant growth; that (ii) nanoparticles would influence plant physiology and that (iii) the concentration of $n\mathrm{CeO_2}$ and $n\mathrm{TiO_2}$ would affect the uptake of Ce and Ti in roots and the translocation of such elements in the vegetative plant fractions and in seeds.

EXPERIMENTAL SECTION

Characterization of $n\mathrm{CeO_2}$ *and* $n\mathrm{TiO_2}$

Cerium (IV) oxide nanopowder and titanium (IV) oxide anatase nanopowder both having a nominal average particle size of 25 nm were purchased from Sigma-Aldrich (St. Louis, MO, USA). Particle characterization was carried out at the Facility for Environmental Nanoscience Analysis and Characterization (FENAC), University of Birmingham (UK).

The specific surface area of the $n\mathrm{CeO_2}$ and $n\mathrm{TiO_2}$ powders was measured by the Brunauer–Emmett–Teller (BET) method by using the Surface Area and Pore Size Analyser SA 3100 plus (Beckman Coulter, Brea, CA, USA). The samples were outgassed at 300 °C for 180 min and the nitrogen adsorption-desorption isotherms were recorded at liquid nitrogen temperature (77K). The BET values were 46.1 $m^2 \cdot g^{-1}$ and 61.6 $m^2 \cdot g^{-1}$ for $n\mathrm{CeO_2}$ and $n\mathrm{TiO_2}$ respectively.

The size distribution of the $n\mathrm{CeO_2}$ and $n\mathrm{TiO_2}$ powders were measured by Atomic Force Microscopy (AFM) method using PSIA XE100 (Park System, Suwon, Korea). The samples were prepared by spreading the $n\mathrm{CeO_2}$ and $n\mathrm{TiO_2}$ powder over a mica sheet pre-treated with poly-L-lysine (Sigma Aldrich). The average height was obtained by measuring at least 100 nanoparticles in non-contact mode. The average height of the $n\mathrm{CeO_2}$ and $n\mathrm{TiO_2}$ powder were 32.6 ± 20.7 nm and 41.8 ± 24.3 nm, respectively. Subsequently, the $n\mathrm{CeO_2}$ and $n\mathrm{TiO_2}$ powders were suspended in deionized water at a concentration of 1000 ppm, sonicated, and simultaneously heated at 60 °C for 30 min. The suspensions were characterized for z-average size, measured as hydrodynamic diameter, and zeta potential, via electrophoretic mobility, by dynamic light scattering (DLS) method using the Nano ZS90 (Malvern Instruments, Malvern, UK). The z-average size of $n\mathrm{CeO_2}$ and $n\mathrm{TiO_2}$ powder was 174 $\pm$ 1.19 and 925 $\pm$ 105 nm, respectively. The zeta potential of the $n\mathrm{CeO_2}$ and $n\mathrm{TiO_2}$ powders were respectively 0.027 ± 0.064 mV and 19.9 ± 0.55 mV. Finally, the $n\mathrm{CeO_2}$ and $n\mathrm{TiO_2}$ powders suspension were characterized also with the Differential Centrifugal Sedimentation (DCS) method using CPS DC24000 UHR (Analytik, Swavesey, UK). The average sizes were 188 $\pm$ 5.9 nm and 690 $\pm$ 17.5 nm, respectively for $n\mathrm{CeO_2}$ and $n\mathrm{TiO_2}$.

Addition of Nanoparticles to Soil

The soil used for this study was collected in Udine, Italy (46°04′52″ N, 13°12′33″ E; top 20 cm) air dried at room temperature and sieved through a 2 mm mesh prior to characterization. The soil was classified as clay soil (sand 26%, silt 6.4% and clay 67.6%) with pH 7.4, cation exchange capacity (CEC) of 13.9 ($cmol \cdot kg^{-1}$ DM), electrical conductivity (EC) of 1235 ($\mu S \cdot m^{-1}$) and organic matter (OM) content of 4.4%.

Eight mixtures of soil and $n\mathrm{CeO_2}$ and $n\mathrm{TiO_2}$ were prepared following the procedure used by Priester *et al.* [36]. The main treatments were made by adding $n\mathrm{CeO_2}$ and $n\mathrm{TiO_2}$ powders directly to the soil and mixing it in a portable concrete mixer previously sealed, to obtain a concentration of 2000 $mg \cdot kg^{-1}$ of either $n\mathrm{CeO_2}$ and $n\mathrm{TiO_2}$. The $n\mathrm{CeO_2}/n\mathrm{TiO_2}$ amended soils were stored in the dark at 10 °C for two weeks. After soil equilibration the final doses of 500 and 1000 $mg \cdot kg^{-1}$, respectively (Ce 500, Ce 1000, Ti 500, Ti 1000) were prepared by serial dilution with soil. Four additional treatments were obtained by combining the stock soils to achieve the following combinations: $n\mathrm{CeO_2}$ 500 $mg \cdot kg^{-1}/n\mathrm{TiO_2}$ 500 $mg \cdot kg^{-1}$ (Ce 500-Ti 500),$n\mathrm{CeO_2}$ 500 $mg \cdot kg^{-1}/n\mathrm{TiO_2}$ 1000 $mg \cdot kg^{-1}$ (Ce 500-Ti 1000), $n\mathrm{CeO_2}$ 1000 $mg \cdot kg^{-1}/n\mathrm{TiO_2}$ 500 $mg \cdot kg^{-1}$ (Ce 1000-Ti 500) and$n\mathrm{CeO_2}$ 1000 $mg \cdot kg^{-1}/n\mathrm{TiO_2}$ 1000 $mg \cdot kg^{-1}$ (Ce 1000-Ti 1000). After another soil equilibration round of three days, five microcosms (4 L polyethylene pots) per treatment were filled with the control and nanoparticles amended soil (n = 45). The control treatment received no nanoparticle amendment.

Plant Growth and Harvest

Eight seeds of spring barley (*Hordeum vulgare* L., *cv.* Tunika) obtained from S.I.S Società Italiana Sementi (San Lazzaro di Savena, Bologna, Italy) were sown in microcosms containing the nCeO$_2$/nTiO$_2$ amended soils. The trial was carried out in a semi-sealed greenhouse under full sunlight. Two weeks after seed planting, the seedlings were thinned to four seedlings per microcosm. At tillering, two plants per pot were removed, therefore 90 plants were observed during the experiment (ten plants per treatment). During the growth period the microcosms were irrigated to maintain the soil at 60% of water holding capacity (WHC). During the barley growth cycle the microcosms were singularly weighed and irrigated to compensate for evapotranspiration. Phenological stages were monitored by adapting the Decimal Growth Scale [37] throughout the growth cycle and were based on 50% of plants within the treatments at each stage. Plants were harvested at physiological maturity. Prior to collecting plants, plant height was measured from soil surface to the flag leaf using a standard meter stick (1 m). Then plant shoots were severed at the collar with a razor blade and then separated into stems, leaves, spikes, and grains. Leaf area was measured using a LI-3100C Area Meter (Li-Cor Corporation, Lincoln, NE, USA). Plant samples were thoroughly washed in tap water and rinsed three times with distilled water. In addition, roots were washed in 400 mL of 0.01 M of nitric acid in a shaker bath at 300 rpm for 5 min to remove metal ions adsorbed at the surface. The plant fractions were oven dried at 105 °C for 24 h and weighed.

Gas Exchange Parameters

The photosynthetic rate at saturating light intensity (A_{max}, μmol $CO_2 \cdot m^{-2} \cdot s^{-1}$), transpiration rate ($T_r$, mmol $H_2O \cdot m^{-2} \cdot s^{-1}$), and stomatal conductance (g_s, mol air $m^{-2} \cdot s^{-1}$) were measured with a portable gas exchange system (LI-6400, LI-COR, Inc. Lincoln, NE, USA). The gas exchanges measurements were carried out on the flag leaf at booting, heading and milk maturity on three individual plants per treatment. The measurements were made after allowing leaves to reach steady-state conditions at saturating photosynthetic active radiation (PAR, 1200 μmol·$m^{-2} \cdot s^{-1}$), at a CO_2 concentration of 400 ppm, and at a temperature of 25 °C, and were collected between 9 and 14 h with evaluation of five measurement periods at intervals of 7–8 days.

TEM Observations

A small leaf portion (2 × 3 mm) was excised close to the central vein of the youngest leaf before the emergence of flag leaves. The fresh samples were fixed for 2 h at 4 °C in 0.1% (wt/vol) buffered sodium phosphate and 3% (wt/vol) glutaraldehyde at pH 7.2. They were then post-fixed with 1% osmium tetroxide (wt/vol) in the same buffer for 2 h, dehydrated in an ethanol series, and embedded in Epon/Araldite epoxy resin (Electron Microscopy Sciences, Fort Washington, PA, USA). Serial ultrathin sections from each sample were cut with a diamond knife, mounted 100/200 folding grids, and then observed under a Philips CM 10 transmission electron microscope (TEM, FEI, Eindhoven, The Netherlands) operating at 80 kV.

Spectroscopy Analysis

Samples of soils were oven-dried (105 °C for 48 h) and digested in 11 mL of a 10 to 1 (*v*/*v*) mixture of 96% (*v*/*v*) sulphuric acid and 30% (*v*/*v*) hydrogen peroxide in Teflon cylinders for 20 min at 200 °C in a microwave oven (MARS Xpress, CEM, Matthews, NC, USA). After digestion, samples were diluted 1 to 20 with milliQ water, filtered through 0.45 μm filters and Ce and Ti were determined with an ICP-OES (Vista MPX, Varian Inc., Palo Alto, CA, USA). Oven-dried plant fractions were digested in 10 mL of a 1 to 4 (*v*/*v*) mixture of 65% (*v*/*v*) nitric acid and 30% (*v*/*v*) hydrogen peroxide in Teflon cylinders for 10 min at 175 °C in microwave oven (CEM MARS Xpress). The plant extracts were filtered with Whatman® PTFE membrane filters (0.45 μm), diluted, and analysed. Total Ce and Ti in roots, stems, and leaves were determined by an ICP-OES using yttrium as internal standard. The Ce and Ti contents in kernels were quantified using an ICP-MS (Aurora M90, Bruker, Bremen, Germany) with an internal standard solution of ^{72}Ge and ^{89}Y. Quality control for both ICP-OES and ICP-MS was carried out using reagent blank samples, and triplicates reading for each sample. Certified standard reference material (NIST 1573a Tomato leaves) was analysed to validate the protocol.

TEM X-ray Microanalysis

Crystal structure and chemistry of the nanoparticles were studied by using a 2010F UHR TEM/STEM (JEOL, Peabody, MA, USA) equipped with a low spherical aberration coefficient (Cs = 0.47 ± 0.01 mm) objective pole piece and energy dispersive X-ray spectrometer (EDXS). The experiments were performed at an accelerating voltage of 200 kV corresponding to an electron wavelength of 2.5 pm. The EDXS spectra were acquired in scanning transmission electron microscopy (STEM) configuration by rasterizing an electron probe of 0.5 nm within the area of interest imaged by a high angle annular dark field (HAADF) detector to accurately determine the chemical assessment of the investigated nanoparticles. Nanodiffraction were acquired in TEM by illuminating the area of interest with a 50 nm parallel electron probe to study the crystal features of individual nanoparticles [38].

Data Analysis

The microcosm study was arranged in a completely randomized factorial design with nine treatments (control soil, two levels of nCeO$_2$, two levels of nTiO$_2$ and four nCeO$_2$/nTiO$_2$ mixtures) and five replicates. Data were tested for homoscedasticity and normality using the Bartlett's test and the Shapiro–Wilk test, respectively. The differences were statistically significant, as determined by one-way and two-way analysis of variance (ANOVA). Tukey's Multiple Comparison test (p = 0.05) in case of significant effects were used to analyse individual effects. Statistical analysis was performed using the SPSS program (ver. 16, SPSS Inc., Chicago, IL, USA).

RESULTS

Phenology and Growth of Barley

A week after sowing, all the plants had germinated, apparently unaffected by the presence of metal oxide nanoparticles in the soil and without early symptoms of phytotoxicity in treated plants. From the 2nd leaf stage we observed that $n\mathrm{CeO_2}$ and$n\mathrm{TiO_2}$ treated plants had a longer vegetative period than the controls (Figure 1).

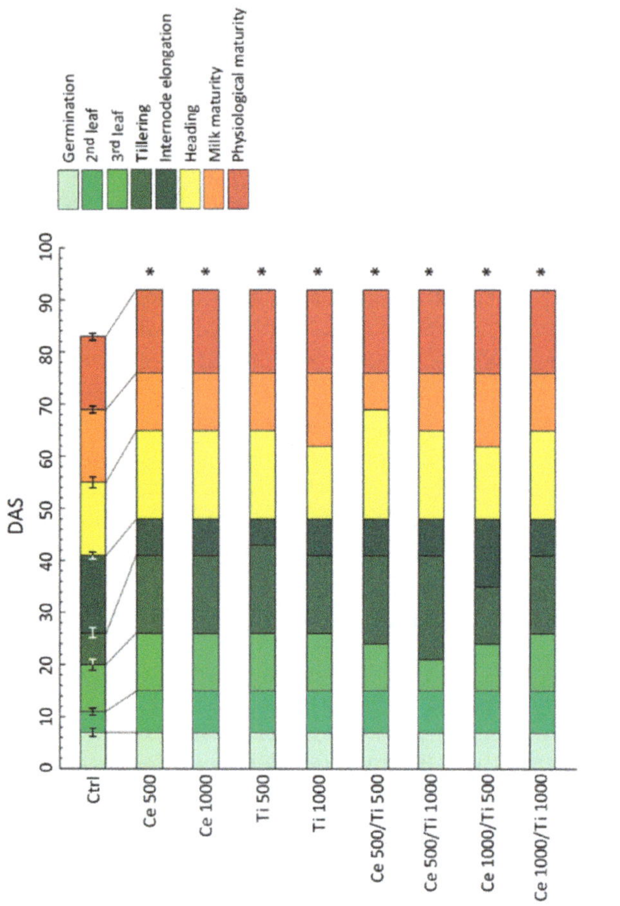

Figure 1: Cumulative contribution of vegetative and reproductive phenophases to the phenology of plants of barley grown in control soil and $n\mathrm{CeO_2}$ and $n\mathrm{TiO_2}$-amended soil. DAS = days after sowing. Error bars represent ± standard error. * denote significant differences with respect to control ($p \leq 0.05$).

This delay was also verified at the 3rd leaf and tillering stages, even if it was less pronounced for the Ce 500/Ti 1000 treatment. The delay in the nanoparticle-treated plants reached its maximum extent (about ten days) at heading and during the ripening stages (Figure 1). Milk maturity was reached on average 65 days after sowing (DAS) in treated plants. Ti 1000 and Ce 1000/Ti 500 plants were the earliest (62 DAS) and the Ce 500/Ti 500 ones were the latest (69 DAS) at entering physiological maturity (Figure 1).

A two-factor analysis of variance (ANOVA) was performed on the following variables: plant height, number of tillers, leaf area per plant and number of spikes. Among the morphological traits that we considered, plant height was the least sensitive to the $n\mathrm{CeO_2}$ and $n\mathrm{TiO_2}$ treatments. The interaction $n\mathrm{CeO_2} \times n\mathrm{TiO_2}$ was not significant (Figure 2A). The plants treated with $n\mathrm{TiO_2}$ were significantly taller than the others were ($p = 0.0035$), whereas there was no statistically significant effect of $n\mathrm{CeO_2}$.

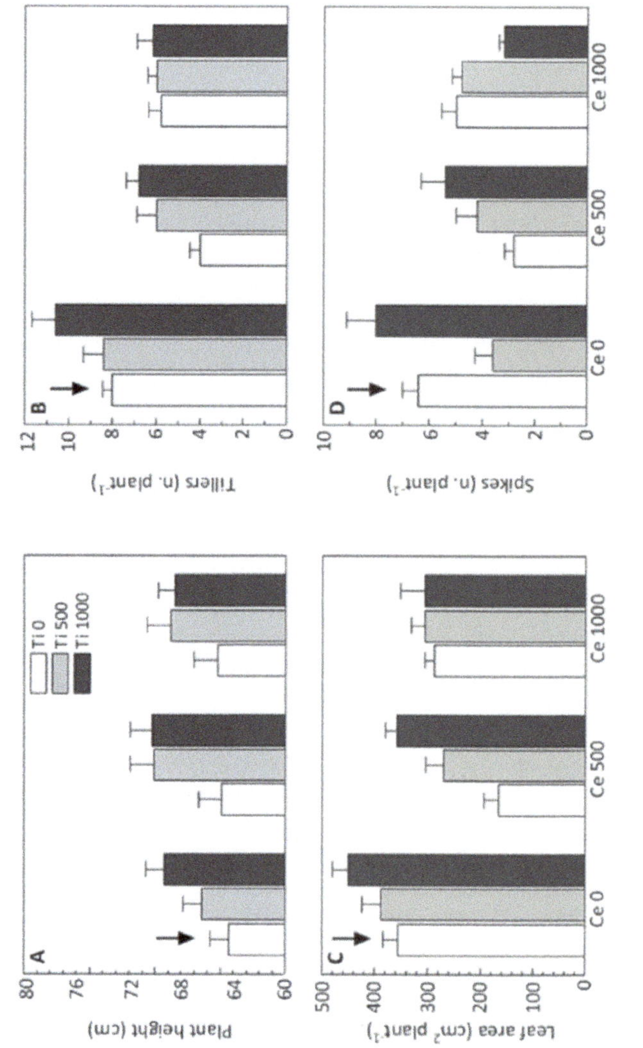

Figure 2: Plant height (**A**); number of tillers (**B**); leaf area (**C**) and number of spikes (**D**) observed in plants of barley grown in control soil and $n\mathrm{CeO_2}$ and $n\mathrm{TiO_2}$-amended soil. Bars are mean standard error ($n = 5$). Arrows indicate the control.

The formation of secondary shoots was significantly influenced, but in an opposite way, by the experimental factors. The interaction effect between $n\mathrm{CeO_2}$ and $n\mathrm{TiO_2}$ on tiller's formation was not statistically significant. The number of tillers per plant was significantly stimulated by $n\mathrm{TiO_2}$, compared to the control ($p = 0.009$). In fact, in Ce 0 plants $n\mathrm{TiO_2}$ 1000 promoted, on average, the formation of 2.60 secondary shoots more (+25%) than the control plants ($p = 0.032$); a more pronounced effect (2.80 tillers more than controls) was recorded in Ce 500 plants ($p = 0.026$) (Figure 2B). In contrast, $n\mathrm{CeO_2}$ had a statistically significant ($p < 0.001$) negative effect on tiller formation (Figure 2B). In particular, on average in Ce 500 and Ce 1000 plants the number of secondary shoots was about 35% lower than both control plants ($p = 0.002$) and Ti treated ones (respectively $p = 0.034$, and $p < 0.001$).

The number of tillers and the plants' total leaf area are closely linked. In fact, the ANOVA showed a significant effect of both $n\mathrm{CeO_2}$ ($p < 0.001$) and $n\mathrm{TiO_2}$ ($p = 0.001$). The interaction $n\mathrm{CeO_2} \times n\mathrm{TiO_2}$ was not significant. Multiple comparisons were run for each simple main effect. As noted earlier, in absence of Ti, Ce 500 had a strong negative effect on plant vegetative growth, in fact the average leaf area per plant was about one-half that of the control (166 *vs.* 356 $\mathrm{cm^2}$ of leaf surface per plant) ($p < 0.001$) (Figure 2C). Such a negative influence of $n\text{CeO}_2$ in Ce 500 plants appeared to have been mitigated by $n\text{TiO}_2$which had a positive significant effect ($p < 0.001$) (Figure 2C). In regards to the plant's response to $n\text{CeO}_2$/$n\text{TiO}_2$ mixtures, although the interaction was not statistically significant ($p = 0.091$), we assumed that $n\text{TiO}_2$ at respectively 500 and 1000 mg per kg of soil was able to remediate the adverse impact on leaf growth of $n\text{CeO}_2$ (Figure 2C).

One of the main yield components in barley is the number of spikes per plant. Two-way ANOVA revealed a statistically significant effect of the main factor $n\text{CeO}_2$ ($p = 0.0116$) and a significant interaction between $n\text{CeO}_2$ and $n\text{TiO}_2$ ($p = 0.0016$). The effect $n\text{TiO}_2$ in this case was not statistically supported, being likely hidden by data variability (Figure 2D). The mean number of spikes in Ce 500 and Ce 1000 plants was 4.2 corresponding to a reduction of 38% than the controls (2D). Regarding interaction, multiple comparisons of the means showed the different main effects of $n\text{CeO}_2$ and $n\text{TiO}_2$. In absence of $n\text{TiO}_2$ (Ti 0), the number of spikes per plant were negatively affected by $n\text{CeO}_2$. There was a statistically significant difference ($p = 0.003$) between Ce 0 and Ce 500 plants (6.4 and 2.8 spikes per plant, respectively; −56%). The negative influence of $n\text{CeO}_2$ was confirmed at the highest dose (Ce 1000) even though it was lighter in magnitude (22% lower than control plants) and not statistically different from Ce 500 (Figure 2D). The same effects were verified also in Ti 1000 plants, where both the levels of $n\text{CeO}_2$ determined a reduction of spike number of 32.5% and 60% respectively for Ce 500 ($p = 0.033$) and Ce 1000 ($p < 0.001$). The intermediate dose of $n\text{TiO}_2$ (500 mg·kg^{-1}) negatively affected the number of spikes in plants (−43.7% than controls, $p = 0.020$). On the contrary, Ti 1000 plants had 8 spikes each (higher that control plants although not statistically significant). In other words, in terms of spike formation our data suggest that, in absence of $n\text{CeO}_2$, the higher dose of $n\text{TiO}_2$ was, at least, harmless to plants. This evidence is in contrast with other growth parameters. Moreover, we detected a negative $n\text{TiO}_2$ dose-effect on plant tillers' number and leaf area (Figure 2B–D).

Gas Exchanges

The leaf photosynthetic rate (A_{max}), stomatal conductance (g_s) and transpiration rate (T_r) at three different growth stages (booting, heading, milk maturity) are shown in Table 1. Because the treatments affected plant development by causing a shift in phenological stages, the gas exchange parameters were evaluated by comparing plants using the same phenological phase.

Table 1: ANOVA p value for the main factors ($n\text{CeO}_2$, $n\text{TiO}_2$) and interaction ($n\text{CeO}_2 \times n\text{TiO}_2$) for leaf photosynthetic rate at saturating light intensity (Amax, $\mu\text{mol CO}_2\text{·m}^{-2}\text{·s}^{-1}$), stomatal conductance (g_s, mol air·$\text{m}^{-2}\text{·s}^{-1}$) and transpiration rate (Tr, mmol $\text{H}_2\text{O·m}^{-2}\text{·s}^{-1}$) recorded at three different phenological stages (booting, heading and milk maturity) in leaves of barley growing in control soil and $n\text{CeO}_2$ and $n\text{TiO}_2$-amended soil. Values are mean ± SE ($n = 5$). Same letters indicated no statistical difference between treatments at Tukey's test. *** $p < 0.001$; ** $p < 0.05$; ns = not significant ($p = 0.05$)

Treatment	Booting			Heading			Milk Maturity		
	Amax	g_s	Tr	Amax	g_s	Tr	Amax	g_s	Tr
$n\text{CeO}_2$	0.0003 ***	0.0335	0.0047 **	ns	ns	ns	ns	ns	ns
$n\text{TiO}_2$	0.0003 ***	0.0105 **	0.0105 **	ns	ns	ns	ns	ns	ns
$n\text{CeO}_2 \times n\text{TiO}_2$	ns	ns	ns	ns	ns	ns	ns	ns	ns
Ctrl	20.4 ± 1.8 b	0.278 ± 0.05 ab	3.06 ± 0.64 b	19.4 ± 1.8 a	0.350 ± 0.06 a	3.24 ± 0.29 a	15.3 ± 1.6 a	0.298 ± 0.05 a	2.95 ± 0.30 a
Ce 500	25.7 ± 1.0 a	0.390 ± 00.5 a	5.36 ± 0.71 a	21.2 ± 1.3 a	0.254 ± 0.02 a	4.33 ± 0.58 a	14.4 ± 1.6 a	0.282 ± 0.04 a	2.53 ± 0.24 a
Ce 1000	19.4 ± 0.9 b	0.249 ± 0.03 b	3.23 ± 0.51 b	19.4 ± 1.7 a	0.220 ± 0.02 a	4.12 ± 0.33 a	17.8 ± 1.1 a	0.349 ± 0.05 a	3.15 ± 0.32 a
Ctrl	17.5 ± 1.8 b	0.205 ± 0.06 b	2.49 ± 0.26 b	21.6 ± 1.4 a	0.339 ± 0.04 a	3.79 ± 0.36 a	16.4 ± 1.3 a	0.229 ± 0.03 a	2.29 ± 0.21 a
Ti 500	22.8 ± 1.5 ab	0.287 ± 00.3 ab	3.73 ± 0.34 ab	18.5 ± 1.6 a	0.229 ± 0.03 a	4.04 ± 0.43 a	15.7 ± 1.4 a	0.327 ± 0.05 a	3.05 ± 0.29 a
Ti 1000	23.9 ± 1.0 a	0.387 ± 0.05 a	3.87 ± 0.45 a	18.6 ± 1.9 a	0.230 ± 0.03 a	3.64 ± 0.62 a	16.1 ± 1.8 a	0.357 ± 0.05 a	3.16 ± 0.30 a

Both $n\text{CeO}_2$ and $n\text{TiO}_2$ treatments had a statistically significant effect on the photosynthetic parameters, whereas their interaction was not significant at any of the growth stages. Table 1 reports the data regarding the main factors. At the booting phase, A_{max} and T_r were positively affected by Ce 500 compared to control plants (respectively +26% and +75%). However, Ce 1000 plants behaved like the controls suggesting that the maximum concentration for a beneficial effect from $n\text{CeO}_2$ had been exceeded (Table 1). As expected, in the subsequent phases of the life cycle of barley A_{max} declined, with no evidence of statistically significant differences between treatments and control. At booting Ti 1000 had an overall positive effect compared to the control plants: A_{max}, gs, and T_r significantly increased by 37, 89, and 92%, respectively. The Ti 500 treatment had an intermediate effect (Table 1).

Plant Uptake and Accumulation of Cerium and Titanium

Table 2 and Table 3 present the concentrations of Ce and Ti in the roots, stems, leaves, and kernels of barley. In general, the concentration of Ce and Ti in the plant tissues showed a dose-response.

Table 2: Ce concentration observed in roots, stems, leaves and kernels of barley grown in control soil and $n\text{CeO}_2$ and $n\text{TiO}_2$-amended soil. Values are mean ± SE. ($n = 5$). Same letters indicated no statistical difference between treatments at Tukey's test ($p \leq 0.05$)

Treatment	Ce Roots (mg·kg^{-1})	Ce Stems (mg·kg^{-1})	Ce Leaves (mg·kg^{-1})	Ce Kernels ($\mu\text{g·kg}^{-1}$)
Ctrl	3.30 ± 0.63 d	0.64 ± 0.12 c	0.73 ± 0.12 c	0.50 ± 0.19 a
Ce 500	45.3 ± 11.6 cd	1.38 ± 0.21 abc	1.62 ± 0.11 bc	0.87 ± 0.57 a
Ce 1000	96.9 ± 1.42 bc	1.75 ± 0.29 ab	3.03 ± 0.46 a	0.69 ± 0.36 a
Ti 500	14.0 ± 0.97 d	0.86 ± 0.29 bc	0.78 ± 0.11 c	0.98 ± 0.59 a
Ti 1000	19.1 ± 1.28 d	0.81 ± 0.21 bc	0.84 ± 0.09 c	1.22 ± 0.73 a
Ce 500/Ti 500	58 ± 9.61 cd	1.39 ± 0.24 abc	1.42 ± 0.15 bc	0.34 ± 0.12 a
Ce 500/Ti 1000	87 ± 5.60 bc	1.69 ± 0.23 ab	1.50 ± 0.08 bc	1.13 ± 0.52 a
Ce 1000/Ti 500	149 ± 19.4 ab	1.98 ± 0.19 a	1.76 ± 0.14 b	0.75 ± 0.41 a
Ce 1000/Ti 1000	164 ± 32 a	2.01 ± 0.17 a	1.79 ± 0.17 b	0.03 ± 0.01 a

Table 3: Ti concentration observed in roots, stems, leaves and kernels of barley grown in control soil and $nCeO_2$ and $nTiO_2$-amended soil. Values are mean ± S.E. ($n = 5$). Same letters indicated no statistical difference between treatments at Tukey's test ($p \leq 0.05$). d.l. = detection limit

Treatment	Ti Roots (mg· kg^{-1})	Ti Stems (mg· kg^{-1})	Ti Leaves (mg· kg^{-1})	Ti Kernels (μg· kg^{-1})
Ctrl	77 ± 3.19 a	0.26 ± 0.043 a	1.03 ± 0.06 a	1.39 ± 1.39 a
Ce 500	66.5 ± 5.15 a	0.19 ± 0.02 a	1.27 ± 0.32 a	0.48 ± 0.48 a
Ce 1000	63.9 ± 2.63 a	0.31 ± 0.02 a	1.31 ± 0.28 a	0.26 ± 0.18 a
Ti 500	66.7 ± 7.49 a	0.28 ± 0.03 a	1.39 ± 0.35 a	0.71 ± 0.61 a
Ti 1000	81.7 ± 4.96 a	0.39 ± 0.06 a	0.96 ± 0.09 a	<d.l.
Ce 500/Ti 500	63.9 ± 4.56 a	0.20 ± 0.09 a	1.33 ± 0.31 a	3.62 ± 2.60 a
Ce 500/Ti 1000	59.4 ± 7 a	0.22 ± 0.16 a	1.15 ± 0.27 a	8.14 ± 4.99 a
Ce 1000/Ti 500	69.1 ± 7.92 a	0.22 ± 0.07 a	1.07 ± 0.22 a	1.34 ± 1.3 a
Ce 1000/Ti 1000	68.4 ± 5.41 a	0.19 ± 0.04 a	0.89 ± 0.11 a	< d.l.

A statistically significant dose-response in Ce concentration was recorded in all plant fractions with the exception of kernels. In Ce 500 and Ce 1000 plants, the mean levels of Ce in the roots were 45.3 mg·kg^{-1} and 96.9 mg·kg^{-1}, respectively (Table 2). The significant dose-dependent response in root Ce accumulation was confirmed also in $nCeO_2/nTiO_2$ treatments.

Ce concentration in stems was significantly different between Ce 500 and Ce 1000 plants both in the case where Ce was individually supplied or when it was associated with the Ti. The Ce root to shoot translocation percentage in treated plants ranged between 1.24 and 9.1, respectively, for Ce 1000 and Ti 500, indicating that Ce accumulation in the aboveground plant fractions occurred at very low magnitude. The highest Ce concentration in leaves was recorded in Ce 1000 plants (3.03 mg·kg^{-1}). As expected, in the leaves of Ctrl plants and Ti 500 and Ti 1000 ones, lower Ce accumulation values (0.73, 0.77 and 0.84 mg·kg^{-1}, respectively) were observed.

Despite the increase in concentration of Ti in the soil ($p = 0.0001$), due to the addition of $nTiO_2$, the Ti uptake and accumulation in the plant fractions did not respond to the treatment. No statistically significant differences in Ti concentration in roots, as well as in stems and leaves, were observed (Table 3). As for Ce, in the case of Ti the root-to-shoot translocation percentage was very low, ranging between 1.42 and 1.91.

Barley grains are used as food for humans and animals, as well as for other applications (e.g., malting and flour). Thus it is appropriate to examine whether nanoparticles are able to reach the kernels during ripening (Table 2 and Table 3). No statistically significant differences among the treatments were recorded. In absolute values, the content of both elements in the kernels was three-four orders of magnitude lower than those recorded in plant leaves (Ce 0.718 μg·kg^{-1} and Ti 1.77 μg·kg^{-1}, respectively in Table 2 and Table 3).

Ultrastructural Analyses

To verify the root uptake and subsequent translocations of $nCeO_2$ and $nTiO_2$ from roots to aerial plant fractions, ultrastructural analyses on plant leaf tissues were carried out. Nanoparticles were not present in untreated control leaf tissues, which presented well preserved ultrastructure and organelles (Figure 3A; Figure 4A; Figure 5A). Rare clusters of nanoparticles were found in leaves sampled from plants grown in soil enriched with the different combinations of $nCeO_2$and $nTiO_2$, at both concentrations (Figure 3).

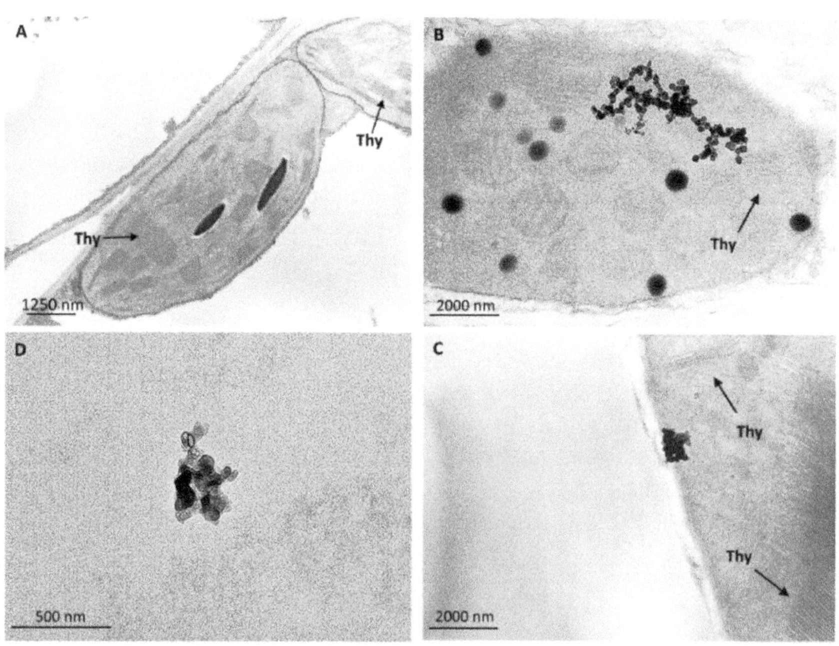

Figure 3: Representative TEM micrographs of leaf tissues from barley plants. In control untreated leaf tissues (**A**) nanoparticles are absent and chloroplast ultrastructure is well preserved. In plants grown in $nCeO_2$ and $nTiO_2$-amended soil clusters of nanoparticles are visible in the stroma of the chloroplasts (**B**, Ce 1000; **C**, Ti 1000) and in the vacuoles of parenchymal cells (**D**, Ce 1000). Chloroplast structure seems not affected by nanoparticle treatment (arrows in **A**, **B** and **C** indicate thylakoids).

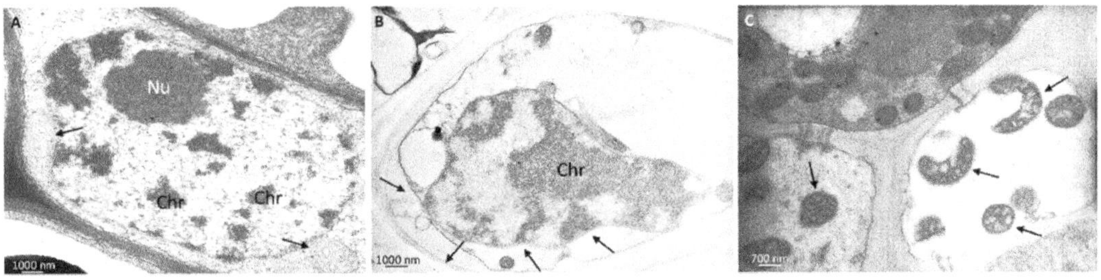

Figure 4: Representative TEM micrographs of leaf tissues from barley plants. (A) In control untreated vascular parenchyma cells mitochondria (arrows) and nuclei appear well preserved with regular shape and intact membranes. Nucleolus (Nu) is recognizable and chromatin (Chr) is normally dispersed. In plants grown in nCeO$_2$and nTiO$_2$-amended soil vascular parenchyma cell has detached plasma membrane (arrows) and nucleus presents lobed shape and condensed chromatin (Chr) (B, Ce 1000). Mitochondria have disorganized, swollen christae (arrows in C, Ce 1000).

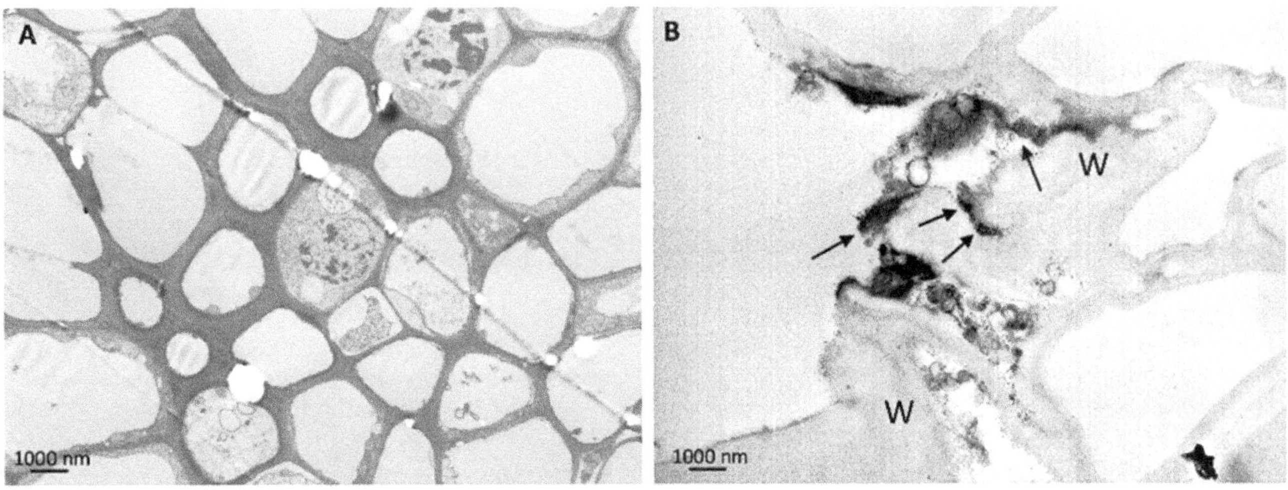

Figure 5. Representative TEM micrographs from secondary vein tissues of barley plants. (A) Control untreated vein tissues appear well structured, cell walls are regular in shape and thickness; (B) Ti 1000: secondary veins showed cells with contorted cell walls (W) associated with little dark precipitates (arrows).

Nanoparticles were observed in parenchyma leaf tissues, in the stroma of the chloroplast and in the vacuoles, (Figure 3B–D). Despite the presence of nCeO$_2$ and nTiO$_2$, the chloroplasts appeared normal, with preserved ultrastructure, and, in general, the cell compartments of the chlorophyll parenchyma did not appear affected by the treatments (Figure 3B). This evidence was in agreement with phenotypical/morphological analyses, as we did not observe macroscopic cell death at the tissue level after the nCeO$_2$ and nTiO$_2$ treatments. Nevertheless, at the vascular tissue, some ultrastructural modifications were visible, especially those affecting the cellular organelles: some nuclei showed condensed chromatin, mitochondria swollen cristae (Figure 4B,C). Only in Ce 1000 and Ti 1000 leaf tissues, few secondary veins showed irregular-shaped cells with contorted walls (Figure 5B).

Nanostructures in Leaf Tissues

After verifying the root-to-leaves translocation of Ce and Ti, STEM EDXS observations were carried out to detect the presence of nCeO$_2$ and nTiO$_2$ within the leaf tissues. Regarding nCeO$_2$, several nanostructure aggregates were observed, with sizes ranging from a few nanometers to some hundreds of nanometers. Interestingly, nanodiffraction measurements revealed an amorphous structure in most of cases. The compositional analysis by EDXS reported in Figure 6 did not show Ce in such aggregates so Ce was unlikely to be present in the form of nanoclusters within the leaf tissues of barley.

In Figure 6A, a brightfield image shows a typical aggregate of clusters with large size dispersion. No Bragg's peaks were observed in the relevant nanodiffraction and the EDXS spectrum showed no evidence of Ce characteristic X-rays, but only those of light elements (Figure 6B). Even though we lack direct experimental evidence, it is very likely that the small amount of Ce in leaves (Table 3) is aggregated in ionic form to organic molecules. It is also worthwhile mentioning that during extensive TEM/STEM sessions, isolated crystalline nCeO$_2$ were observed in only two cases. Therefore, we conclude that$n$CeO$_2$ induces a massive formation of amorphous clusters of light elements rather than nanometer-scale clusters of CeO$_2$. Such aggregates were abundant in the leaves of Ce 1000 plants but absent in control ones (Figure 5).

Several aggregates of nanoparticles were observed in the leaves of plants exposed to nTiO$_2$. In Figure 6C–E, a representative result of the chemical and structural analysis performed on a Ti aggregate is shown. Figure 6D shows a brightfield image of several dark aggregates. The nanodiffraction pattern acquired demonstrates the crystalline nature of the aggregates (Figure 6D). The measured lattice spacing is compatible with TiO$_2$ crystals. Indeed, the diffraction intensities belong to TiO_2 nanocrystallites with different orientation with respect to the primary electron beam. As an example, the arrow in the pattern points to the systematic reflection of a TiO_2 anatase particle oriented close to the [312] zone axis (Figure 6D). To identify the chemical signature of the element contained in the nanostructures, the EDXS spectrum was acquired in the same area of the nanodiffraction, and the emission of characteristic fluorescence X-rays of Ti atoms was recorded (Figure 6E). In conclusion, we report the presence of TiO_2 nanoclusters in parenchyma leaf tissues—particularly inside the chloroplasts—of barley plants exposed to $n$$TiO_2$.

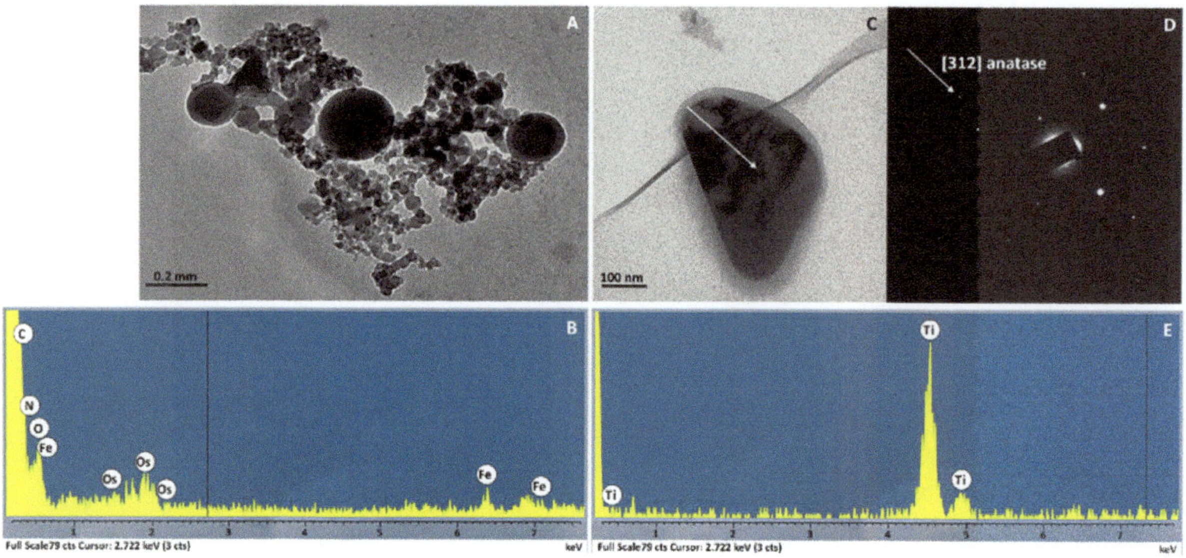

Figure 6: (A) Brightfield TEM image of representative nanostructures observed in Ce 1000 leaves; (B) EDXS spectrum as acquired in the particle area shown in brightfield mode; (C) Brightfield TEM image of Ti nanostructures in Ti 1000 leaves; (D) Nanodiffraction pattern as acquired on individual nanostructures indicated by the arrow in the brightfield image (see text); (E) EDXS spectrum as acquired on the region indicated in the brightfield image.

DISCUSSION

Model simulations have demonstrated that flows of ENMs are currently able to reach natural ecosystems [39]. For this reason, questions are rising about the consequences of the interaction of ENMs with biota. Plant phenological traits and growth parameters always respond to the physical environment, so they can be used to assess the adaptive behaviours of the plants and evaluate their relationships to the ecosystem where they grow.

It was reported that ENMs have the potential to influence the growth of some crops with enhancing or inhibitory effects on plant growth according to their composition, size and physical and chemical properties [10,36,40].

Studying the phenological stages of barley plants, according to the findings of Rico *et al.* [20] and Yoon *et al.* [40], we verified that $n$$CeO_2$ and $n$$TiO_2$ treated plants had a longer vegetative period than the controls. This fact *per se* may not be negative. In fact, a longer vegetative phase may promote higher biomass and grain yield because plants are allowed to produce more photosynthetically active leaves and therefore more photosynthates [41].

Our data showed that $n$$TiO_2$ were associated with effects opposite to those induced by $n$$CeO_2$. First, we observed an$n$$TiO_2$ dose-response effect on vegetative growth. Second, the compensation of the adverse effects of $n$$CeO_2$ observed in plants grown in $n$$CeO_2$/$n$$TiO_2$-treated soils was probably due to the beneficial effects of $n$$TiO_2$ on plant metabolism. Several evidences obtained in different experimental conditions support this hypothesis. Studies carried out on *Spinacia oleracea*have demonstrated that $n$$TiO_2$ promote plant photosynthesis increasing light absorbance and transformation of light energy [42] and enhancing Rubisco activity [22]. Also, it was demonstrated that $n$$TiO_2$, could significantly improve CO_2 fixation by plants, where it enhances absorption and transmission of the solar energy into the chain electron transport in chloroplasts [43]. Our data support such evidences, even though we cannot provide a physiological explanation. On the other hand, working with *Cucurbita pepo* grown in soil containing 400–800 $mg{\cdot}kg^{-1}$ of $n$$CeO_2$, Zhao *et al.* [44] did not observe any differences in the photosynthetic rate between treated plants and controls. Such different responses suggest that, at least at the intermediate level of concentration, different interactions with metabolism occur in various species exposed to metal nanoparticles. However, we still lack a systematic study on the effects on basic metabolism of plants induced by different nanomaterials and different levels of exposition.

At the end of the plant life cycle we studied a number of biometric variables observing some differences between treated plants and the controls. Recently, observations made on plants of barley grown for the entire life-cycle in a soil amended with 125–250–500 $mg{\cdot}kg^{-1}$ $n$$CeO_2$ were reported [20]. Thus, we can compare part of our data with those presented in that paper. In general, our results regarding the effects induced by $n$$CeO_2$ on plant growth are in contrast with Rico *et al.* [20]. One of the most relevant differences was that in our case all the plants were able to reach the reproductive stage, whereas Rico *et al.* [19] did not observe the formation of spikes in $n$$CeO_2$ 500 $mg{\cdot}kg^{-1}$ treated plants. Moreover, they observed that$n$$CeO_2$ 500 $mg{\cdot}kg^{-1}$ increased the height of barley plants and the accumulation of dry matter in the shoots, whereas in our case the vegetative growth was not stimulated by any level of $nCeO_2$. If we consider studies carried out on other crop species, our data partially agree with the findings of Priester *et al.* [36] in *Glycine max*, whereas Zhao *et al.* [44] did not observe statistically significant differences in biometric parameters in plants of *Cucumis sativum* between control plants and those treated with $nCeO_2$.

With regard to Ce and Ti accumulation in the plant tissues, it must be emphasized that both DLS and DCS analyses, carried out for the higher $nCeO_2$ and $nTiO_2$ test suspension (1000 ppm), indicate agglomeration of nanoparticles. Therefore, the actual bioavailability of nanomaterials may be lower than expected. For this reason, we believe that the concentrations of Ce and Ti in the plant tissues were underestimated. In addition, similarly to what happens in the case of conventional contaminants, soil pH, OM, texture and CEC have an important influence on the fate of nanomaterials in soil, [45,46] and particularly with respect to those, which release ions, such as $nCeO_2$ [47].

It was demonstrated that after root uptake, Ce was able to reach the plant leaves by moving through the vascular system [44]. We verified that Ce did not translocate easily since only a small fraction of the element moved from the roots to the aerial biomass of plants. Our findings and data agree with those obtained in similar studies respectively on soybean, cucumber, and wheat [37,44,48].

The increase in Ti concentration in the roots of Ce 500 and Ce 1000 plants could be due to a stimulatory effect of Ce on root growth and the formation of adventitious roots [49]. Therefore, $nCeO_2$ treated plants were able to explore a greater volume of soil, accumulating more Ti and Ce compared to the controls. This raises interesting questions about the bioavailability of Ti, which occurs in the soil as $nTiO_2$ from anthropogenic sources. It was previously observed that an increase in root assimilation and translocation of Ti after exposure to $nTiO_2$ would mean that the nano formulation of Ti makes it more bioavailable [50]. Other studies have shown that plants could translocate $nTiO_2$ into their aerial fractions. Titanium nanoparticles of 5 nm in diameter were found in leaves of *Arabidopsis thaliana* [51], in addition both the uptake and translocation of 100 nm $nTiO_2$ to the leaves of *Nicotiana tabacum* were observed [52]. However, the soil environment is very different and much more complex than artificial conditions. In our case, we observed very different results.

Macro- and micro-morphological observations indicate that under our conditions $nCeO_2$ and $nTiO_2$ induced limited cell injuries, at least in the parenchyma tissues. Inside the tissues, metal nanoparticles, depending on their type, shape and concentration may cause either cell death or other side effects [53]. Alternatively, metal nanoparticles can be well tolerated by the cells; however, this does not mean that nanoparticles do not affect cellular pathways [54]. The primary ultrastructural alteration we detected in leaf tissues of plants treated with metal oxide nanoparticles was the condensed chromatin in the nuclei and swollen mitochondria of vascular parenchyma cells. Condensed chromatin and fragmented nuclei, as well as swollen mitochondria, are described in programmed cell death (PCD), reported in cell response to different environmental stimuli and stresses, induced by pathogens [55] and abiotic factors as salinity, cold stress, waterlogging, or hypoxia [56,57].

Referring to the literature findings [26,58,59] we would have expected to find crystalline forms of both elements in the plant tissues. Several nanostructures were observed in the leaf tissues of $nCeO_2$ treated plants; however, microanalysis did not confirm the presence of Ce in such aggregates. This is in contrast with Zhao *et al.* [58] which verified the presence of$nCeO_2$ aggregates within vascular tissues of corn, thus demonstrating that Ce nanoparticles migrate through the xylem under transpiration. In studies on soybean plants grown in $nCeO_2$ amended soil, it was observed that most of the Ce stored in the pods was in the form of $nCeO_2$ [59].

We have not investigated the speciation of Ce and Ti in plant tissues, however this evidence constitutes a rather strong indication of Ce biotransformation. The biotransformation of ceria nanoparticles in plant tissues was demonstrated in soybean [16] and in cucumber [60,61]. In our case we hypothesized that the formation of the amorphous clusters could be related to the presence of intracellular Ce and, a defensive mechanism against Ce-cytotoxicity, as previously demonstrated [62]. To the contrary, Ti nanoparticles were able to cross biological barriers in plant tissues. In fact, we report the presence of Ti nanoclusters in parenchymatic cells of barley leaves. Early evidence of root to shoot translocation of $nTiO_2$ was found in hydroponically grown cucumber seedlings, using micro X-ray fluorescence (μ-XRF) and micro X-ray absorption spectroscopy (μ-XANES) [63]. Subsequently, the same evidence was confirmed in older cucumber plants growing in $nTiO_2$-enriched soil, as well [64]. Our work undertaken with the same method agrees with these findings. Furthermore, unlike what was observed for Ce, we did not observe evidence of biotransformation, confirming the literature findings [64].

CONCLUSIONS

ENMs are currently considered as an emerging class of environmental contaminants and thus, as mentioned above, neither (i) potential interactions with other pollutants [65,66] and/or (ii) the possibility of co-occurrence of ENMs in the environment can be excluded [67]. In other words, there exists a chance of simultaneously exposure of target organisms to different types of ENMs. This implies that the horizon of knowledge gaps on the relationships between ENM and biota has to be moved farther. According to Kumar *et al.* [67] "to properly incorporate exposure of more than one type of ENM, data on toxicity due to mixture of ENMs for a given target organ are required".

With regard to plants, it is very likely that several research groups are currently working on this topic. However, in literature we found only one paper reporting evidence of stimulating effects on germination and early growth of *Glycine max*induced by a mixture of $nSiO_2$ and $nTiO_2$ [68]. Therefore, we reiterate that this paper is the first to report data on the effects of a co-exposure to different metal oxide nanoparticles on a worldwide important crop. Moreover, our data were collected at the end of the life-cycle of soil grown barley plants.

Data available on the effects of ENMs in humans, crop plants, and livestock are not enough to allow for a thorough evaluation of their potential and of their safety. With respect to the aims of this study, although no visual symptoms of toxicity have been detected in plants, we demonstrated that the phenology and growth of barley were affected by $n\text{CeO}_2$and $n\text{TiO}_2$. All plants concluded their life cycle producing seeds. However, in treated plants we verified differences in some biometric parameters compared to the control ones. In particular, $n\text{CeO}_2$ at the lower concentration were associated with a reduction in the leaf area, the number of tillers and spikes per plant, and for this reason the number of kernels per plant. It can be assumed that this will lead to a reduction in crop production. An attenuation of such adverse effects was observed in plants treated with the higher dose of $n\text{CeO}_2$. Titanium nanoparticles were associated with positive effects on plants. First, we observed $n\text{TiO}_2$ positive dose-response effect on vegetative growth. Second, in plants co-exposed to $n\text{CeO}_2$ and $n\text{TiO}_2$, it is likely that the beneficial effects of Ti on plant metabolism have more than compensated for the adverse effects of Ce. Lacking literature data, at this moment we cannot discuss further these results and this part of our experiments should be considered simply exploratory. However, we demonstrated that the co-occurrence of $n\text{CeO}_2$ or $n\text{TiO}_2$ in soil determined in barley plants effects other than those observed in plants exposed separately to nanomaterials. It is likely that the study on this issue will be further dealt in the near future, through developing appropriate experimental protocols to study the physiological bases of plant response to ENMS co-occurrence. From an ecological point of view, our data suggest that the fate of $n\text{CeO}_2$ and $n\text{TiO}_2$ could be different. In both cases, their bioaccumulation in plants is minute. Cerium nanoparticles inside plant tissues seem to dissolve into the ionic form that most likely undergoes a subsequent biotransformation. Titanium oxide nanoparticles are found in crystalline form in the leaves of barley and also in the seeds, although in small concentrations, so in this form the $n\text{TiO}_2$ may be able to continue on through the food chain. Further research should be carried out on the intricate relationships that exist in the soil-plant system with respect the fate of nanomaterials.

ACKNOWLEDGMENTS

This work was in part supported by a project funded by DISA—Department of Agriculture and Environmental Sciences, University of Udine, through Grant n.64 dd. 08-09.2014 (RANDOLPH—Relazioni tra nanoparticelle metalliche e piante superiori). STEM EDXS experiments were partially supported by PRIN 2012 NOXSS and Progetto premiale MIUR 2013 USCEF. The authors acknowledge technical assistance in the use of BET by Anastasios Papadiamantis and access to the lab facilities of the Environmental Nanoscience research group at Birmingham University. We also acknowledge Giorgio Angelo Lucchini (Dipartimento di Scienze Agrarie e Ambientali—Produzione, Territorio, Agroenergia, Università degli Studi di Milano, Italy) for ICP-MS analyses.

AUTHOR CONTRIBUTIONS

Marchiol and Guido Fellet conceived and designed the experiments; Alessandro Mattiello and Filip Pošćić performed the experiments; Costanza Zavalloni, Rita Musetti and Elvio Carlino respectively carried out gas exchanges measurements, TEM observations and STEM EDX analysis; Luca Marchiol analyzed the data and wrote the paper.

REFERENCES

1. Tourinho, P.S.; Van Gestel, C.A.M.; Lofts, S.; Svedsen, C.; Soares, A.M.V.M.; Loureiro, S. Metal-based nanoparticles in soil: Fate, behavior, and effects on soil invertebrates. *Environ. Toxicol. Chem.* 2012, *31*, 1679–1692.
2. Batley, G.; Kirby, J.K.; McClaughlin, M.J. Fate and risks of nanomaterials in aquatic and terrestrial environments. *Acc. Chem. Res.* 2013, *46*, 854–862.
3. Piccinno, F.; Gottschalk, F.; Seeger, S.; Nowack, B. Industrial production quantities and uses of ten engineered nanomaterials in Europe and the world. *J. Nanopart. Res.* 2012.
4. Gogos, A.; Knauer, K.; Bucheli, T.D. Nanomaterials in plant protection and fertilization: Current state, foreseen applications, and research priorities. *J. Agric. Food Chem.* 2012, *60*, 9781–9792.
5. Liu, H.H.; Cohen, Y. Multimedia environmental distribution of engineered nanomaterials. *Environ. Sci. Technol.* 2014,*48*, 3281–3292.
6. Zhang, Z.; He, X.; Zhang, H.; Ma, Y.; Zhang, P.; Ding, Y.; Zhao, Y. Uptake and distribution of ceria nanoparticles in cucumber plants. *Metallomics* 2011, *3*, 816–822.
7. OECD. List of Manufactured Nanomaterials and List of Endpoints for Phase One of the Sponsorship Programme of Testing on Manufactured Nanomaterials: Revision. OECD Environment, Health and Safety Publications Series on the Safety of Manufactured Nanomaterials, No. 27, Paris, France. 2010. Available online: http://www.oecd.org/officialdocuments/publicdisplaydocumentpdf/?cote=env/jm/mono(2010)46&doclanguage=en (accessed on 30 December 2015).
8. Rico, C.M.; Majumdar, S.; Duarte-Gardea, M.; Peralta-Videa, J.R.; Gardea-Torresdey, J.L. Interaction of nanoparticles with edible plants and their possible implications in the food chain. *J. Agric. Food Chem.* 2011, *59*, 3485–3498.
9. Miralles, P.; Church, T.L.; Harris, A.T. Toxicity, uptake, and translocation of engineered nanomaterials in vascular plants. *Environ. Sci. Technol.* 2012, *46*, 9224–9239.
10. Gardea-Torresdey, J.L.; Rico, C.M.; White, J.C. Trophic transfer, transformation, and impact of engineered nanomaterials in terrestrial environments. *Environ. Sci. Technol.* 2014, *48*, 2526–2540.
11. Capaldi Arruda, S.C.; Silva, A.L.; Galazzi, R.M.; Azevedo, R.A.; Arruda, M.A. Nanoparticles applied to plant science: A review. *Talanta* 2015, *131*, 693–705.
12. López-Moreno, M.L.; de La Rosa, G.; Hernández-Viezcas, J.A.; Peralta-Videa, J.R.; Gardea-Torresdey, J.L. X-ray Absorption Spectroscopy (XAS) corroboration of the uptake and storage of CeO_2 nanoparticles and assessment of their differential toxicity in four edible plant species. *J. Agric. Food Chem.* 2010, *58*, 3689–3693.

13. Larue, C.; Castillo-Michel, H.; Sobanska, S.; Trcera, N.; Sorieul, S.; Cécillon, L.; Ouerdane, L.; Legros, S.; Sarret, S. Fate of pristine TiO_2 nanoparticles and aged paint-containing TiO_2 nanoparticles in lettuce crop after foliar exposure. *J. Hazard. Mater.* 2014, *273*, 17–26.
14. Liang, Y.; Bradford, S.A.; Simunek, J.; Heggen, M.; Vereeken, H.; Klumpp, E. Retention and remobilization of stabilized silver nanoparticles in an undisturbed loamy sand soil. *Environ. Sci. Technol.* 2013, *47*, 12229–12237.
15. Ma, Y.; Kuang, L.; He, X.; Bai, W.; Ding, Y.; Zhang, Z.; Zhao, Y.; Chai, Z. Effects of rare earth oxide nanoparticles on root elongation of plants. *Chemosphere* 2010, *78*, 273–279.
16. López-Moreno, M.L.; de la Rosa, G.; Hernández-Viezcas, J.A.; Castillo-Michel, H.; Botez, C.E.; Peralta-Videa, J.R.; Gardea-Torresdey, J.L. Evidence of the differential biotransformation and genotoxicity of ZnO and CeO_2nanoparticles on soybean (*Glycine max*) plants. *Environ. Sci. Technol.* 2010, *44*, 7315–7320.
17. Ma, Y.; Zhang, P.; Zhang, Z.; He, X.; Zhang, J.; Ding, Y.; Zhang, J.; Zheng, L.; Guo, Z.; Zhang, L.; *et al.* Where does the transformation of precipitated ceria nanoparticles in hydroponic plants take place? *Environ. Sci. Technol.* 2015, *49*, 10667–10674.
18. Majumdar, S.; Peralta-Videa, J.R.; Bandyopadhyay, S.; Castillo-Michel, H.; Hernandez-Viezcas, J.-A.; Sahi, S.; Gardea-Torresdey, J.L. Exposure of cerium oxide nanoparticles to kidney bean shows disturbance in the plant defense mechanisms. *J. Hazard. Mater.* 2014, *278*, 279–287.
19. Rico, C.M.; Peralta-Videa, J.R.; Gardea-Torresdey, J.L. Differential effects of cerium oxide nanoparticles on rice, wheat, and barley roots: A Fourier Transform Infrared (FT-IR) microspectroscopy study. *Appl. Spectrosc.* 2015, *69*, 287–295.
20. Rico, C.M.; Barrios, A.C.; Tan, W.; Rubenecia, R.; Lee, S.C.; Varela-Ramirez, A.; Peralta-Videa, J.-R.; Gardea Torresdey, J.L. Physiological and biochemical response of soil-grown barley (*Hordeum vulgare* L.) to cerium oxide nanoparticles.*Environ. Sci. Pollut. Res.* 2015, *22*, 10551–10558.
21. Hong, F.; Zhou, J.; Liu, C.; Yang, F.; Wu, C.; Zheng, L.; Yang, P. Effect of nano-TiO2 on photochemical reaction of chloroplasts of spinach. *Biol. Trace Elem. Res.* 2005, *105*, 269–279.
22. Gao, F.; Hong, F.; Liu, C.; Zheng, L.; Su, M.; Wu, X.; Yang, F.; Wu, C.; Yang, P. Mechanism of nano-anatase TiO_2 on promoting photosynthetic carbon reaction of spinach. *Biol. Trace Elem. Res.* 2006, *111*, 239–253.
23. Yang, F.; Liu, C.; Gao, F.; Su, M.; Wu, X.; Zheng, L.; Hong, F.; Yang, P. The improvement of spinach growth by nano-anatase TiO_2 treatment is related to nitrogen photoreduction. *Biol. Trace Elem. Res.* 2007, *119*, 77–88.
24. Linglan, M.; Chao, L.; Chunxiang, Q.; Sitao, Y.; Jie, L.; Fengqing, G.; Fashui, H. Rubisco activase mRNA expression in spinach: Modulation by nanoanatase treatment. *Biol. Trace Elem. Res.* 2008, *122*, 168–178.
25. Qi, M.; Liu, Y.; Li, T. Nano-TiO_2 improve the photosynthesis of tomato leaves under mild heat stress. *Biol. Trace Elem. Res.* 2013, *156*, 323–328.
26. Asli, S.; Neumann, M. Colloidal suspensions of clay or titanium dioxide nanoparticles can inhibit leaf growth and transpiration via physical effects on root water transport. *Plant Cell Environ.* 2009, *32*, 577–584.
27. Ruffini Castiglione, M.; Giorgetti, L.; Geri, C.; Cremonini, R. The effects of nano-TiO_2 on seed germination, development and mitosis of root tip cells of *Vicia narbonensis* L. and *Zea mays* L. *J. Nanopart. Res.* 2011, *1*, 2443–2449.
28. Song, U.; Shin, M.; Lee, G.; Roh, J.; Kim, Y.; Lee, E.J. Functional analysis of TiO_2 nanoparticle toxicity in three plant species. *Biol. Trace Elem. Res.* 2013, *155*, 93–103.
29. Frazier, T.P.; Burklew, C.E.; Zhang, B. Titanium dioxide nanoparticles affect the growth and microRNA expression of tobacco (*Nicotiana tabacum*). *Funct. Integr. Genom.* 2014, *14*, 75–83.
30. Pakrashi, S.; Jain, N.; Dalai, S.; Jayakumar, J.; Chandrasekaran, P.T.; Raichur, A.M.; Chandrasekaran, N.; Mukherjee, A. *In vivo* genotoxicity assessment of titanium dioxide nanoparticles by *Allium cepa* root tip assay at high exposure concentrations. *PLoS ONE* 2013.
31. Keller, A.K.; McFerran, S.; Lazareva, A.; Suh, S. Global life cycle releases of engineered nanomaterials. *J. Nanopart. Res.*2013, *15*, 1692.
32. Liu, R.; Lal, R. Potentials of engineered nanoparticles as fertilizers for increasing agronomic productions. *Sci. Total Environ.* 2015, *514*, 131–139.
33. Singh, D.; Kumar, A. Human exposure of engineered nanoparticles from plants irrigated with contaminated waters: Mixture toxicity issues and challenges ahead. *Adv. Sci. Lett.* 2014, *20*, 1204–1207.
34. Servin, A.D.; White, J.C. Nanotechnology in agriculture: Next steps for understanding engineered nanoparticle exposure and risk. *NanoImpact* 2016.
35. Food and Agriculture Organization of the United Nations. FAOSTAT. Available online: http://faostat3.fao.org/download/Q/QC/E (accessed on 20 February 2016).
36. Priester, J.H.; Ge, Y.; Mielke, R.E.; Horst, A.M.; Moritz, S.C.; Espinosa, K.; Gelb, J.; Walker, S.L.; Nisbet, R.M.; An, Y.J.;*et al.* Soybean susceptibility to manufactured nanomaterials with evidence for food quality and soil fertility interruption. *Proc. Natl. Acad. Sci. USA* 2012.
37. Zadoks, J.C.; Chang, T.T.; Konzak, C.F. A Decimal code for the growth stages of cereals. *Weed Res.* 1974, *14*, 415–421.
38. Carlino, E. TEM for Characterization of semiconductor nanomaterials. In *Transmission Electron Microscopy Characterization of Nanomaterials*; Kumar, C.S.S.R., Ed.; Springer: Berlin, Germany, 2014; pp. 89–138.
39. Colman, B.P.; Arnaout, C.L.; Anciaux, S.; Gunsch, C.K.; Hochella, M.F., Jr.; Kim, B.; Lowry, G.V.; McGill, B.M.; Reinsch, B.C.; Richardson, C.J.; *et al.* Low concentrations of silver nanoparticles in biosolids cause adverse ecosystem responses under realistic field scenario. *PLoS ONE* 2013.
40. Yoon, S.-J.; Kwaka, J.I.; Lee, W.-M.; Holden, P.A.; Ana, Y.-J. Zinc oxide nanoparticles delay

soybean development: A standard soil microcosm study. *Ecotoxicol. Environ. Saf.* 2014, *100*, 131–137.

41. Dofing, S.M. Phenological development-yield relationships in spring barley in a subarctic environment. *Can. J. Plant Sci.* 1995, *75*, 93–97.
42. Yang, F.; Hong, F.; You, W.; Liu, C.; Gao, F.; Wu, C.; Yang, P. Influences of nano-anatase TiO_2 on the nitrogen metabolism of growing spinach. *Biol. Trace Elem. Res.* 2006, *110*, 179–190.
43. Morteza, E.; Moaveni, P.; Farahanim, H.A.; Kiyani, M. Study of photosynthetic pigments changes of maize (*Zea mays*L.) under nano TiO_2 spraying at various growth stages. *SpringerPlus* 2013.
44. Zhao, L.; Sun, Y.; Hernandez-Viezcas, J.A.; Servin, A.D.; Hong, J.; Niu, G.; Peralta-Videa, J.R.; Duarte-Gardea, M.; Gardea-Torresdey, J.L. Influence of CeO_2 and ZnO Nanoparticles on cucumber physiological markers and bioaccumulation of Ce and Zn: A life cycle study. *J. Agric. Food Chem.* 2013, *61*, 11945–11951.
45. Cornelis, G.; Hund-Rinke, K.; Kuhlbusch, T.; van den Brink, N.; Nickel, C. Fate and bioavailability of engineered nanoparticles in Soils: A review, critical reviews in environmental science and technology. 2014, *44*, 2720–2764.
46. Schlich, K.; Hund-Rinke, K. Influence of soil properties on the effect of silver nanomaterials on microbial activity in five soils. *Environ. Pollut.* 2015, *196*, 321–330.
47. Schwabe, F.; Tanner, S.; Schulin, R.; Rotzetter, A.; Stark, W.; von Quadt, A.; Nowack, B. Dissolved cerium contributes to uptake of Ce in the presence of differently sized CeO_2-nanoparticles by three crop plants. *Metallomics* 2015, *7*, 466–477.
48. Rico, C.M.; Lee, S.C.; Rubenecia, R.; Mukherjee, A.; Hong, J.; Peralta-Videa, J.R.; Gardea-Torresdey, J.L. Cerium oxide nanoparticles impact yield and modify nutritional parameters in wheat (*Triticum aestivum* L.). *J. Agric. Food. Chem.*2014, *62*, 9669–9675.
49. Zhang, P.; Ma, Y.H.; Zhang, Z.Y.; He, X.; Li, Y.Y.; Zhang, J.; Zheng, L.R.; Zhao, Y.L. Species-specific toxicity of ceria nanoparticles to Lactuca plants. *Nanotoxicology* 2013, *10*, 1–8.
50. Foltête, A.-S.; Masfaraud, J.-F.; Bigorgne, E.; Nahmani, J.; Chaurand, P.; Botta, C.; Labille, J.; Rose, J.; Férard, J.-F.; Cotelle, S. Environmental impact of sunscreen nanomaterials: Ecotoxicity and genotoxicity of altered TiO_2nanocomposites on *Vicia Faba. Environ. Pollut.* 2001, *159*, 2515–2522.
51. Kurepa, J.; Paunesku, T.; Vogt, S.; Arora, H.; Rabatic, B.M.; Lu, J.; Wanzer, M.B.; Woloschak, G.E.; Smalle, J.A. Uptake and distribution of ultrasmall anatase TiO_2 alizarin red S nanoconjugates in *Arabidopsis thaliana*. *Nano Lett.* 2010, *10*, 2296–2302.
52. Ghosh, M.; Bandyopadhyay, M.; Mukherjee, A. Genotoxicity of titanium dioxide (TiO_2) nanoparticles at two trophic levels: Plant and human lymphocytes. *Chemosphere* 2010, *81*, 1253–1262.
53. Sohaebuddin, S.K.; Thevenot, P.T.; Baker, D.; Eaton, J.W.; Tang, L. Nanomaterial cytotoxicity is composition, size, and cell type dependent. *Part. Fibre Toxicol.* 2010, *7*, 22–39.
54. Panariti, A.; Miserocchi, G.; Rivolta, I. The effect of nanoparticle uptake on cellular behavior: disrupting or enabling functions? *Nanotechnol. Sci. Appl.* 2012, *5*, 87–100.
55. Lam, E.; Kato, N.; Lawton, M. Programmed cell death, mitochondria and the plant hypersensitive response. *Nature*2012, *411*, 848–853.
56. White, E. Life, death, and the pursuit of apoptosis. *Genes Dev.* 1996, *10*, 1–15.
57. Kratsch, H.A.; Wise, R.R. The ultrastructure of chilling stress. *Plant Cell Environ.* 2000, *23*, 337–350.
58. Zhao, H.; Zhou, Q.; Zhou, M.; Li, C.; Gong, X.; Liu, C.; Qu, C.; Wang, L.; Si, W.; Hong, F. Magnesium deficiency results in damage of nitrogen and carbon cross-talk of maize and improvement by cerium addition. *Biol. Trace Elem. Res.*2012, *148*, 102–109.
59. Hernandez-Viezcas, J.A.; Castillo-Michel, H.; Cooke Andrews, J.; Cotte, M.; Rico, C.; Peralta-Videa, J.R.; Ge, Y.; Priester, J.H.; Holden, P.A.; Gardea-Torresdey, J.L. *In situ* synchrotron X-ray fluorescence mapping and speciation of CeO_2 and ZnO nanoparticles in soil cultivated soybean (*Glycine max*). *ACS Nano* 2013, *7*, 1415–1423.
60. Zhang, P.; Ma, Y.; Zhang, Z.; He, X.; Zhang, J.; Guo, Z.; Tai, R.; Zhao, Y.; Chai, Z. Biotransformation of ceria nanoparticles in cucumber plants. *ACS Nano* 2012, *6*, 9943–9950.
61. Ma, Y.; Zhang, P.; Zhang, Z.; He, X.; Li, Y.; Zhang, J.; Zheng, L.; Chu, S.; Yang, K.; Zhao, Y.; *et al.* Origin of the different phytotoxicity and biotransformation of cerium and lanthanum oxide nanoparticle in cucumber. *Nanotoxicology* 2015,*30*, 262–270.
62. Horie, M.; Nishio, K.; Fujita, K.V.; Endoh, S.; Miyauchi, A.; Saito, Y.; Iwahashi, H.; Yamamoto, K.; Murayama, H.; Nakano, H.; *et al.* Protein adsorption of ultrafine metal oxide and its influence on cytotoxicity toward cultured cells.*Chem. Res. Toxicol.* 2009, *22*, 5431–5453.
63. Servin, A.D.; Castillo-Michel, H.; Hernandez-Viezcas, J.A.; Corral Diaz, B.; Peralta-Videa, J.R.; Gardea-Torresdey, J.L. Synchrotron micro-XRF and micro-XANES confirmation of the uptake and translocation of TiO_2 nanoparticles in cucumber (*Cucumis sativus*) plants. *Environ. Sci. Technol.* 2012, *46*, 7637–7643.
64. Servin, A.D.; Morales, M.I.; Castillo-Michel, H.; Hernandez-Viezcas, J.A.; Munoz, B.; Zhao, L.; Nunez, J.E.; Peralta-Videa, J.R.; Gardea-Torresdey, J.L. Synchrotron verification of TiO_2 accumulation in cucumber fruit: A possible pathway of TiO_2 nanoparticle transfer from soil into the food chain. *Environ. Sci. Technol.* 2013, *47*, 11592–11598.
65. Balbi, T.; Smerilli, A.; Fabbri, R.; Ciacci, C.; Montagna, M.; Grasselli, E.; Brunelli, A.; Pojana, G.; Marcomini, A.; Gallo, G.; *et al.* Co-exposure to n-TiO_2 and Cd^{2+} results in interactive effects on biomarker responses but not in increased toxicity in the marine bivalve *M. galloprovincialis*. *Sci. Total Environ.* 2014, *493*, 355–364.

66. Ribas Ferreira, J.L.; Lonné, M.N.; Franc, T.A.; Maximilla, N.M.; Lugokenski, T.H.; Costa, P.G.; Fillmann, G.; Antunes Soares, F.A.; de la Torre, F.R.; Monserrat, J.M. Co-exposure of the organic nanomaterial fullerene C60 with benzo[a]pyrene in *Danio rerio* (zebrafish) hepatocytes: Evidence of toxicological interactions. *Aquat. Toxicol.* 2014, *147*, 76–83.
67. Kumar, A.; Kumar, P.; Anandan, A.; Fernandes, T.F.; Ayoko, G.A.; Biskos, G. Engineered nanomaterials: Knowledge gaps in fate, exposure, toxicity, and future directions. *J. Nanomater.* 2014.
68. Lu, C.M.; Zhang, C.Y.; Wen, J.Q.; Wu, G.R.; Tao, M.X. Research of the effect of nanometer materials on germination and growth enhancement of *Glycine max* and its mechanism. *Soybean Sci.* 2002, *21*, 168–172.

CHAPTER 4 ALANINE AMINOTRANSFERASE CONTROLS SEED DORMANCY IN BARLEY

Kazuhiro Sato, Miki Yamane, Nami Yamaji, Hiroyuki Kanamori, Akemi Tagiri, Julian G. Schwerdt, Geoffrey B. Fincher, Takashi Matsumoto, Kazuyoshi Takeda, and Takao Komatsuda

INTRODUCTION

Barley (*Hordeum vulgare*) was domesticated *ca.* 10,000 years ago from its wild progenitor *H. vulgare* subsp. *spontaneum* in the Fertile Crescent[1,2]. Wild barley is characterized by long grain dormancy, which lasts for several months after grain maturation[3]. Dormancy can be defined as the failure of an intact, viable seed to germinate under conditions that would normally be favourable for germination[4]. The wild barley's long dormancy means that, initially, the grain will not germinate in response to transient moisture availability and will therefore survive hot, dry summers. However, after some months of higher temperatures in the summer, grain dormancy is broken and this allows the grains to germinate after rains in the autumn. The selection pressure for shortened seed dormancy has been driven by the post domestication sowing activities of barley[5]. The process of malting barley kernels for brewing beer or distilling whisky involves controlled germination. Pedigrees of cultivars show that during the 19th century, malting barleys from the UK and Moravia spread to neighbouring countries and also to East Asia and the Americas[6]. Long periods of seed dormancy became undesirable because they required a longer storage period prior to malting. Thus, the control of dormancy is of great practical importance in agricultural production of crop species and in grain utilization by food and beverage industries.

Two major quantitative trait loci (QTLs) on chromosome 5H, one near the centromere (*Qsd1*) and another in the telomeric region of the long arm (*Qsd2*) have been identified[7]. Of the two QTLs, *Qsd2* was isolated and proved to encode a mitogen-activated protein kinase kinase 3 (ref.[8]). The *Qsd1* QTL mapped in the cross of wild barley H602 and malting barley cv. Haruna Nijo[9]. Haruna Nijo carries a dominant short dormancy allele whereas the wild barley carries a recessive long dormancy allele. There are some reports of other genes that control seed dormancy in wheat[10] and rice[11,12], but none are orthologous with *Qsd1*. Furthermore, seed dormancy in wild barley is much longer than seed dormancy observed in these other cereal species.

We have identified a gene that defines a major dormancy trait in wild ancestral forms of barley and show that it encodes alanine aminotransferase, which has not previously been linked with dormancy in plants. We show that a single amino acid substitution in this enzyme reduces the dormancy period of mature barley grain. Our data suggest that the *qsd1* gene for long dormancy appears to be a mutable gene that is used for adaptation to different environments and this is consistent with the large variation in *qsd1* sequence that exists in nature.

RESULTS

Qsd1 encodes Alanine Aminotransferase

For the initial map-based cloning of *Qsd1*, we identified a genetic substitution line carrying the long dormancy *qsd1* allele of H602, together with Haruna Nijo›s short dormancy alleles for the other three QTLs[13,14]. A total of 910 F_2 plants were produced to genetically map *Qsd1* (ref.[14]). The flanking *EST1* and *EST12* markers were used to select recombinants. All 13 expressed sequence tag (EST) markers were assayed in the recombinants. *EST4R* co-segregated with *Qsd1* in a total of 4,792 F_3 plants (Fig. 1a). The *EST4R* marker was used for clone selection from genomic libraries of Haruna Nijo[15] and H602 (ref.[2]). One and three

Citation: Kazuhiro Sato , Miki Yamane , Nami Yamaji , Hiroyuki Kanamori, Akemi Tagiri, Julian G. Schwerdt, Geoffrey B. Fincher, Takashi Matsumoto, Kazuyoshi Takeda & Takao Komatsuda, Alanine aminotransferase controls seed dormancy in barley, DOI: 10.1038/ncomms11625.

clones were selected for Haruna Nijo and H602 libraries, respectively. Of these, one clone from each library comprised a contig. An alignment of the genomic sequences from the two genotypes showed that these contigs shared 100,231 bp of sequence. On these contigs a full-length complementary DNA clone[16]*LC054176*maps at the*EST4R*marker and another full-length cDNA[17]*AK372829*maps, 345 bp away, at the*EST5*marker next to*EST4R*. Among the 4,792 F_3plants, one plant F_3#2,253, recombined between*Qsd1*and*EST5*and another plant F_3#2,372 recombined between*Qsd1*and*EST4F*(Fig. 1a), indicating that*Qsd1*is located in a 9,467 bp region between*EST5*and*EST4F*and thereby*Qsd1*is located in either*LC054176*,*AK372829*or the intergenic region. Homozygous F_4plants from F_3#2,372 discretely segregated*qsd1*/*qsd1*(long dormancy) and*Qsd1*/*Qsd1*(short dormancy) in a Mendelian fashion (Fig. 1b).

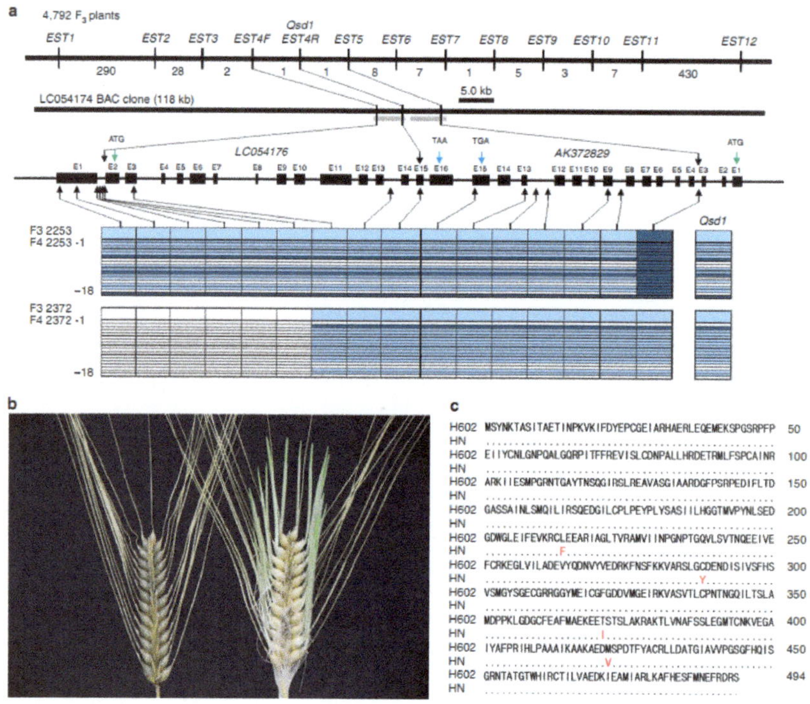

Figure 1: The isolation of*Qsd1*.

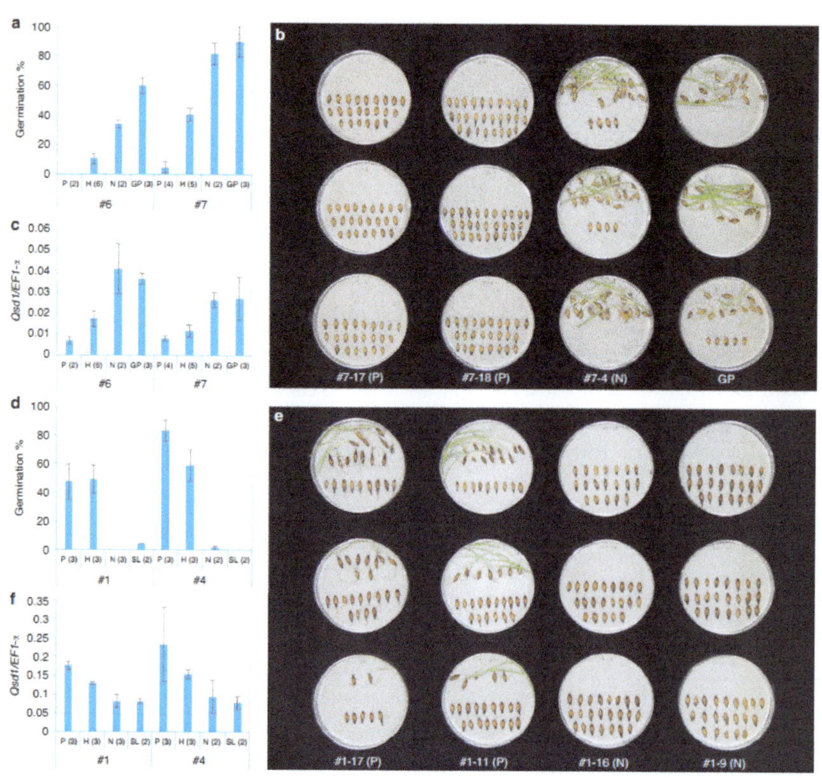

Figure 2: Effects of RNAi knockdown and complementation for*Qsd1*.

Within the 9,467 bp region, there were only four non-synonymous single nucleotide polymorphisms (SNPs) and all of them were located in the coding region of*AK372829*, indicating that one or more of these were likely to be responsible for the*Qsd1*phenotype. The*AK372829*gene encodes an alanine aminotransferase (AlaAT), a protein of 495 amino acid residues. For proof-of-function experiments, the AlaAT gene was used to transform barley. However, the Golden Promise variety used in our standard*Agrobacterium*-mediated transformation experiments carries the dominant short dormancy*Qsd1*allele, so we developed knockdown plants by RNA interference (RNAi) with the pANDA vector[18] with the objective of developing long dormancy lines.Figure 2ashows the effects of*Qsd1*RNAi knockdown on seed germination. Of the two events, #7 T_2transgenic homozygous positive plants showed <5% germination after 21 days at 21 °C (Fig. 2a,b), while heterozygous plants showed intermediate levels of about 40% germination under the same conditions. These values may be the compared with values of 80–90% obtained for negative control lines and the wild type Golden Promise. Both the homozygous and heterozygous transgenic T_2plants showed decreased transcript levels for the*AK372829*AlaAT gene (Fig. 2c).

We also conducted complementation tests using genetic substitution lines that contained the recessive long dormancy*qsd1*allele of wild barley H602 in a background of Golden Promise (id. B1H602GP20-7) by transforming with a construct carrying the cloned Haruna Nijo short dormancy*Qsd1*genomic DNA.Figure 2dshows the effects of the transgene*Qsd1*on seed germination. The T_1plants carrying the*Qsd1*transgene showed vigorous seed germination, which was significantly higher than both the transgene negative plants and the parent substitution line (*qsd1*/*qsd1*) (Fig. 2d,e). Transcript levels in the embryos increased in transgene positive T_1plants (Fig. 2f). These transgenic data clearly demonstrated that the*Qsd1*gene is responsible for shortened dormancy in barley grain. Given the complexity of the mechanisms of seed dormancy and germination and their operation at many levels in the plant[19,20], the large effect of the*Qsd1*gene on grain dormancy in barley indicates that the encoded AlaAT enzyme plays a central role in the release of dormancy in this species.

Association between SNPs and Dormancy

To identify any association between dormancy and the four SNPs identified in the long and short dormancy alleles of the AlaAT gene, the additive effects of individual*Qsd1*alleles were compared with the parents used for*Qsd1*mapping[9]. Among the four non-synonymous SNPs in the delimited region, SNP E9 matched the*Qsd1*/*qsd1*alleles. This SNP is associated with a change from a leucine (L) residue at amino acid position 214 in wild barley H602 to a phenylalanine (F) residue in cultivar Haruna Nijo (Fig. 1c). The SNPs E11 and E14 showed a much lower level of correlation with the dormancy phenotype, but it is not possible to exclude the possibility that they also have effects on the dormancy level observed.

To further confirm that the SNP alleles were correlated with dormancy in barley grain more generally, the germination scores of 353 cultivated and 14 wild barley accessions were analysed. The germination (low-dormancy) scores were compared among haplotypes of non-synonymous SNPs in*AK372829*. Average germination scores of cultivated barleys were lowest in haplotype TACC (non-synonymous SNP haplotype at E14, E13, E11, E9) and highest in CATG (Haruna Nijo haplotype). There was a clear difference in average germination between E9 alleles. Variation at the other three SNP alleles showed relatively smaller effects on germination. The number of haplotypes having allele G in E9 (n=46) was much lower than haplotypes having allele C (n=307). Most of the barleys having the allele G have pedigrees that track malting and brewing applications[6], indicating that the allele G might have resulted from the specific selection of reduced dormancy for the malting process.

To track the origin of the*Qsd1*mutation that resulted in the reduced dormancy time, the genomic regions of*Qsd1*in 27 wild and 210 cultivated barley accessions were sequenced and aligned. Eight sub-clusters (clades) of two wild barleys (W), three cultivated types (C) and mixtures of wild and cultivated types (WC) were evident (Fig. 3).

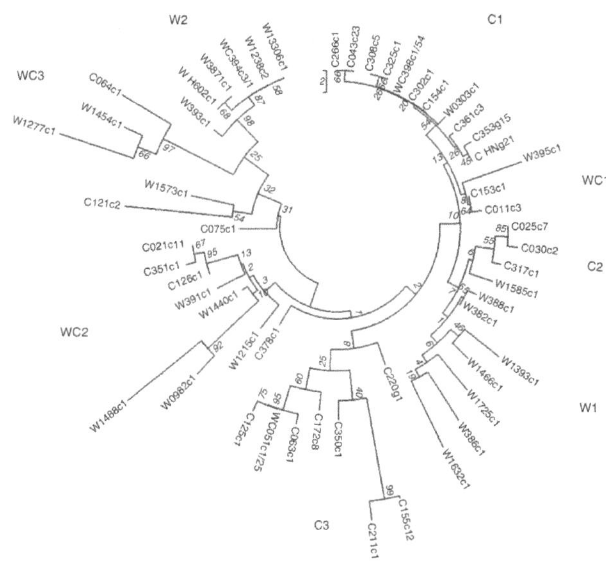

Figure 3: Molecular relationships of haplotypes for*Qsd1* **in wild and cultivated barleys.** W and C represent wild and cultivated barleys, respectively, and WC means mixtures of wild and cultivated barleys. Haplotype name is composed of type of barley (W, C or WC), representative accession number, non-synonymous SNP genotype at exon 9 and number of accessions in the haplotype (order of W/C in case of mixture).

Structure and Diversity of QSD1 Protein

The question now arises as to how the AlaAT enzyme might be a major determinant of the length of barley grain dormancy. There are five AlaAT genes in barley. The short dormancy gene*Qsd1*is located on chromosome 5HL.*Hv2*is located on chromosome 2HS, while*Hv3*is likely to be on chromosome 1H,*Hv4*is on chromosome 5HS and*Hv5*is on chromosome 2HL (ref.[21]). Estimated domains of AlaAT class I and class II subfamilies (IPR004839) are highly conserved. The amino acid sequence of QSD1=Hv1 are highly similar (E-value <E−145) for AlaATs in other plant species and for four barley sequences (Hv2, Hv3, Hv4 and Hv5part) of Haruna Nijo (Fig. 4a,b). None of these sequences have been previously reported to be related to dormancy. Motif analysis of QSD1 and its homologues using the SALAD database[22]showed that Hv3 and Hv4 shared the same motifs with QSD1 (Fig. 4b). One of the non-synonymous amino acid changes from leucine (L) of*qsd1*to phenylalanine (F) of*Qsd1*, corresponding to E9, was located on motif 11 (Fig. 4c). QSD1 has the only haplotype that shows phenylalanine (F) at E9 among the sequences examined, whereas L is the most common in other species. The result implies that the evolution of*Qsd1*represented the dominant mutation for reduced dormancy in barley.

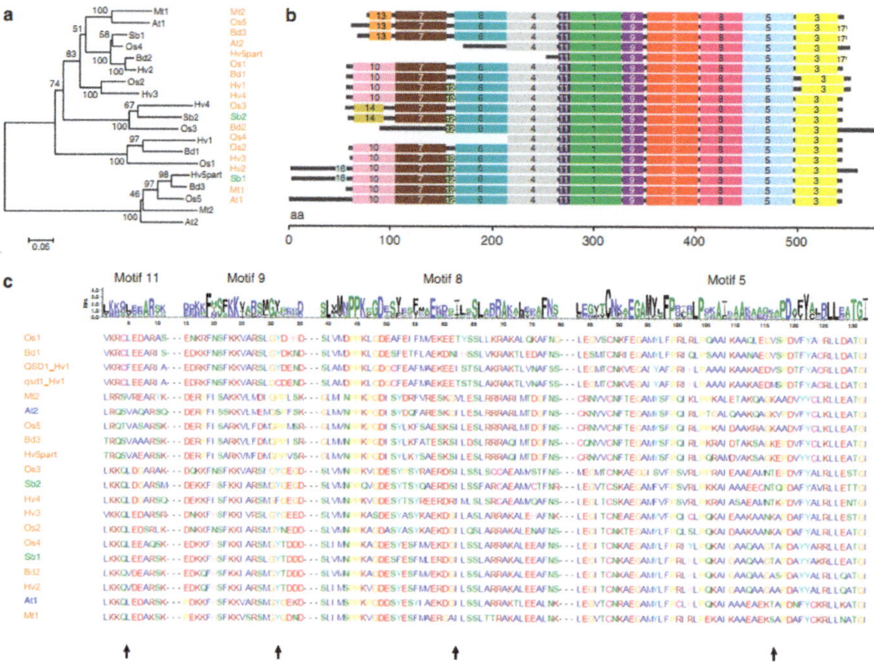

Figure 4: Alignment of QSD1 homologues in plant species. (a) Multiple alignment of QSD1 amino acid sequences (Hv1: Haruna Nijo). Sequences are listed from accessions showing similarity to QSD1 (E-value oE 145) by the blastp analysis in NCBI nr. Species and homologues with accession numbers are Hordeum vulgare: Hv1 (BAK04026.1), Hv2 (BAK07780.1), Hv3 (P52894.1), Hv4 (BAK05632.1), Hv5part (BAJ90574.1); Oryza sativa: Os1 (NP_001063248.1), Os2 (NP_001064504.1) Os3 (NP_001064505.2), Os4 (NP_001060284.1), Os5 (NP_001058716.1): Sorghum bicolor: Sb1 (XP_002463187.1), Sb2 (XP_002467302.1), Brachypodium distachyon: Bd1 (XP_003578159.1), Bd2 (XP_010235387.1), Bd3 (XP_003557680.1), Arabidopsis thaliana: At1 (NP_173173.3), At2 (AAK68842.1), Medicago truncatula: Mt1 (XP_003627448.1), Mt2 (XP_003613139.1). Scale bar indicates Poisson Correction distance. (b) Motif analysis of Qsd1 homologues using SALAD database22. (c) Multiple alignment of motifs which show non-synonymous substitutions between QSD1 (Hv1: Haruna Nijo) and qsd1 (Hv1: H602). Arrows indicate positions of amino-acid substitutions located in motifs 11 (exon9), 9 (exon11), 8 (exon13) and 5 (exon14).

*Qsd1*expression profiles

In addressing the question posed above regarding AlaAT function in dormancy, transcriptional profiling was used to first define the timing of the appearance of enzyme in the grain.

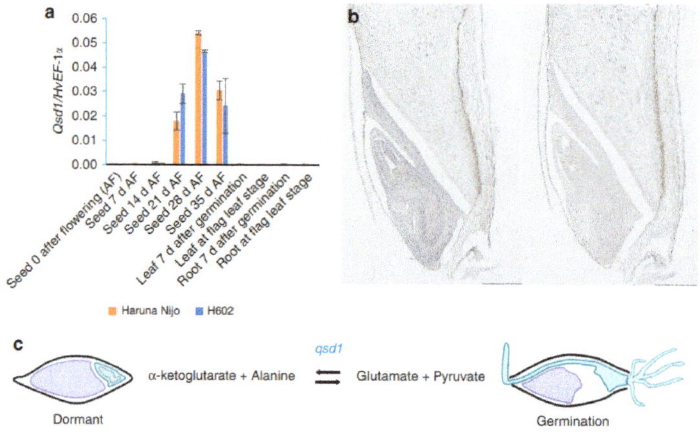

Figure 5: Expression of*Qsd1***and an estimated model of seed germination.** . (a) Relative transcript levels of copy numbers on Qsd1/ EF-1a in different organs and growth stages in the less-dormant cultivar Haruna Nijo and dormant wild barley H602 (for primer sequences). Difference of averages between Haruna Nijo and H602 is not significantly different (t-test, P40.05) in every pair of samples. Error bars represent s.e., n ¼ 3. (b) Qsd1 expression in an embryo of 19 days after flowering in H602 which is detected by in situ hybridization with an anti-sense (left) and sense (right) Qsd1 30 -UTR transcript. Scale bars represent 500 mm. (c)

A suggested model of seed dormancy controlled by qsd1 (derived from H602). Possible role of qsd1 is in the building up of alanine in the dormant grain. Other constitutive alanine aminotransferases catalyse aketoglutarate þ alanine2glutamate þ pyruvate pathway ubiquitously. Embryo is indicated in green.

The transcription of*Qsd1*/*qsd1*was detected in developing grains at 21, 28 and 35 days after flowering (Fig. 5a) but transcript levels were very low in grains at 0, 7 and 14 days after flowering. Very low transcript levels were also detected in young or older leaves and roots. The grain-specific expression in the long dormancy haplotype (*qsd1*: H602) was further analysed by RNA*in situ* hybridization 19 days after flowering. Transcripts were detected in the embryo but not in the endosperm (Fig. 5b). The microarray-based analysis of the*Qsd1*homologue in rice also showed that transcription was embro-specific[23], which is quite different from the wider expression patterns of paralogous rice genes. The embryo-specific expression in*Qsd1*and*Os1*increases during grain maturation, whereas expression levels of other (*Os2-Os5*,Fig. 4a) rice genes in the embryo decrease during the maturation process. It was possible that the phenotype is caused by temporal differences in expression of the two genes, but no major temporal differences could be detected between AlaAT transcript abundance in Haruna Nijo and H602 (Fig. 5a).

Structure and Diversity of QSD1 protein

Next, we investigated the possibility that the L214F substitution in short dormancy barleys might affect AlaAT activity. To identify the spatial disposition of L214, we constructed a homology model based on the crystal structure of Hv3 (ref.[24]), which is a dimeric, pyridoxal 5′-phosphate-dependent enzyme. Thus, the Hv3 crystal structure has been used here to locate the position of the L214F substitution that defines the difference between the enzymes encoded by the wild type*qsd1*(L) and mutant*Qsd1*(F) alleles, where L is conserved in most cereal orthologs (Fig. 4c). The location of L214 but it is not close to the active site, to the cofactor binding site or to the junction zone of the two proteins of the dimer. The L214F substitution has not yet been observed in other plant species (Fig. 4c). So, at this stage one would suspect that the conservative L214F substitution would not have a dramatic effect on enzymatic activity; we believe it is more likely that the region of the substitution is involved in homo-dimer or -trimer formation, or in binding of another, regulatory protein onto the surface of the AlaAT enzyme.

DISCUSSION

The fact that the qsd1 protein is encoded by the*qsd1*allele in wild barley implies that qsd1 has an essential function in seed dormancy, through a mechanism that is potentially common with other plant species. The qsd1 enzyme has a L residue at amino acid site 214 and this is conserved in orthologous cereal proteins (motif 11,Fig. 4c). The re-sequencing of wild and cultivated barley germplasm representing the global genetic diversity of barley did not show a single accession carrying the lesion allele without qsd1 function. This strongly supports the notion that qsd1, but not QSD1, represents the fundamental function of seed dormancy. The haplotype analyses showed that QSD1 protein structures may vary, even among wild barleys. The wild barley H602 (Fig. 3, W_H602: W2 clade) and the cultivar Haruna Nijo (C_HN: C1 clade) were located most distantly among the haplotypes. Of the haplotypes observed in the C1 clade, only haplotype C353 shares the SNP allele G at exon 9 with C_HN. The dominant*Qsd1*allele is carried by the cultivars with C_HN and C353 haplotypes, indicating that the dominant mutation in the*Qsd1*caused the reduced dormancy. The C_HN haplotype is distributed mainly in the Mediterranean and European areas, but is also found in elite barleys in Japan and Korea. The C353 haplotype is found only in European pedigrees of malting barleys. On the basis of the substitution at the E9 SNP, WC398 is a direct ancestral descendant of haplotype C_HN and a Palestinian wild barley; the latter could be the wild ancestral type of*Qsd1*. Another Palestinian wild haplotype, W0303, was also located in the C1 clade. Although it is not possible to exclude minor cultivars of clade C as candidate immediate ancestors, the dominant*Qsd1*allele appears to have originated from the*qsd1*allele of domesticated barley, rather than directly from wild barley. This notion is consistent with the conclusion that*Qsd1*was not responsible for barley domestication, but that its importance for the transition of barley from a primary food source to a useful raw material for malting and beer production was quickly recognized by farmers and early barley breeders, soon after the domestication events. The large C1 clade involves wild barleys not only from Palestine (WC398), but also from Lebanon (W1585) and Jordan (W1215). The result indicates that the*Qsd1*allele is in the lineage from wild barleys of the southern Levant, where the first barley domestication also occurred[1,2].

It is likely that the activity of AlaAT in the embryo of grains will influence the length of dormancy at the biochemical level[25]. In plants, AlaAT is a pivotal enzyme in several key metabolic pathways that involve nitrogen assimilation, carbon metabolism and protein synthesis through its reversible transfer of an amino group from glutamate to pyruvate, to form oxaloacetate and alanine[26]. Certain isoenzymes can act as glutamate-glyoxylate amino transferases[27]and many of the enzymes can transfer amino groups from other amino acids, at lower rates. AlaAT has also been implicated in stress responses, in anaerobic glycolysis, in branched chain amino acid synthesis[28]and in nitrogen use efficiency[24,29]. In barley, amino acids are necessary for the synthesis of cellular proteins and secreted hydrolytic enzymes after the initiation of germination[30]and it is possible that AlaAT plays a necessary and central role in linking nitrogen and carbon metabolism with protein synthesis as sugars and amino acids are released from storage proteins and starch during the germination process. The barley AlaAT encoded by the*qsd1*gene is located almost exclusively in the embryo (Fig. 5b), where one can envisage not only that the enzyme needs to be active in order for germination to proceed, but also that it could be a key target enzyme for the control of metabolic flux through the nitrogen and carbon metabolism pathways.

As shown inFig. 5a,*Qsd1*and*qsd1*are expressed at similar levels but differ by a non-synonymous amino acid change on the external face of the protein that is linked with the alternative phenotypes. Knockdown of the*Qsd1*allele in cv. Golden Promise (Fig. 2c) causes a reduction in*Qsd1*expression and an increase in the dormancy level. We suggest that this is due to the lower abundance of*Qsd1*in these lines leading to lower AlaAT activity (Fig. 2a). However, complementation of*qsd1*by overexpression of*Qsd1*, led to a significant increase in germination (Fig. 2d). In these plants the total *Qsd1* expression level is increased (considering both alleles together; Fig 2f). However because the non-transgenic control and the transgenic negative plants still showed reasonable levels of *qsd1* expression, we believe that it is the specific expression of the *Qsd1* allele (that is, the difference between the proteins derived from the non-synonymous amino acid change) rather than the cumulative increase in expression that causes different levels of dormancy.

As mentioned above, dormancy and germination are complex processes and dormancy has been linked with redox signalling by reactive oxygen species[31], levels of the phytohormones GA, ethylene and ABA, and anoxic conditions[30]. Indeed, alanine accumulates during periods of hypoxia, as do some isoforms of the AlaAT[32,33], while some isoforms of AlaAT are involved in the conversion of alanine to pyruvate during recovery from hypoxia[20]. Thus, AlaAT isoenzymes might be important both in the accumulation of alanine in hypoxic conditions and in its removal during the recovery from hypoxia (Fig. 5c). Low oxygen potentials are observed in developing grains and seeds from several species and could result in the accumulation of alanine[34]. It is known that the hard seed coat of legumes prevents germination by restricting the supply of oxygen in the embryo. The yellow seed coat locus ECY1 in *Brassica napus* also controls dormancy and has been implicated in alanine metabolism[35]. This is consistent with the use of hydrogen peroxide (H_2O_2) to break dormancy in barley seeds[36] but little is known about the oxygen status of mature barley grain and whether the release of low oxygen levels is an important early event in germination or whether the various isoforms of barley AlaAT play a direct role in breaking dormancy is not clear at this stage.

Despite these interpretative constraints, it is clear from the genetic data presented here that AlaAT is centrally important in the release of dormancy in barley grain. In practical terms, our data suggest that the *qsd1* gene for long dormancy appears to be a mutable gene that might be used for adaptation to different environments and this is consistent with the large variation in *qsd1* sequence that exists in nature. For example, the long dormancy *qsd1* gene could be used to control pre-harvest sprouting in higher rainfall areas to enhance global adaptation of barley. On the other hand, certain haplotypes which carry the *Qsd1* mutation are associated with barley lines that have been developed for industrial uses, such as in the malting and brewing industries. The *Qsd1* mutation that was discovered soon after barley domestication has proved to be essential for the transition of barley utilization from food to beverage in human diets and contributes further to the debate as to what extent the development of ancient agrarian societies was driven by the human appetite for flour and bread, or for beer and alcohol[37].

METHODS

Scoring of Seed Dormancy

Spike samples are harvested at physiological maturity (when the colour of first internode turns to yellow). Spikes are dried 48 h by the dehumidifier at the condition of 30 °C and 10% RH, and then stored at −20 °C until use. Fifty seeds of each sample were exposed to 25 °C for dormancy reduction treatment and germinated for 4 days on moistened filter paper in Petri dishes at 25 °C. Seed non-dormancy was scored as the percentage of germinated seeds. Each germination test was replicated twice.

Population Development and Genetic Mapping

Qsd1 was initially mapped with the 93 doubled haploid phenotype data on a map derived from the cross between the cultivar Haruna Nijo and the wild barley line H602 consisting of 2,890 EST markers[13]. Of the four short dormancy QTLs, *Qsd1* mapped at the centromeric region of the long arm of chromosome 5H. The method of population development is described in Sato *et al.*[14] In brief, BC_3F_1 recombinant chromosome substitution lines (RCSLs) derived from the cross between H602 (donor parent) and Haruna Nijo (recurrent parent) were genotyped and plants heterozygous for a chromosome segment with *Qsd1* were selected. A RSCL carrying the long dormancy allele *qsd1* and short dormancy alleles for other three QTLs was identified[13,14]. Using the initial back-crossed population of this line, a total of 910 F_2 plants and 4,792 F_3 plants derived from the heterozygous F_2 plants were produced to genetically map *Qsd1* (ref.[14]). The segregation of the plants for seed germination (%) fitted a mono-factorial 1:2:1 Mendelian ratio. The direction of dominance was towards reduced dormancy (allele of Haruna Nijo) and the degree of dominance was 0.93, indicating that *Qsd1* shows nearly complete dominance[14]. *Qsd1*/*qsd1* genotype was determined during 35 days of 25 °C dormancy reduction treatment. Because the heterozygotes and non-dormant homozygotes were not clearly separated, the plants showing intermediate dormancy levels were further scored for segregation in F_4 to estimate the genotype of *Qsd1* in F_3. *EST1* and *EST12* markers, which flanked *Qsd1*, were used to select recombinants. The alignment of these two marker sequences with the rice genome revealed a segment carrying 26 rice genes (IRGSP/RAP build 5) that showed high sequence co-linearity with the other 10 barley genes (*EST2* to *EST11*). By scoring dormancy, *Qsd1* was narrowed down to the region of three EST markers (*EST4F*, *EST4R* and *EST5*). A co-segregated marker with an estimated genotype of *Qsd1* was identified in two plants from the F_3 population. From each of these two plants, 18 F_4 plants were used to confirm the respective genotype of F_3 by segregation in the progeny.

BAC Sequencing and Sequence Annotation

The primer set of co-segregated markers with *Qsd1* was used to select BAC clones of the less-dormant parent Haruna Nijo[15] and the dormant parent H602 (ref.[2]). The BAC library of H602 was developed by custom order (Amplicon Express Co.). The inserts of H602 were cloned into the vector pCC1 (Epicenter vector in Invitrogen DH10B T1 resistant) in a *Hind*III cloning site. The average insert size was 180 kb and the total clone number was 172,800 (6 ×). From the BAC library of Haruna Nijo[14], a clone (GenBank ID:LC054174) was sequenced and a contig of 117,759 bp containing the marker sequence was obtained. Another BAC library of H602 was also screened with the *EST4R* marker and a contig sequence of 121,395 bp was obtained from another BAC clone (GenBank ID:LC054175). Each shotgun library from the selected clone was sequenced using a 3730xl sequencer (Applied Biosystems). Reads were assembled by phred/phrap software into contigs. The EST sequence was mapped on contig sequences to confirm the target sequence on the BAC clone. The Rice Genome Automated Annotation System (http://ricegaas.dna.affrc.go.jp/) was used for gene prediction on the contig sequence. Sequence polymorphisms between Haruna Nijo and H602 were detected by the alignment of contig sequences by CLUSTALX (http://www.clustal.org/).

Development of Knockdown Transgenic Plants

A 509-bp cDNA region from*AK372829*was amplified using the forward primer 5′-caccaaagctaggagatgg-3′ and reverse primer 5′-gaacacactgccccaaaagt-3′and cloned into the pENTR/D topo vector (Invitrogen). The construct was recombined (according to the manufacturer›s instructions) into the binary vector pANDA modified to contain a GatewayTM cassette downstream of the maize Ubi-1 promoter[18]. The construct was introduced into the*Agrobacterium*strain EHA101. Immature embryos of cv. Golden Promise were used for*Agrobacterium*-mediated transformation. Transformation was performed by the protocol of Hensel*et al.*[38]In addition, pre-treatment with heat for immature embryos[39]and the dividing method for immature embryos before selection with hygromycin[40]were employed for more efficient transformation. The T_0plants which had single copy inserts, as shown by Southern analysis, were examined further. We obtained several independent transgenic lines, and two (ids. #6 and #7) were used for further analysis. Genotypes of T_1plants were determined by the inserts of segregating T_2progeny and each T_2plant from the T_1plants were further genotyped by segregation in the T_3progeny.

Development of Complementation Transgenic Plants

The recombinant chromosome substitution line was used as a donor of dormant*qsd1*/*qsd1*allele in the genetic background of cv. Golden Promise. Twenty-three B_1F_1plants were genotyped by the Golden Gate Assay (Illumina Co.), using selected 384 genome-wide SNP markers from barley OPA1 (ref.[41]) to identify the substituted segment from H602. Selected B_1F_1plants with the substituted segment were self-pollinated to generate homozygous*qsd1*/*qsd1*substitution lines. *Agrobacterium*-mediated transformation of the substituted line (id. B1H602GP20-7), using a vector construct carrying a cloned Haruna Nijo*Qsd1*genomic DNA, was conducted to complement the*qsd1*allele of the recipient line. The*Qsd1*open reading frame was amplified using the forward primer 5′-cccg ggctcctctgcgccaccttagc-3′ and the reverse primer 5′-ggtacccag cgccaccagagttgaga-3′ and KOD -Plus- DNA polymerase from Haruna Nijo genomic DNA. This fragment was cloned into the pTA vector (TOYOBO) and sequenced. The open reading frame region was digested by*Sac*I and*Hind*III and ligated into binary vector pBUH3 (ref.[40], provided by Dr Y. Ito) and driven by the maize Ubi-1 promoter. We obtained several independent T_0transgenic plants. Segregating T_1plants (ids #1 and #4) were produced from single T_0plants and further genotyped for the*Qsd1*transgene in the T_2progeny.

Phenotypic Analysis of Transgenic Barley

To measure dormancy levels, transgenic barley lines were grown in a growth chamber with 12 h day length and at 15 °C. Expression levels were measured with the same method described below. Dormancy levels were measured at the time when the difference of long dormancy and short dormancy samples were maximized.

AlaAT Homology Model

The barley AlaAT amino acid sequences were aligned to the barley AlaAT crystal structure[24]and manually edited to ensure alignment of homologous catalytic motifs. Gap regions longer than 11 residues were modelled*de novo*using I-TASSER[42]. Modeller[43]was used to construct homology models of AlaAT using the barley AlaAT[24]and*de novo*predictions as templates. Modeller DOPE functions, ProSA[44]and Procheck[45]were used to assess the reliability of candidate models. High-scoring models were passed through a Modeller loop refinement script and final models chosen using Modeller DOPE, Modeller GA32, ProSA and Procheck assessments.

Plant Materials for allelic Diversity Analysis

A total of 365 barley accessions preserved at Okayama University were evaluated for seed dormancy for two growing seasons. Nucleotide polymorphisms were detected by sequencing PCR amplicons from target regions..

Phylogenic Analysis of*Qsd1*

The genomic regions of*Qsd1*in 27 wild and 210 cultivated barley accessions are sequenced. Non-redundant haplotypes were collected and aligned according to their similarities with the haplotypes of H602. The similarities of 54 sequenced haplotypes were calculated by the neighbour joining method of MEGA6 (ref.[46]).

Expression analyses of target genes

Transcript levels of Haruna Nijo and H602 were evaluated. Samples from seedling root, seedling leaf, adult root, adult leaf and maturing seeds (ovules) at 0, 1, 2, 3, 4 and 5 weeks after flowering were measured for*Qsd1*/*qsd1*transcripts and replicated three times. The quantitative PCR analysis was conducted using THUNDERBIRD SYBR qPCR Mix (TOYOBO Co.) on a Step One thermal cycler (Applied Biosystems Co.) with initial conditions of 95 °C, 60 s, 40 cycles of denaturing of 95 °C, 60 s and extension at 60 °C for 60 s. The*qsd1*gene segment comprising part of the 3′-UTR (300 bp) was amplified from cDNA isolated from immature barley spike using specific primers

. The PCR product was cloned into the pBluescript IIKS (+) vector (Stratagene, La Jolla, CA, USA). Two clones with different insert orientations were linearized with*Not*I or*EcoR*I and were used as templates to generate antisense and sense probes using T3 or T7 RNA polymerase.*In situ*hybridization based on a digoxigenin-labelled RNA probe and subsequent immunological detection followed methods described elsewhere[47].

Data availability

BAC clone sequences described in this paper have been deposited in the DDBJ nucleotide database with accession codesLC054174: cv. Haruna Nijo BAC cloneHNB550I12andLC054175: wild barley H602 BAC cloneHSP216G11. The data that support the findings of this study are available from the corresponding author upon request.

ACKNOWLEDGMENTS

We thank Ms N. Ooe and Y. Motoi for their excellent technical assistance, Drs S. Nakamura, M. Pourkheirandish, S. Toki for providing various analytical methods and Dr S. Ito for providing the pBUH3 vector for transformation. Barley seed samples and the Haruna Nijo BAC library were provided through the National Bioresource Project of Barley, MEXT of Japan. We acknowledge financial supports given by the Ministry of Agriculture, Forestry and Fisheries of Japan GD 3004, TRC1002, TRG1002 and TRS1001 to K.S., T.M. and T.K., and by the Japan Science and Technology Agency (CREST) to K.S. and K.T.

REFERENCES

1. von Bothmer R. & Komatsuda T.Barley: Improvement, Production, and Usesed. Ullrich S. E. 14–62John Wiley and Sons, Inc. (2011).
2. Pourkheirandish M.*et al.*Evolution of the seed dispersal system in barley.Cell162, 527–539 (2015).
3. Takeda K. & Hori K.Geographical differentiation and diallel analysis of seed dormancy in barley.Euphytica153, 249–256 (2007).
4. Bewley J. D.Seed germination and dormancy.Plant Cell9, 1055–1066 (1997).
5. Fuller D. Q. & Allaby R. G.Seed dispersal and crop domestication: shattering, germination and seasonality in evolution under cultivation in fruit development and seed dispersal.Ann. Plant Rev.38, 238–295 (2009).
6. Fischbeck G.Diversity in Barley (Hordeum vulgare)eds Bothmer R., von Hintum T., van Knüpffer H., Sato K. 29–52Elsevier Science B.V. (2003).
7. Han F.*et al.*Verification of barley seed dormancy loci via linked molecular markers.Theor. Appl. Genet.92, 87–91 (1996).
8. Nakamura S.*et al.*Mitogen-activated protein kinase kinase 3 regulates seed dormancy in barley.Curr. Biol.26, 775–781 (2016).
9. Hori K., Sato K. & Takeda K.Detection of seed dormancy QTL in multiple mapping populations derived from crosses involving novel barley germplasm.Theor. Appl. Genet.115, 869–876 (2007).
10. Nakamura S.*et al.*A wheat homolog of MOTHER OF FT AND TFL1 acts in the regulation of germination.Plant Cell23, 3215–3229 (2011).
11. Sugimoto K.*et al.*Molecular cloning of Sdr4, a regulator involved in seed dormancy and domestication of rice. Proc. Natl Acad. Sci. USA107, 5792–5797 (2010).
12. Gu X. Y., Liu T., Feng J., Suttle J. C. & Gibbons J.The qSD12 underlying gene promotes abscisic acid accumulation in early developing seeds to induce primary dormancy in rice.Plant Mol. Biol.73, 97–104 (2010).
13. Sato K., Nankaku N. & Takeda K.A high-density transcript linkage map of barley derived from a single population.Heredity103, 110–117 (2009).
14. Sato K., Matsumoto T., Ooe N. & Takeda K.Genetic analysis of seed dormancy QTL in barley.Breed. Sci.59, 645–650 (2009).
15. Saisho D., Myoraku E., Kawasaki S., Sato K. & Takeda K.Construction and characterization of a bacterial artificial chromosome (BAC) library for Japanese malting barley ‹Haruna Nijo›.Breed. Sci.57, 29–38 (2007).
16. Sato K.*et al.*Development of 5006 full-length cDNAs in barley: a tool for accessing cereal genomics resources. DNA Res.16, 81–89 (2009).
17. Matsumoto T.*et al.*Construction and sequence analysis of barley 24,783 full-length cDNAs.Plant Physiol.156, 20–28 (2011).
18. Miki D. & Shimamoto K.Simple RNAi vectors for stable and transient suppression of gene function in rice.Plant Cell Physiol.45, 490–495 (2004).
19. Bentsink L. & Koorneef M.Seed dormancy and germination.Arabidopsis Book6, e0119 (2008).
20. Nonagaki H.Seed dormancy and germination-emerging mechanisms and new hypotheses.Front. Plant Sci.5, 233–246 (2014).
21. The International Barley Genome Sequencing Consortium.A physical, genetic and functional sequence assembly of the barley genome.Nature491, 711–716 (2012).
22. Mihara M., Itoh T. & Izawa T.SALAD database: a motif-based database of protein annotations for plant comparative genomics.Nucleic Acids Res.38, D835–D842 (2010).
23. Sato Y.*et al.*RiceXPro Version 3.0: expanding the informatics resource for rice transcriptome.Nucleic Acids Res.41, D1206–D1213 (2013).
24. Duff S. M.*et al.*The enzymology of alanine aminotransferase (AlaAT) isoforms from*Hordeum vulgare*and other organisms, and the HvAlaAT crystal structure.Arch. Biochem. Biophys.528, 90–101 (2012).
25. Weitbrecht K., Müller K. & Leubner-Metzger G.First off the mark: early seed germination.J. Exp. Bot.62, 3289–3309 (2011).
26. Miyashita Y., Dolferus R., Ismond K. P. & Good A. G.Alanine aminotransferase catalyses the breakdown of alanine after hypoxia in*Arabidopsis thaliana*.Plant J.49, 1108–1121 (2007).
27. Liepman A. H. & Olsen L. J.Alanine aminotransferase homologs catalyze the glutamate:glyoxylate aminotransferase reaction in peroxisomes of Arabidopsis. Plant Physiol.131, 215–227 (2003).
28. Ricoult C., Echeverria L. O., Cliquet J. & Limami A. M.Characterization of alanine aminotransferase (AlaAT) multigene family and hypoxic response in young seedlings of the model legume*Medicago truncatula*.J. Exp. Bot.57, 3079–3089 (2006).
29. Ashok K.*et al.*Genetic engineering of improved nitrogen use efficiency in rice by the tissue-specific expression of alanine aminotransferase.Plant Biotechnol. J.6, 722–732 (2008).
30. Fincher G.Barley: Improvement, Production, and

Usesed. Ullrich S. E. 449–477John Wiley and Sons, Inc. (2011).

31. Bykova N. V.*et al.*Redox-sensitive proteome and antioxidant strategies in wheat seed dormancy control. Proteomics11, 865–882 (2011).
32. Good A. G. & Crosby W. L.Anaerobic induction of alanine aminotransferase in barley root tissue.Plant Physiol.90, 1305–1309 (1989).
33. Rocha M.*et al.*Analysis of alanine aminotransferase in various organs of soybean (*Glycine max*) and in dependence of different nitrogen fertilisers during hypoxic stress.Amino Acids39, 1043–1053 (2010).
34. Borisjuk L. & Rolletschek H.The oxygen status of the developing seed.New Phytol.182, 17–30 (2009).
35. Wang F.*et al.*Embryonal control of yellow seed coat locus ECY1 is related to alanine and phenylalanine metabolism in the seed embryo of*Brassica napus*.Genes Genome Genet.6, 1073–1081 (2016).
36. Fontaine O., Huault C., Pavis N. & Billard J. P.Dormancy breakage of*Hordeum vulgare*seeds: effects of hydrogen peroxide and scarification on glutathione level and glutathione reductase activity.Plant Physiol. Biochem.32, 677–683 (1994).
37. Dietrich O., Heun M., Notroff J., Schmidt K. & Zarnkow M.The role of cult and feasting in the emergence of Neolithic communities. New evidence from Göbekli Tepe, south-eastern Turkey.Antiquity86, 674–695 (2012).
38. Hensel G., Valkov V., Middlefell-Williams J. & Kumlehn J.Efficient generation of transgenic barley: the way forward to modulate plant–microbe interactions.J. Plant Physiol.165, 71–82 (2008).
39. Hiei Y., Ishida Y., Kasaoka K. & Komari T.Improved frequency of transformation in rice and maize by treatment of immature embryos with centrifugation and heat prior to infection with*Agrobacterium tumefaciens*. Plant Cell Tiss. Organ Cult.87, 233–243 (2006).
40. Nigorikawa M.*et al.*Enhanced saccharification of rice straw by overexpression of rice exo-glucanase.Rice5, 14 (2012).
41. Sato K. & Takeda K.An application of high-throughput SNP genotyping for barley genome mapping and characterization of recombinant chromosome substitution lines.Theor. Appl. Genet.119, 613–619 (2009).
42. Roy A., Kucukural A. & Zhang Y.I-TASSER: a unified platform for automated protein structure and function prediction.Nat. Protoc.5, 725–738 (2010).
43. Martí-Renom M. A.*et al.*Comparative protein structure modeling of genes and genomes.Ann. Rev. Biophys. Biomol. Struct.29, 291–325 (2000).
44. Wiederstein M. & Sippl M. J.ProSA-web: interactive web service for the recognition of errors in three-dimensional structures of proteins.Nucleic Acids Res.35, W407–W410 (2007).
45. Laskowski R. A., MacArthur M. W., Moss D. S. & Thornton J. M.PROCHECK - a program to check the stereochemical quality of protein structures.J. Appl. Cryst.26, 283–291 (1993).
46. Tamura K., Stecher G., Peterson D., Filipski A. & Kumar S.MEGA6: molecular evolutionary genetics analysis version 6.0.Mol. Biol. Evol.30, 2725–2729 (2013).
47. Komatsuda T.*et al.*Six-rowed barley originated from a mutation in a homeodomain-leucine zipper I-class homeobox gene.Proc. Natl Acad. Sci. USA104, 1424–1429 (2007).

CHAPTER 5

INVOLVEMENT OF ALTERNATIVE SPLICING IN BARLEY SEED GERMINATION

Qisen Zhang, Xiaoqi Zhang, Songbo Wang, Cong Tan, Gaofeng Zhou, Chengdao Li

INTRODUCTION

Seed germination initiates a transition of biological activities from quiescent to active statuses. Most of the machineries for energy metabolism and protein synthesis are present in dehydrated seeds, but integrity and physiochemical status are different from living cells. One of biological activities quickly activated after seed imbibition is the resumption of respiration. An increase in gas exchanges was observed within minutes upon seed imbibition in maize embryos [1]. Membrane repair is another important biological activity observed in cotton cotyledons [2]. Membrane phospholipid is in a gel status in dry seed [3]. Influx of water onto cells during imbibition disturbs the membrane structure. Solute leakages from cells were observed [4]. Mitochondrial repair is important in the initiation of seed germinations, as shown by the increased in respiration and biosynthesis of cytochrome c oxidase [5]. Cell wall modification was also very active in the germinating embryos [6]. However, limited information is available on whether these processes involved alternatively spliced transcripts and which AS genes played important roles in germinating embryos.

Alternative splicing (AS) increases transcriptomic plasticity and proteomic diversity, which extends gene-coding capacities in eukaryotic genomes. It regulates cell differentiation and many developmental processes [7,8,9]. It generates more than one mRNA variant from precursor mRNA transcripts and regulates transcript levels by introducing premature termination [10,11]. It produces transcript isoforms with altered sequences which target different cellular locations or have different functionalities [12]. It plays important roles in co- and post-transcriptional regulations in *Arabidopsis* [7,8,9] and in soybean embryos [13]. It is involved in many biological processes including nutrient uptake [14], circadian clock, organ differentiation, biotic and abiotic stress responses in *Arabidopsis*; oil formation in soybean; and temperature and light responses in *Physcomitrella patens* [7,8,9,15,16].

Barley is an important cereal crop ranking fourth in worldwide cereal production after maize, rice and wheat. It has been used as animal feed and for health food ingredients, but good quality barley is mainly used for malt in the brewing industry, which is a controlled seed germinating process. Thus, understanding of seed germination will provide valuable information for controlling malting and brewing qualities and also for understanding of seed germination properties such as germination rates and seed vigour. We conducted RNA sequencing (RNA-seq) experiments to dissect common AS events and transcriptional activities in germinating embryos of four barley varieties. Between 552 and 669 common AS transcripts were identified. The AS transcripts were predominantly mapped to pathways/ complexes of ribosome, RNA transporting, spliceosome, oxidative phosphorylation, mRNA surveillance, carbon metabolism and plant hormone signal transduction pathways. Furthermore, since AS transcripts were predominantly found not only in protein and energy metabolic pathways, but also in regulatory pathways including spliceosome and hormone signal transduction, they could play very important roles in seed germination.

MATERIALS AND METHODS

Plant Materials

Four barley varieties of *Hordeum vulgare* L Bass, Baudin, Harrington and Stirling were selected for this study, since we wanted to dissect common AS events. Seeds of the four varieties were obtained from Department of Agriculture and Food, Western Australia. These seeds were harvested from the same field trial in Western Australia. The seeds were tested without pre-harvest sprouting damages and dormancy before

Citation: Zhang Q, Zhang X, Wang S, Tan C, Zhou G, Li C (2016) Involvement of Alternative Splicing in Barley Seed Germination. PLoS ONE 11(3): e0152824. doi:10.1371/journal.pone.0152824.

they were used for germination to extract RNA. For RNA preparation, about 400 seeds were sterilized with 1% (w/v) hypochlorite solution for 15 min, rinsed with running water and transfered them to 10 cm Petri dishes with two layers of filter paper and 16 ml distilled water each. After imbibing for 24 h or 48 at 21 °C in dark, germinating embryos were separated from endosperms and placed in a microfuge tube on dry ice. The embryos were stored at –80 °C before RNA extraction. Total RNA was extracted with phenol-SDS reagents (http://onlinelibrary.wiley.com/doi/10.1002/0471142727.mb0403s09/pdf).

Sequencing and Bioinformatics Analysis

Details of RNA sequencing and bioinformatics analysis methodologies were published previously [6]. A brief summary was described below.

RNA-seq Library Preparation and Sequencing.

Total RNA was extracted and treated with DNase to remove genomic DNA contamination. The quality of the RNA extract was assessed using the Agilent 2100 Bioanalyzer system with a minimum RNA integrated number value of 8. Messenger RNA was enriched by Magnetic beads with Oligo (dT) according to the manufacturer's protocol and fragmented into short fragments. The mRNA products were used as templates to synthesize the first-strand cDNA using reverse transcriptase together with random hexamer primers. The second-strand cDNA was synthesized after adding reaction buffer, dNTPs, RNase H, DNA polymerase and $MgCl_2$ with the remaining mRNA completely removed. Double stranded cDNA was purified using the QIAquick PCR Purification Kit according to the manufacturer's protocol and end-repaired with T4 DNA polymerase and Klenow DNA polymerase. An "A" base was added to the 3' end of DNA fragments. The cDNA was ligated to sequencing adapters with a 'T' base overhang at the 3' end. DNA fragments with lengths from 250 bps to 500 bps were selected from electrophoresis gels and purified before PCR amplification of 15 cycles. The cDNA libraries were validated through the Agilent 2100 Bioanalyzer and ABI StepOnePlus Real-Time PCR system. A paired-end sequencing protocol was performed on the platform of HiSeq2000 (Illumina) [17].

Sequencing data Quality Control.

In order to ensure valid biological conclusions, quality control (QC) and data filtering on the raw data were performed strictly prior to further processes and bioinformatics analysis. As an evaluation of raw data quality, we calculated the composition and distribution of bases (A, T, C, G, N) by using the FastQC program (version 0.11.2,http://www.bioinformatics.babraham.ac.uk/projects/fastqc/). The percentage of A/G should almost equal that of T/C and the rate of N should be less than one per cent. The proportion of low quality (Qs < 20) bases should be less than 10 per cent. If not, a re-sequencing process was activated. In the data filter process, dirty reads with sequencing adapters, reads with N rates greater than 10 per cent and those with more than 50 per cent low quality (Qs < 20) were discarded.

Alignment with Reference Sequences.

In order to select a program for alignment, we compared SOAPaligner/Soap2 with other aligners including Bowtie, SOAP and MAQ. We found SOAPaligner/Soap2 performed better than others. SOAPaligner/Soap2 increased aligning speed dramatically and achieved better alignment rates for total and unique mapping [18]. Therefore, SOAPaligner/Soap2 (version 2.21) was chosen to align clean reads of each sample against the Morex genome (IBSC) [19] with a seed length of 32, allowing at most five mismatches in a read and maximal insert size of 1000 bp.

QC of Alignments.

SOAPcoverage was used for QC of the alignments. Coverage, depth and mapped rates (including unique mapped reads, total mapped read and unmapped reads) were assessed.

Differential expression analysis.

The mRNA levels of genes were expressed by RPKM (Reads Per Kilo base-pair per Million mapped reads), which can eliminate the effects of gene length and data amount.

$$\mathrm{RPKM} = \frac{10^6 C}{NL/10^3}$$

Provided RPKM is the expression level of gene X, C is the number of reads mapped perfectly on gene X, N is the total number of reads uniquely mapped on the reference genes and L is the length of gene X. A strict algorithm—developed on the theoretical basis of the Poisson distribution and Multiple Testing [20]—was used to calculate FDR and screen DEGs between different groups [21]. The significance of differentially expressed genes under different conditions was judged by a stringent threshold of FDR ≤ 0.001 and an absolute value of $\log_2$ratio ≥ 1.

Alternative Splicing Transcript Identification.

In order to detect alternative splicing evens, we had evaluated a few software for detecting splicing junctions, including TopHat, SpliceMap and MapSplice and SOAPsplice (version 1.10) [22]. We selected SOAPsplice for this work, since it produced the highest call rate (detected true junction number/true junction in total) and the lowest false positive rate (detected false junction number/the number of junctions all detected). Firstly, SOAPsplice mapped complete reads to the reference genome. Secondly, the unmapped (IUM) reads were mapped with the spliced alignment algorithm. Basically, SOAPsplice divided the IUM reads into two segments, which were derived from different exons in the premature mRNA. SOAPsplice first detected the longest 5' end segment of an IUM read that was mapped to the reference, then aligned the remaining segment to the reference sequences. Major criteria were:

1. Each segment should be longer than 8 bp.
2. No mismatch and no gaps were allowed in the alignment of each segment.
3. Distance of two segments was expected to range between 50 bp and 50,000 bp. Majority of known intron size was in this range in Eukaryote [23].
4. The boundary nucleotides of an intron should be in

the form of "GT-AG", "GC-AG" or "AT-AC". When spliced alignment produced multiple hits, a splice junction candidate with "GT-AG" boundary was given the highest priority, followed by candidates with "GC-AG" and "AT-AC" boundaries [24].

5. When the segments had multiple hits to the reference, consideration was given to the cases where one segment had a unique hit while the other had multiple hits. The closest pair of hit was reported for this read. SOAPsplice ignored the other cases.

6. when the reads spanned more than 2 exons, an additional step was applied to detect junctions with reads longer than 50 bp. For the reads between 50 bp—100 bp, SOAPsplice spliced the read into two equal size segments. With reads longer than 100 bp, SOAPsplice spliced the reads into multiple segments of 50 bp from the 5' end until the remaining segment was in length between 50 and 100 bp, which was then spliced into two equal segments. SOAPsplice considered each segment as a sub-read, and treated it with the above alignment protocol.

7. Finally, SOAPsplice checked and concatenated the separated alignment hits for sub-reads to build the alignment for the original read [22].

Functional and Pathway Analysis.

To investigate the biological implications, AS transcripts were mapped onto biological pathways using the blast tool in http://www.genome.jp/kaas-bin/kaas_main [25]. The AS transcripts were those with the same type of AS occurred in at least three out of four barley varieties at 24 h and 48 h. The protein sequences were downloaded from IBSC database (ftp://ftpmips.helmholtz-muenchen.de/plants/barley/public_data/) and used for the mapping.

The RNA-seq data have been submitted to NCBI GEO database (accession number GSE66024).

RESULTS

RNA-seq read stat istics and alignments to barley reference genome

RNA-seq generated about 26 million reads which accounted for at least 2.3 G base pairs for each barley library. More than 67% and 45% of the reads were mapped to genomic and annotated gene sequences, respectively (International Barley Sequencing Consortium, IBSC) [19]. Between 7800 and 8700 novel transcripts were detected in the eight libraries, which were equivalent to about 30% of high confident genes. Gene coverage was used to calculate the percentage of a gene covered by the reads. The RNA-seq data showed that the majority of the genes (60%) had 90–100% gene coverage and another 25% of genes had at least 50% gene coverage.

Numbers and types of AS events in germinating embryos

Large numbers of AS events (2635–3901) or genes (2229–3201) were detected in the germinating barley embryo (Table 1). About 1.2 AS events per AS gene were estimated (Table 1). The current gene annotation in barley has 15754 intron-containing genes [19]. Thus, 14–20% of intron-containing genes underwent AS during in germinating barley embryos. Moreover, alternative 3' splicing (A3S) was the major AS event in the germinating barley embryo, accounting for 39–46% of total AS events, while intron retention (IR) and alternative 3' splicing (A5S) accounted for 29–36% and 18–23% of total AS events, respectively. Of other AS event types, 2–3% was exon skipping, while alternative first exon and alternative last exon were negligible, and the mutually-exclusive exon was not detectable. When we calculated the numbers of AS transcripts at 24 h and 48 h germination, we have found that a half or over a half AS transcripts detected at 24 h had disappeared at 48 h (45–74%, Table 1). Almost same number of new AS transcripts were detected at 48 h germination. When analyses for common AS transcripts detected in at least three out of four barley varieties at 24 h and 48 h, the number of AS genes were 669 and 549 at 24 h and 48 h, respectively (Table 2). The percentages of A3S, IR and A5S transcripts over total AS transcripts were 34–45%, 32–34% and 16–21%, respectively. Once again, about a half of commonly detected AS transcripts identified at 24 h were not present at 48 h (Table 2).

Table 1: Alternative splicing events and genes

Varieties	AS types	Events		Genes		Turnover
		24 h	48 h	24 h	48 h	
Bass	Alternative 3' splicing	1710 (43)	1203 (42)	1388	1037	888 (64)
	Intron retention	1166 (29)	1033 (36)	927	849	501 (54)
	Alternative 5' splicing	823 (21)	521 (18)	725	462	537 (74)
	Exon skipping	189 (4)	81 (2)	148	68	
	Alternative first exon	12 (0)	8 (0)	12	8	
	Alternative last exon	1 (0)	3 (0)	1	3	
	Total	**3901**	**2849**	**3201**	**2427**	
Harrington	Alternative 3' splicing	1217 (46)	1148 (43)	1022	979	603 (59)
	Intron Retention	781 (29)	836 (31)	635	737	286 (45)
	Alternative 5' splicing	547 (20)	559 (21)	490	511	319 (65)
	Exon skipping	74 (2)	87 (3)	66	72	
	Alternative first exon	14 (0)	10 (0)	14	10	
	Alternative last exon	2 (0)	1 (0)	2	1	
	Total	**2635**	**2641**	**2229**	**2310**	
Stirling	Alternative 3' splicing	1447 (40)	1180 (39)	1213	1004	716 (59)
	Intron Retention	1199 (33)	1046 (35)	965	833	521 (54)
	Alternative 5' splicing	746 (21)	627 (21)	649	565	422 (65)
	Exon skipping	130 (3)	92 (3)	101	77	
	Alternative first exon	12 (0)	3 (0)	12	3	
	Alternative last exon	3 (0)	3 (0)	3	3	
	Total	**3537**	**2951**	**2943**	**2485**	
Baudin	Alternative 3' splicing	1504 (43)	1287 (43)	1216	1083	717 (59)
	Intron Retention	1015 (29)	897 (30)	823	728	453 (55)
	Alternative 5' splicing	795 (23)	646 (21)	698	575	475 (68)
	Exon skipping	116 (3)	107 (3)	101	91	
	Alternative first exon	8 (0)	14 (0)	8	14	
	Alternative last exon	3 (0)	5 (0)	3	5	
	Total	**3441**	**2956**	**2849**	**2496**	

Table 2: Number of AS transcripts commonly occurring in at least three out of four barley varieties

AS types	24 h	48 h	Turnover
A3S	299 (45)	228 (34)	155 (52)
IR	229 (34)	213 (32)	105 (46)
A5S	141 (21)	108 (16)	80 (57)
Total	669	549	

Mapping AS Transcripts onto Biological Pathways

The AS transcripts were mapped on to KEGG pathways of ribosome, RNA transport, spliceosome, oxidative phosphorylation, mRNA surveillance and over 100 other biological pathways (S4–S10 Tables). The top 20 pathways with most AS transcript numbers were mainly responsible for protein synthesis (ribosome, RNA transport, protein processing in endoplasmic reticulum and biosynthesis of amino acids), energy and carbon metabolism (glycolysis, starch and sugar metabolism and carbon metabolism) and post-transcriptional regulation (spliceosome, and ubiquitin mediated proteolysis). The genes undergone AS in germination embryos also included those coding for membrane repair, phytohormone responsive (S4–S10Tables) and cell wall modification [6]. An auxin efflux carrier (MLOC_12686), three AUX/IAA responsive transcripts (MLOC_ *MLOC_54787*, *MLOC_75842* and *MLOC_65332*) and a serine/threonine protein kinase SRK2 (SnRK2) (*MLOC_22145*) gene were among them.

Transcription of AS genes

The mRAN levels of AS genes were shown in S11–S16 Tables. Some of the AS genes could be highly expressed in the germinating embryos (>1000 RPKM). The mRNA levels were low to intermediate for most of the genes (10–1000 RPKM). There was not significant difference in transcript levels for most genes between 24 h and 48 h.

Gene Expression Correlation Analysis

Coordinative regulation of phytohormones play key roles in seed germination [26]. After identifications of several AS hormone responsive transcripts (Fig 1), we conducted gene expression correlation analyses to investigate co-expressed genes in germinating embryos. The co-expressed genes with high correlation values (> = 0.9 or < = -0.9) were then mapped to KEGG pathways (http://www.genome.jp/kaas-bin/kaas_main). These co-expressed genes were mapped onto 50 to 120 KEGG pathways (Table 3, S17–S21 Tables). Genes coding for Auxin efflux carrier and Auxin responsive proteins (AUX/IAA) had many co-expressed genes, while Serine/threonine protein kinase SRK2 gene (SnRK2) had fewer co-expressed genes in protein synthesis and energy metabolism pathways (Table 3).

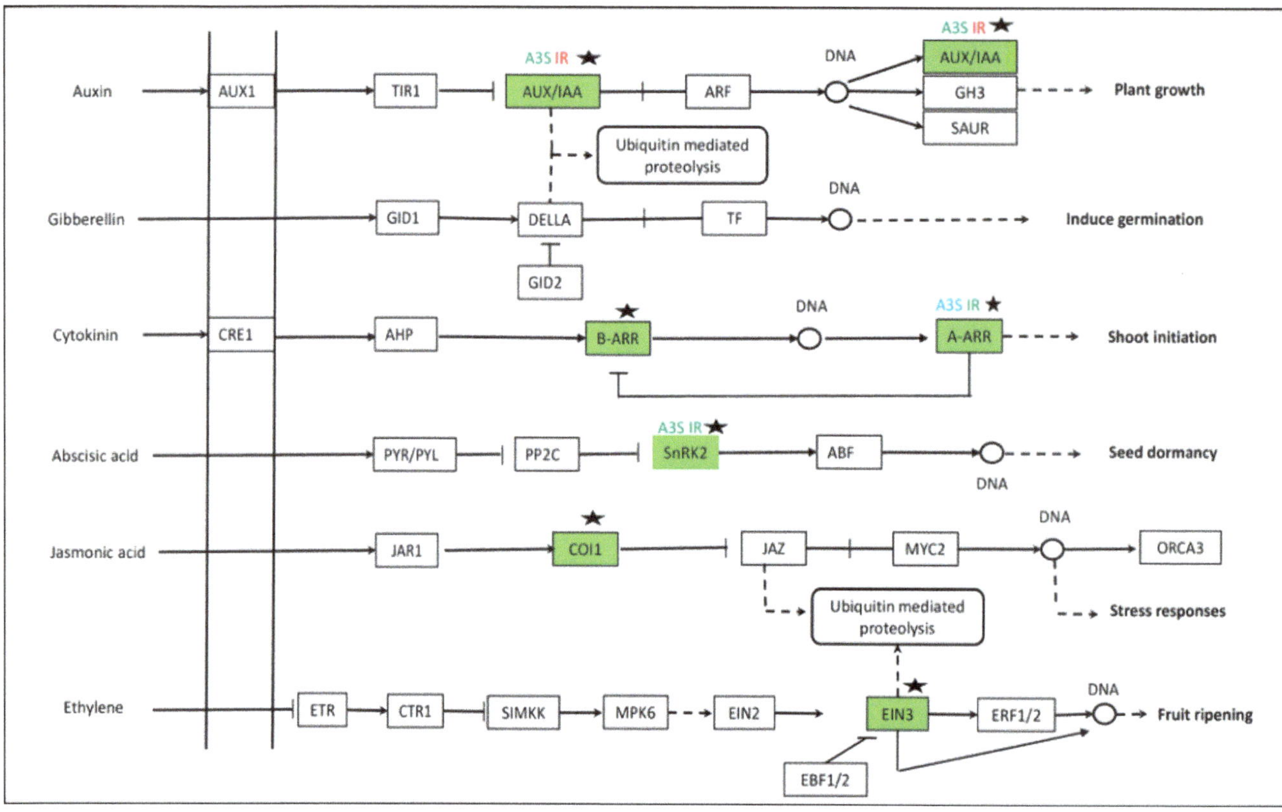

Figure 1: AS transcripts on plant hormone signal transduction pathway.

AS transcripts were indicated by A3S, IR or A5S on tops of the corresponding genes. Blue and red colors indicated AS at 24 h and 48 h, respectively, while green color indicated that AS was detected at both 24 h and 48 h. AUX/IAA: Auxin-responsive protein IAA (MLOC_65332, A3S, MLOC_54787, IR and MLOC_75842, IR); JAZ: Jasmonate ZIM domain-containing protein (MLOC_12120, IR); ARR-A: Two-component response regulator A family (MLOC_64100, A3S, IR, MLOC_14492, A3S); SnRK2: Serine/threonine-protein kinase SRK2 (MLOC_69212,A3S, IR; MLOC_3013, IR; MLOC_63787, A3S, MLOC_22145, A3S, IR). The black stars indicated mRNA levels at 54–100 RPKM. AUX1: Auxin influx carrier; TIR1: Transport inhibitor response 1; AUX/IAA: Auxin-responsive protein IAA (MLOC_58812, MLOC_60624, MLOC_54255); ARF: Auxin response factor; GH3: Auxin response GH3 gene family; SAUR: SAUR family protein; GID1: Gibberellin receptor GID1; DELLA: DELLA protein; GID2: F-box protein GID2; TF: Phytochrome-interacting factor 4; CRE1:Cytokinin receptor; AHP: Histidine-containing phosphor-transfer protein; B-ARR: Two-component response regulator ARR-B (MLOC_3955); ARR-A: Two-component response regulator A family (MLOC_64100); PYR/PYL: Abscisic acid receptor PYR/PYL family; PP2C: Protein phosphatase 2C; SnRK2: Serine/threonine-protein kinase SRK2 (MLOC_62759, MLOC_69212); ABF: ABA response element binding factor; JAR1: Jasmonic acid-amino synthetase; COI1: Coronatine-insensitive protein1 (MLOC_4800); JAZ: Jasmonate ZIM domain-containing protein (MLOC_12120); MYC2: Transcription factor MYC2; ETR: Ethylene receptor; CTR1: Serine/threonine-protein kinase; SIMKK: Mitogen-activated protein kinase kinase; MPK6: Mitogen-activated protein kinase C; EIN2: Ethylene-insensitive protein 2; EIN3: Ethylene-insensive protein 3 (MLOC_14619); EBF1/2: EIN3-binding F-box protein; ERF1: Ethylene-responsive transcription factor 1.

Table 3: Co-expression of AS hormone responsive genes with genes responsible for protein synthesis and energy metabolism

Pathways	PIN *MLOC_12686*	AUX/IAA *MLOC_54787*	AUX/IAA *MLOC_75842*	SnRK2 *MLOC_22145*
Ribosome	5	2	12	0
Biosyn AA	8	3	2	2
RNA transport	9	7	1	0
Carbon Met	9	6	3	2
Oxidative phos	4	2	0	0
Spliceosome	1	6	0	0
Plant h s trans	1	2	1	3

Co-expression of AS auxin efflux carrier (PIN), auxin responsive protein (AUX/IAA) and serine/threonine protein kinase SRK2 (SnRK2) genes with other genes were analysed. The co-expressed genes with high correlation values (> = 0.9 or < = -0.9) were mapped onto KEGG pathways (http://www.genome.jp/kaas-bin/kaas_main). The number of co-expressed genes for each hormone responsive protein were listed for ribosome, biosynthesis of amino acid (Biosy AA), RNA transport, carbon metabolism (Carbon Met), oxidative phosphorylation (Oxidative phos), spliceosome and plant hormone signal transduction (Plant h s trans) pathways.

doi:10.1371/journal.pone.0152824.t003

Auxin Efflux Carrier.

Auxin efflux carrier transcript (*MLOC_12686*) underwent IR type of AS (S10 Table). It had over 354 co-expressed genes (S17 Table). They were mapped onto over 280 KEGG pathways (S17 Table). Pathways of protein synthesis (ribosome, biosynthesis of amino acid, RNA transport) and energy metabolism (carbon metabolism, oxidative phosphorylation) were on the top of the list with most gene numbers. It also co-expressed with many genes functioning in DNA replication, cell cycle, purine metabolism and nucleotide excision repair pathways. Expression of auxin efflux carrier gene was negatively correlated with expressions of several metabolic genes in glycolysis and TCA cycle (Table 4).

Table 4: Co-expression of AS hormone responsive genes with genes functioning in energy metabolism pathways

AS hormone responsive proteins	Co-expressed genes	Corr values
Auxin efflux carrier (MLOC_12686)	Hexokinase (MLOC_53317)	-0.93
	Glyceraldehyde-3-phosphate dehydrosenase (MLOC_18233)	-9.97
	Pyruvate dehydrogenase E1 (MLOC_53947)	0.92
	Malate dehydrogenase (MLOC_61949)	-0.92
	Citrate synthase (MLOC_66594)	-0.92
AUX/IAA (MLOC_54787)	Hexokinase (MLOC_53317)	0.94
	Glucose-6-phosphate isomerase (MLOC_1497)	0.92
	6-Phosphofructokinase (MLOC_67234)	0.94
	6-Phosphofructokinase (MLOC_11145)	0.93
	Pyruvate dehydrogenase E1 (MLOC_63372)	0.93
	Isocitrate dehydrogenase (MLOC_69600)	0.9
AUX/IAA (MLOC_75842)	Glyceraldehyde-3-phosphate dehydrosenase (MLOC_52515)	0.94
SnRK2 (MLOC_22145)	2,3-Bisphophoglycerate-independent phosphoglycerate mutase (MLOC_52687)	-0.96

Co-expression of AS auxin efflux carrier, auxin responsive protein (AUX/IAA) and serine/threonine protein kinase SRK2 (SnRK2) genes with energy metabolism genes were analysed. The co-expressed genes with high correlation values (> = 0.9 or < = -0.9) were listed. Corr: correlation

doi:10.1371/journal.pone.0152824.t004

AUX/IAA

We identified three AS *AUX/IAA* transcripts (*MLOC_54787*, *MLOC_75842* and *MLOC_65332*). Two of them (*MLOC_54787* and *MLOC_75842*) underwent IR type of AS (S10 Table), while the third (*MLOC_65332*) underwent A3S (S5 and S6 Tables). Both IR *AUX/IAA* transcripts have over 320 co-expressed genes (S18 and S19 Tables). These co-expressed genes were mapped onto over 100 KEGG pathways. The genes co-expressed with AS *AUX/IAA* gene (*MLOC_75842*) was preferably mapped onto ribosome pathway, but the genes co-expressed with other *AUX/IAA* gene *(MLOC_54787)* were preferably mapped on to carbon metabolism and oxidative phosphorylation pathways (Table 3). The AS *AUX/IAA* gene (*MLOC_54787*) could play more important roles in energy metabolism, while the other (*MLOC_75842*) was likely related to protein synthesis as shown by the number of co-expressed genes in corresponding pathways (Table 3). Unlike AS auxin efflux carrier, expressions of both IR *AUX/IAA* genes were positively correlated with expressions of several genes in glycolysis and TCA cycle (Table 4). The third A3S *AUX/IAA* gene (*MLOC_65332*) had only nine co-expressed genes (S21 Table).

SnRK2.

SnRK2 gene (SnRK2, *MLOC_22145*) underwent both A3S and IR type of AS (Fig 1, S6 Table,S10 Table). It had 138 co-expressed genes (S20 Table). Two co-expressed genes were mapped to carbon metabolism pathways and three co-expressed genes were mapped onto spliceosome. It was negatively correlated with the expression of a gene in glycolysis (*MLOC_52687*, Table 4).

DISCUSSION

Number of AS Transcripts in Germinating Embryos

We have identified a large number of AS transcripts in geminating embryos (Tables 1 and 2). The total number of AS transcripts is roughly consistent at a certain time point in each barley variety (2200–3900). However, because AS is dynamically regulated and half turnover time is approximately 24 hours, i. e. about a half of the AS transcripts detected at 24 h have disappeared at 48 h and new AS transcripts are detected (Tables 1 and 2), number of genes undergoing AS could be very large during seed germination. Furthermore, the major AS type in barley embryos is the same as in various maize tissues [27], but differs from other plant species. In germinating barley embryos, the major AS type is A3S (34–45%), while in rice,*Sorghum bicolor*, maize, soybean and *Arabidopsis*, IR is the major AS type [28,29,30,31].

Potential Roles of AS in Seed Germination

AS regulates many different biological processes. It plays key roles in abiotic stress tolerance in plants[11]. Two AS variants of heat shock transcription factors are generated under heat stress in *Arabidopsis*—one targets for nonsense-mediated decay while the other short form binds to heat shock transcription factor promoter and activates self-transcription. Other stress-induced AS transcripts code for rice drought-regulated dehydration-responsive element binding protein 2 [32], potato cold-regulated invertase [33], invertase inhibitor [34] and indeterminate domain 14 transcription factor [35]. All AS variants of these transcripts have different biological functions and add additional capacities for environmental adaptations. Several splicing regulators also undergo AS in stress conditions including splicing regulators 30 and 45a in*Arabidopsis* [9]. Variants of AS transcripts also regulate dormancy and germination in plants. Transcripts of Viviparous 1 (VP1) [36] and ABI3 [37], have been detected to undergo AS. VP1 is a regulator of seed maturation. Mutation of VP1 gene disrupted seed maturation and promoted germination in maize [38,39]. Orthologues of maize VP1 genes were *ABI3* in*Arabidopsis* [40], *PtABI3* in poplar [41], *vp-1* in wheat [36], *Osvp1* in rice [42] and *Afvp1* in wild oat [43]. All variants of AS *VP1* orthologue transcripts had different effects on seed germination. For example, several variants of AS *VP1* transcripts were detected in wheat embryos [36]. Some of them causes pre-harvest sprouting [36]. In this study, we have identified AS transcripts commonly detected in four barley varieties. These AS transcripts could possibly affect protein synthesis, energy metabolism and post-transcriptional regulations, since they participated in the activities of these pathways as shown that they were predominately mapped on to ribosome, RNA transport, biosynthesis of amino acid, protein processing in endoplasmic reticulum, glycolysis, oxidative phosphorylation, spliceosome, ubiquitin-mediated proteolysis and plant signal transduction pathways (S4 Table).

Since VP1 gene played very important roles in both seed maturation and seed germination, we have closely examined the changes in its transcript level and AS patterns. Barley genome contains one annotated *VP1* gene (MLOC_69727) (IBSC, [19]). The transcript levels were consistent among the four barley varieties ranging at 20–32 RPKM (S11 Table). However, this gene underwent A3S at 24 h and 48 h, A5S at 48 h and IR type of AS at 48 h (S5 and S6Tables, S8 and S10 Tables), implicating that AS variants of VP1 could be involved in seed germination.

Furthermore, synthesis of auxin was detected as early as the second day in germinating common bean seeds [44]. Auxin also plays key roles in control seed dormancy in a coordinative network of auxin and ABA signalling [45]. Gene expression correlation analysis showed that AS auxin efflux carrier and AUX/IAA transcripts could play key roles in regulating biological processes in germinating embryos, since expressions of these hormone responsive genes was correlated with the expression of genes in protein synthesis and energy metabolism pathways (Fig 1, Tables 3 and 4). Barley genome contained 12 annotated Auxin efflux carrier genes in chromosome 1H, 3H, 5H, 6H and 7H (IBSC). Only one of them from chromosome 7H (MLOC_12686) underwent IR type of AS at both 24 h and 48 h germination. This gene was expressed in all four barley varieties with mRNA levels ranging from 50–158 RPKM. Furthermore, this gene was co-expressed with 355 other genes. Many of them functioned in protein synthesis and energy metabolism (S17 Table).

CONCLUSIONS

Alternative splicing could play important roles in seed germination. It was involved in most active biological processes including protein synthesis, carbon and energy metabolism in the germinating embryos.

AUTHOR CONTRIBUTIONS

Conceived and designed the experiments: QZ CL. Performed the experiments: QZ XZ. Analyzed the data: QZ CT GZ. Contributed reagents/materials/analysis tools: SW. Wrote the paper: QZ.

REFERENCES

1. Botha FC, Potgieter GP, Botha AM (1992) Respiratory metabolism and gene expression during seed germination. Plant Growth Regulation 11: 211–224. doi: 10. 1007/bf00024560
2. Sandoval JA, Huang ZH, Garrett DC, Gage DA, Chapman KD (1995) N-acylphosphatidylethanolamine in dry and imbibing cottonseeds. Amounts, molecular species, and enzymatic synthesis. Plant Physiology 109: 269–275. pmid:7480326 doi: 10. 1104/pp. 109. 1. 269
3. Crowe JH, Crowe LM (1992) Membrane integrity in anhydrobiotic organisms: Toward a mechanism for stabilizing dry cells. In: Somero G, Osmond C, Bolis C, editors. Water and Life: Springer Berlin Heidelberg. pp. 87–103.
4. Bewley J (1997) Seed germination and dormancy. Plant Cell 9: 1055–1066. pmid:12237375 doi: 10. 1105/tpc. 9. 7. 1055
5. Ehrenshaft M, Brambl R (1990) Respiration and mitochondrial biogenesis in germinating embryos of maize. Plant Physiology 93: 295–304. pmid:16667450 doi: 10. 1104/pp. 93. 1. 295
6. Zhang Q, Zhang X, Pettolino F, Zhou G, Li C (2016) Changes in cell wall polysaccharide composition, gene transcription and alternative splicing in germinating barley embryos. J Plant Physiol 191: 127–139. doi: 10. 1016/j. jplph. 2015. 12. 007. pmid:26788957
7. Eckardt NA (2013) The plant cell reviews alternative splicing. Plant Cell 25: 3639. doi: 10. 1105/tpc. 113. 251013. pmid:24179133
8. Reddy ASN, Marquez Y, Kalyna M, Barta A (2013) Complexity of the alternative splicing landscape in plants. Plant Cell 25: 3657–3683. doi: 10. 1105/tpc. 113. 117523. pmid:24179125
9. Staiger D, Brown JW (2013) Alternative splicing at the intersection of biological timing, development, and stress responses. Plant Cell 25: 3640–3656. doi: 10. 1105/tpc. 113. 113803. pmid:24179132
10. Kalyna M, Simpson CG, Syed NH, Lewandowska D, Marquez Y, Kusenda B, et al. (2012) Alternative splicing and nonsense-mediated decay modulate expression of important regulatory genes in *Arabidopsis*. Nucleic Acids Research 40: 2454–2469. doi: 10. 1093/nar/gkr932. pmid:22127866
11. Liu J, Sun N, Liu M, Liu J, Du B, Wang X, et al. (2013) An autoregulatory loop controlling *Arabidopsis* HsfA2 expression: Role of heat shock-induced alternative splicing. Plant Physiology 162: 512–521. doi: 10. 1104/pp. 112. 205864. pmid:23503691
12. Syed NH, Kalyna M, Marquez Y, Barta A, Brown JWS (2012) Alternative splicing in plants coming of age. Trends in plant science 17: 616–623. doi: 10. 1016/j. tplants. 2012. 06. 001. pmid:22743067
13. Aghamirzaie D, Nabiyouni M, Fang Y, Klumas C, Heath L, Grene R, et al. (2013) Changes in RNA splicing in developing soybean (Glycine max) embryos. Biology 2: 1311–1337. doi: 10. 3390/biology2041311. pmid:24833227
14. Li W, Lin W-D, Ray P, Lan P, Schmidt W (2013) Genome-wide detection of condition-sensitive alternative splicing in *Arabidopsis* roots. Plant Physiology 162: 1750–1763. doi: 10. 1104/pp. 113. 217778. pmid:23735510
15. Wu HP, Su YS, Chen HC, Chen YR, Wu CC, Lin WD, et al. (2014) Genome-wide analysis of light-regulated alternative splicing mediated by photoreceptors in*Physcomitrella patens*. Genome Biology 15: 2014–2015. doi: 10. 1186/gb-2014-15-1-r10
16. Chang C-Y, Lin W-D, Tu S-L (2014) Genome-wide analysis of heat-sensitive alternative splicing in *Physcomitrella patens*. Plant Physiology 165: 826–840. pmid:24777346 doi: 10. 1104/pp. 113. 230540
17. Pease J, Sooknanan R (2012) A rapid, directional RNA-seq library preparation workflow for Illumina[reg] sequencing. Nat Meth 9.
18. Li R, Yu C, Li Y, Lam T-W, Yiu S-M, Kristiansen K, et al. (2009) SOAP2: an improved ultrafast tool for short read alignment. Bioinformatics 25: 1966–1967. doi: 10. 1093/bioinformatics/btp336. pmid:19497933
19. Mayer K, Waugh R, Langridge P, Close TJ, Wise RP, Graner A, et al. (2012) A physical, genetic and functional sequence assembly of the barley genome. Nature 491: 711–716. doi: 10. 1038/nature11543. pmid:23075845
20. Benjamini Y, Yekutieli D (2001) The control of the false discovery rate in multiple testing under dependency. The Annals of Statistics 29: 1165–1188. doi: 10. 1214/aos/1013699998
21. Audic S, Claverie J-M (1997) The significance of digital gene expression profiles. Genome Research 7: 986–995. pmid:9331369
22. Huang S, Zhang J, Li R, Zhang W, He Z, Lam T-W, et al. (2011) SOAPsplice: Genome-wide detection of splice junctions from RNA-Seq data. Frontiers in Genetics 2: 46. doi: 10. 3389/fgene. 2011. 00046. pmid:22303342
23. Trapnell C, Pachter L, Salzberg SL (2009) TopHat: discovering splice junctions with RNA-Seq. Bioinformatics 25: 1105–1111. doi: 10. 1093/bioinformatics/btp120. pmid:19289445
24. Burset M, Seledtsov IA, Solovyev VV (2000) Analysis of canonical and non-canonical splice sites in mammalian genomes. Nucleic Acids Research 28: 4364–4375. pmid:11058137 doi: 10. 1093/nar/28. 21. 4364
25. Kanehisa M, Goto S, Sato Y, Kawashima M, Furumichi M, Tanabe M (2014) Data, information, knowledge and principle: back to metabolism in KEGG. Nucleic Acids Research 42: D199–D205. doi: 10. 1093/nar/gkt1076. pmid:24214961
26. Vanstraelen M, Benkova E (2012) Hormonal interactions

in the regulation of plant development. Annu Rev Cell Dev Biol 28: 463–487. doi: 10. 1146/annurev-cellbio-101011-155741. pmid:22856461

27. Thatcher SR, Danilevskaya ON, Meng X, Beatty M, Zastrow-Hayes G, Harris C, et al. (2016) Genome-Wide Analysis of Alternative Splicing during Development and Drought Stress in Maize. Plant Physiology 170: 586–599. doi: 10. 1104/pp. 15. 01267. pmid:26582726
28. Marquez Y, Brown JW, Simpson C, Barta A, Kalyna M (2012) Transcriptome survey reveals increased complexity of the alternative splicing landscape in *Arabidopsis*. Genome Research 22: 1184–1195. doi: 10. 1101/gr. 134106. 111. pmid:22391557
29. Shen Y, Zhou Z, Wang Z, Li W, Fang C, Wu M, et al. (2014) Global dissection of alternative splicing in paleopolyploid soybean. The Plant Cell 26: 996–1008. doi: 10. 1105/tpc. 114. 122739. pmid:24681622
30. Thatcher SR, Zhou W, Leonard A, Wang B- B, Beatty M, Zastrow-Hayes G, et al. (2014) Genome-wide analysis of alternative splicing in Zea mays: Landscape and genetic regulation. The Plant Cell Online 26: 3472–3487. doi: 10. 1105/tpc. 114. 130773
31. Min X, Powell B, Braessler J, Meinken J, Yu F, Sablok G (2015) Genome-wide cataloging and analysis of alternatively spliced genes in cereal crops. BMC Genomics 16: 721. doi: 10. 1186/s12864-015-1914-5. pmid:26391769
32. Matsukura S, Mizoi J, Yoshida T, Todaka D, Ito Y, Maruyama K, et al. (2010) Comprehensive analysis of rice DREB2-type genes that encode transcription factors involved in the expression of abiotic stress-responsive genes. Molecular Genetics and Genomics 283: 185–196. doi: 10. 1007/s00438-009-0506-y. pmid:20049613
33. Bournay AS, Hedley PE, Maddison A, Waugh R, Machray GC (1996) Exon skipping induced by cold stress in a potato invertase gene transcript. Nucleic Acids Research 24: 2347–2351. pmid:8710506 doi: 10. 1093/nar/24. 12. 2347
34. Brummell DA, Chen RK, Harris JC, Zhang H, Hamiaux C, Kralicek AV, et al. (2011) Induction of vacuolar invertase inhibitor mRNA in potato tubers contributes to cold-induced sweetening resistance and includes spliced hybrid mRNA variants. Journal of Experimental Botany 62: 3519–3534. doi: 10. 1093/jxb/err043. pmid:21393382
35. Seo PJ, Kim MJ, Ryu JY, Jeong EY, Park CM (2011) Two splice variants of the IDD14 transcription factor competitively form nonfunctional heterodimers which may regulate starch metabolism. Nat Commun 2: 303. doi: 10. 1038/ncomms1303. pmid:21556057
36. McKibbin RS, Wilkinson MD, Bailey PC, Flintham JE, Andrew LM, Lazzeri PA, et al. (2002) Transcripts of Vp-1 homeologues are misspliced in modern wheat and ancestral species. Proceedings of the National Academy of Sciences 99: 10203–10208. doi: 10. 1073/pnas. 152318599
37. Sugliani M, Brambilla V, Clerkx EJM, Koornneef M, Soppe WJJ (2010) The Conserved splicing factor SUA controls alternative splicing of the developmental regulator ABI3 in*Arabidopsis*. Plant Cell 22: 1936–1946. doi: 10. 1105/tpc. 110. 074674. pmid:20525852
38. McCarty DR, Carson CB, Stinard PS, Robertson DS (1989) Molecular analysis of viviparous-1: An abscisic acid-insensitive mutant of maize. Plant Cell 1: 523–532. pmid:12359900 doi: 10. 2307/3868973
39. McCarty DR, Hattori T, Carson CB, Vasil V, Lazar M, Vasil IK (1991) The Viviparous-1 developmental gene of maize encodes a novel transcriptional activator. Cell 66: 895–905. pmid:1889090 doi: 10. 1016/0092-8674(91)90436-3
40. Giraudat J, Hauge BM, Valon C, Smalle J, Parcy F, Goodman HM (1992) Isolation of the *Arabidopsis* ABI3 gene by positional cloning. The Plant Cell 4: 1251–1261. pmid:1359917 doi: 10. 2307/3869411
41. Rohde A, Adiles-Diaz W, Montagu MV, Boerjan W (1998) Isolation and expression analysis of an ABSCISIC ACID-INSENSITIVE 3 (ABI3) homologue from Populus trichocarpa. Journal of Experimental Botany 49: 1059–1060. doi: 10. 1093/jxb/49. 323. 1059
42. Hattori T, Terada T, Hamasuna S (1994) Sequence and functional analyses of the rice gene homologous to the maize Vp1. Plant Molecular Biology 24: 805–810. pmid:8193305 doi: 10. 1007/bf00029862
43. Jones HD, Peters NCB, Holdsworth MJ (1997) Genotype and environment interact to control dormancy and differential expression of the VIVIPAROUS 1 homologue in embryos of Avena fatua. The Plant Journal 12: 911–920. pmid:9375401 doi: 10. 1046/j. 1365-313x. 1997. 12040911. x
44. Bialek K, Michalczuk L, Cohen JD (1992) Auxin biosynthesis during seed germination in *Phaseolus vulgaris*. Plant Physiol 100: 509–517. pmid:16652991 doi: 10. 1104/pp. 100. 1. 509
45. Liu X, Zhang H, Zhao Y, Feng Z, Li Q, Yang HQ, et al. (2013) Auxin controls seed dormancy through stimulation of abscisic acid signaling by inducing ARF-mediated ABI3 activation

CHAPTER 6

NITRIC OXIDE AND REACTIVE OXYGEN SPECIES MEDIATE METABOLIC CHANGES IN BARLEY SEED EMBRYO DURING GERMINATION

Zhenguo Ma, Frédéric Marsolais, Natalia V. Bykova, and Abir U. Igamberdiev

Keywords: barley (*Hordeum vulgare* L.), dormancy, germination, nitric oxide, reactive nitrogen species, reactive oxygen species

INTRODUCTION

The early stage of seed germination starting from the uptake of water by dry seeds is a challenge for plants, in particular because seed coat represents a robust barrier for oxygen (Bewley, 1997). Only after radicle protrusion the mobilization of storage reserves becomes not significantly limited by oxygen supply (Taylorson and Hendricks, 1977; Bewley and Black, 1994; Bewley, 1997). As a consequence, after imbibition seeds develop anaerobic conditions limiting oxidative respiration (reviewed in Bykova et al., 2015). Despite this, seed ATP levels and energy charge remain high in early germination (Duff et al., 1998; Logan et al., 2001; Benamar et al., 2003). The increase of reduction level of NAD, NADP, and other electron transferring compounds upon oxygen depletion immediately after imbibition results in the formation of reactive species of oxygen (ROS) and of nitrogen (RNS) (Bethke, 2009).

It has been suggested that generation of nitric oxide (NO) under seed coat is increased upon depletion of oxygen representing a mechanism of supporting redox and energy balance in germinating seeds together with activation of fermentation processes (Debska et al., 2013). In germination, after the initial depletion of oxygen within several hours, the transition to the mainly anaerobic step takes place, characterized by NO production. These phases correspond to the stages I and II described by Bewley and Black (1994). While the generation of NO is an unavoidable consequence of low oxygen and high reduction level of electron transferring compounds (Igamberdiev and Hill, 2004), the turnover of NO that leads to balancing of redox and energy levels is achieved via the systems participating in scavenging of RNS such as class 1 plant hemoglobin (phytoglobin; Igamberdiev et al., 2006) and S-nitrosoglutathione reductase, which detoxifies nitrosylated glutathione returning it to the active glutathione pool and represents one of the reactions catalyzed by the class III alcohol dehydrogenase (Corpas et al., 2013). This results in the oxidation of NAD(P)H and possibly in the support of ATP formation by mitochondria (Stoimenova et al., 2007). However, the particular aspects of NO and RNS metabolism in germinating seeds are not investigated in detail and need further clarification.

It has been suggested that NO and ROS are crucial for dormancy breaking, while generation of NO under seed coat is increased upon depletion of oxygen representing a mechanism of supporting redox and energy balance in germinating seeds together with the activation of fermentation processes (Bethke et al., 2004, 2007; Igamberdiev and Hill, 2004). While ROS formation starts upon imbibition, the NO production is triggered by the progressing depletion of oxygen under seed coat upon germination. NO nitrosylates proteins and glutathione, while ROS can interact with NO and facilitate its scavenging (Wulff et al., 2009). The interplay between NO and ROS is important for understanding how the seed breaks dormancy and maintains its energy status during germination (Bykova et al., 2015).

In this paper we have presented a wide-range study of the development of oxidative metabolism in embryo during

Citation: Zhenguo Ma, Frédéric Marsolais, Natalia V. Bykova, and Abir U. Igamberdiev, Nitric Oxide and Reactive Oxygen Species Mediate Metabolic Changes in Barley Seed Embryo during Germination, doi: 10.3389/fpls.2016.00138.

the first 2 days of germination of the cereal (barley) seeds of two cultivars with different germination rates. The main part of embryo in early germination is scutellum which supplies intermediates for carbohydrate and amino acid synthesis and contributes to acidification of endosperm. We have traced the development of production of NO, ROS and their scavenging systems in the course of germination of two barley cultivars differing in dormancy level. It is concluded that NO production results in the increase of protein nitrosylation. The mechanisms involved in scavenging of NO and RNS include the expression of phytoglobin and GSNO-reductase, as well as the interaction of NO and superoxide anion with the following degradation of peroxynitrite. The role of NO turnover during the early stage of plant development is discussed in relation to the maintenance of redox and energy levels in cells, which supports the development of germinating seeds and facilitates the efficiency of responses to biotic and abiotic stress factors in the course of germination process and dormancy breaking.

MATERIALS AND METHODS

Barley Seeds

Seeds of two barley (*Hordeum vulgare* L.) cultivars, Harrington (non-dormant, 96% of seeds germinate after 24 h) and Sundre [more dormant, its germination resistance is about 1.2 days determined by the method of Gordon (1971)] were soaked with sterile deionized water from 0 (dry seeds) to 48 h on filter paper in Petri dishes. Embryos (usually 100 mg) were isolated and ground in liquid nitrogen with mortar and pestle. The Sundre embryos analyzed from 0 to 9 h after imbibition could not be distinguished into dormant and non-dormant and therefore represented a mixture of seeds with different germination potential.

Measurement of Protein Concentration

The method of measuring concentration of proteins was based on standard protocol of Bradford reagent (Sigma–Aldrich). Bovine serum albumin was used as a standard.

ATP and ADP Measurement

Extraction of ATP and ADP was conducted according to Joshi et al. (1989) and Yuroff et al. (2003). The tissue powder was lysed in ice-cold 2.4 M perchloric acid (1 ml to 25 mg) for 60 min on ice, centrifuged 5 min at 20,000 × *g*, and supernatant (0.5 ml) was neutralized by 4 M KOH. Quantification of ATP in the neutralized solution was performed by chemiluminescent analysis using the protocol of ATP Detection Kit (Thermo Fisher Scientific). Content of ADP was determined using EnzyLightTM ADP Assay Kit (EADP-100, BioAssay Systems). Fine tissue powder (25 mg) was homogenized in 200 µl of ice-cold 50 mM potassium phosphate buffer (pH 7.0) and centrifuged at 12,000 ×*g* for 5 min. Supernatant was used for ADP analysis, followed the standard protocol of the kit.

Measurement of NO Content

Preparation of hemoglobin (Hb) solution and NO measurement were conducted according toMurphy and Noack (1994). Hb (50 mg) was dissolved in 10 ml of 50 mM Tris-HCl, pH 7.0. Sodium dithionite ($Na_2S_2O_4$, 15 mg) was added directly into fresh Hb solution (which contained mainly MetHb) and reacted for 3 min to reduce MetHb (Fe^{3+}) to Hb (Fe^{2+}), which immediately became HbO_2. The prepared HbO_2 solution was desalted through PD-10 column (GE Healthcare Life Sciences). The final concentration of HbO_2 was determined by spectrophotometry at 415 nm ($\varepsilon = 131$ mM^{-1} cm^{-1}), adjusted to ~10 µM and stored in dim light.

Nitric oxide was extracted from tissue powder with 50 mM Tris-HCl buffer, pH 7.0, containing 0.6% (w/v) PVP at 4°C. Supernatant was collected after centrifugation for 10 min at 15000 ×*g*. To 0.5 ml of supernatant 12.5 µl superoxide dismutase (SOD; 4,000 U/ml) and 12.5 µl catalase (10,000 U/ml) was added and incubated for 2 min at room temperature to remove ROS. A volume of 0.475 ml of the desalted HbO_2 solution was added and the tubes were incubated for 5 min. Absorbance at 401 and 421 nm was recorded to calculate the concentration of NO in tissue using the extinction coefficient $\varepsilon = 77$ mM^{-1} cm^{-1}.

It is recommended to use the two methods for measuring NO (Gupta and Igamberdiev, 2013). The measurement of NO content by the hemoglobin method was in agreement with the method using chemiluminescent NO analyzer CLD 88 p (Eco-Physics, Dürnten, Switzerland; not shown). However, it was not possible to compare these two methods before radicle protrusion when NO is trapped inside the seed coat, therefore we present here the data obtained by the hemoglobin method.

Measurement of Protein SH-Groups and *S*–Nitrosylation

Measurement of protein *S*–nitrosylation was conducted according to Jaffrey et al. (2001) by reduction of R-SNO to R-SH in the presence of ascorbate followed by the assay of free thiol groups using 5,5′–dithiol–bis (2–nitrobenzoic acid; DTNB; Riddles et al., 1983). Proteins were extracted from tissue powder with 50 mM HEPES, pH 8.0, containing 1 mM EDTA, 0.1 mM neocuproine, 0.2% (w/v) SDS and 0.5% (w/v) CHAPS. The samples were centrifuged at 4°C for 10 min at 15,000 ×*g* and proteins were precipitated from supernatant by two volumes of acetone (-20°C) overnight. After centrifuging at 4°C and 15,000 ×*g* for 10 min, protein precipitate was washed four times with chilled 70% acetone. The protein samples were re-suspended in the same volume of the extraction buffer. The clear protein solution was collected to measure the quantity of R-SNO. Ascorbate (50 µl of 100 mM solution) was added in 0.9 ml samples, the same volume of H_2O was added to control. After incubation for 1 h at 25°C, 50 µl of 10 mM DTNB in 75 mM phosphate buffer, pH 7, was added, followed by measurement of absorbance at 412 nm against control samples. The mixture of ascorbate and DTNB in extraction buffer, and DTNB in the same buffer were set up as blank for treatment and control groups, respectively. The difference of R-SH quantity between sample and control groups was used to calculate quantity of R-SNO. Namely, the quantity of R-SH generated by ascorbate treatment corresponded to that of R-SNO in proteins. The measurement of free SH-groups in proteins was conducted without ascorbate treatment.

Measurement of Superoxide Anion and Hydrogen Peroxide

To extract O·−2, the fine powder of fresh tissue was homogenized in 8 M KOH and centrifuged for 10 min at 15000 *g* at 4°C. The superoxide anion level was measured in supernatant by the method modified from Sun and Trumpower (2003) and Dahlgren et al. (2007) at 550 nm by reduction of cytochrome *c*. Cytochrome *c* (0.1 ml, 5 mg ml^{-1}) was dissolved in 1.397 ml of 0.2 M potassium phosphate buffer, pH 8.6, and 3 µl extract solution was added finally, homogenized immediately and incubated for 15 min at 25°C. In the reference set 50 units of SOD and 100 units of catalase were added to the solution. The value of $\varepsilon = 21.5\ mM^{-1}\ cm^{-1}$ for the reduced cytochrome c (Fe^{2+}) was used, and based on a one-to-one molar stoichiometry between O·−2produced and cytochrome *c* molecules reduced, the actual amount of O·−2 produced was quantified.

Concentration of hydrogen peroxide was measured according to the method of Lu et al. (2009). Fine powder of isolated scutella was homogenized in 6% trichloroacetic acid (TCA) for 30 min at 4°C, centrifuged at 15000 ×*g* for 10 min, and then insoluble polyvinylpyrrolidone (PVPP; 50 mg/ml) was added. The samples were centrifuged at 15000 ×*g* for additional 3 min. The extract was diluted 1000 fold with 0.1 M carbonate buffer (pH 10.2). The preparation of reagents followed the method of Pérez and Rubio (2006) and Lu et al. (2009). Ten ml of 6.5 mM luminol and 2 ml of 3 mM $CoCl_2$ in 0.1 M sodium carbonate buffer (pH 10.2) were mixed, diluted to 100 ml in the same buffer and stored for at least 1 h in dark. The solution was further diluted 10 times in the same buffer and stored at 4°C in dark overnight before use. The 40 µl sample was mixed with 10 µl of the sodium carbonate buffer and the mixture was incubated at 30°C for 15 min. Catalase (500 units) was added and incubated at the same condition as a control. Ten µl of each sample and 200 µl of reaction reagent were added into 5 ml SARSTEDT tubes to measure chemiluminescence (CL) with FB 12 Luminometer (Berthold Detection Systems GmbH, Germany). The difference of CL response between each treatment and corresponding background was considered as CL specific for H_2O_2 in samples. The amount of H_2O_2 produced per gram tissue was calculated using the standard curve.

Expression of the Genes Encoding *S*–Nitrosoglutathione Reductase and Class 1 Phytoglobin

Primers of *ADH3* (GenBank: X12734.1), *Hb* (GenBank: U94968.1), and *Mub1* (GenBank:M60175.1) for Real-Time Polymerase Chain Reaction (RT PCR) were designed using NCBI/Primer-BLAST according to known cDNA sequence of the *ADH3* and *Hb* genes encoding correspondingly GSNOR (which is a class III alcohol dehydrogenase) and class 1 phytoglobin, where *Mub1* was set up as a reference gene. Specific primers of *ADH3*, *Hb*, and *Mub1* were: ADH3-forward: 5′ - GTCTCTCAACTGGACTTGGTG - 3′ and ADH3-reverse: 5′ - CTTAGCTTGTTCGTATTTTGCAGG - 3′; Hb forward: 5′ - ACCAACCCCAAGCTCAAGAC - 3′ and Hb reverse: 5′ - CTGCCACGCCGTATTTCAAG - 3′; and Mub1-forward: 5′ - CACCGGCAAGGTAACCAG - 3′ and Mub1-reverse: 5′ - GACATAGGTGAGTCCGCAC - 3′, respectively, which were synthesized by Eurofins. Extraction of total RNA was carried out using the RNeasy Plant Mini Kit (QIAGEN) and all of the steps followed the standard protocol of its manufacturer. Concentration of total RNA was measured using NanoDrop 1000 instrument (Thermo Fisher Scientific) and 1.5 µg RNA was loaded onto 1% agarose gel to observe the quality of the extracted RNA. RNase-free DNase I treatment of RNA followed the protocol of Ambion (Thermo Fisher Scientific). Reverse transcription of RNA followed the manufacturer's protocol for the qScript™ cDNA SuperMix (Quanta Biosciences). The single strand cDNA was used as template in the following PCR. The cDNA synthesized in this step was diluted into a series of cDNA solutions to determine efficiency of the primers. PCR of *ADH3, Hb*, and *Mub1* followed the manufacturer's protocol for the SsoFastTM EvaGreen® Supermix (Bio-Rad Laboratories): 0.5 µl 10 µM forward primer, 0.5 µl 10 µM reverse primer, 1 µl cDNA (diluted eightfold) and 5 µl SsoFast™ EvaGreen® Supermix were mixed and adjusted to 10 µl using nuclease-free water. Samples were transferred to Hard-Shell 96-well plate (Bio-Rad Laboratories) and sealed tightly using a transparent membrane. No template control (NTC) was set up for each biological sample and run in duplicate. Biological replicates corresponded to independent RNA extracts, and three technical replications were run for each biological replicate. The program for the RT PCR reactions was set up as enzyme activation at 95°C for 30 s, followed by 40 cycles of 95°C for 10 s and 58°C for 30 s. PCR products were viewed by electrophoresis on 2% agarose gel. PCR was performed on a CFX96 Real-Time PCR Detection System and results analyzed with Bio-Rad CFX Manager. Data was expressed as the cycle number required for reaching a threshold fluorescence value (C_q). Data was normalized to the mean C_q of the reference gene, for which variation between samples was <1. The specificity of primer pairs was confirmed by melt curve analysis in comparison with controls without template. PCR efficiency was calculated from a standard curve of C_q versus the logarithm of starting template quantity. Each assay was optimized so that efficiency ranged between 97 and 101%, with a coefficient of determination (R^2) > 0.99.

Determination of *S*-nitrosoglutathione reductase (GSNOR, EC 1.2.1.46) was modified from the method of Wünsche et al. (2011). GSNOR was extracted from embryo with 50 mM Tris-HCl buffer, pH 8.0, containing 100 mM NaCl and 0.1 mM EDTA. Supernatant was collected after centrifuging for 15 min at 18,000 ×*g* and passed through Sephadex G-10 column. Fifty µl of the filtered solution was taken to measure enzymatic activity at 340 nm in final 1 ml of 50 mM Tris-HCl (pH 8.0) containing 0.4 mM GSNO and 0.2 mM NADH ($\varepsilon = 6.22\ mM^{-1}cm^{-1}$). The reaction was initiated by adding GSNO and change of absorbance was recorded for 3 min.

Measurement of Ascorbate and Glutathione

Ascorbate and glutathione were extracted with 6% TCA from tissue powder. The homogenate was centrifuged at 12,000 ×*g* for 20 min at 4°C. Supernatant was used for measurement of reduced and oxidized ascorbate and glutathione. Ascorbate (ASC) and dehydroascorbate (DHA) were determined according to Kampfenkel et al. (1995), and the absorbance was recorded at 525 nm using spectrophotometer. Reduced glutathione (GSH) and oxidized glutathione (GSSG) were determined according to Zaharieva and Abadía (2003). The method is based on the reaction of DTNB with GSH forming

5-thionitrobenzoic acid (TNB) recorded at 412 nm. The oxidized glutathione (GSSG) was measured after its reduction by glutathione reductase (Sigma).

Measurement of Enzymes Involved in Oxidative Metabolism

The extraction method was modified from Murshed et al. (2008). All extraction steps were performed on ice. Catalase, SOD, alcohol and lactate dehydrogenases were extracted from embryo with 50 mM Tris-HCl buffer, pH 8.0. The enzymes of ascorbate-glutathione cycle were extracted from fine powder of isolated embryo with 50 mM MES-KOH buffer, pH 6.5, containing 40 mM KCl, 2 mM $CaCl_2$, 1 mM ascorbate (for ascorbate peroxidase, added freshly). The ratio of extraction buffer to the scutellum tissue was 1 ml to 20 mg powder of tissue in fresh weight (FW). The samples were vortexed and then centrifuged at 15000 g for 10 min. Supernatant was collected as a crude enzyme solution.

Measurement of SOD (EC 1.15.1.1) was performed according to Kuthan et al. (1986) and Gupta et al. (1993) by inhibition of reduction of cytochrome *c*. SOD was extracted by 50 mM potassium phosphate buffer, pH 7, containing 0.1 mM EDTA and 1% (w/v) polyvinylpolypyrrolidone (PVP). The spectrophotometric buffer contained 0.05 M potassium phosphate (pH 7.8), 0.1 mM EDTA, and 0.05 mM xanthine. Fifty μl of crude SOD solution and 50 μl 5 mg/ml cytochrome *c* on ice were mixed with 0.85 ml of the buffer until absorbance at 550 nm was constant and finally 0.05 ml of xanthine oxidase (0.02 U/ml) was added and mixed. The increase of absorbance at 550 nm was recorded for 2 min. KCN (2 mM) was added to SOD extract to measure activity of MnSOD. The activity of Cu/Zn SOD was calculated as a difference of total SOD and MnSOD. The activity of Fe SOD, which is mainly chloroplastic, was negligible and not considered.

The methods for measuring activities of the ascorbate-glutathione cycle enzymes were modified from Nakano and Asada (1987). The assay medium for ascorbate peroxidase (APX, EC 1.11.1.11) was 50 mM potassium phosphate buffer (pH 7.0) containing 0.25 mM sodium ascorbate and 50 μl sample extract. The reaction was started by adding H_2O_2 (final concentration 0.25 mM) and the reaction rate was determined spectrophotometrically by absorbance change at 290 nm (coefficient of absorbance, $\varepsilon = 2.8\ mM^{-1}\ cm^{-1}$). Dehydroascorbate reductase (DHAR, EC 1.8.5.1) activity was measured at 265 nm ($\varepsilon = 14\ mM^{-1}\ cm^{-1}$). The assay buffer contained 50 mM HEPES buffer (pH 7.0), 0.1 mM EDTA, 2.5 mM GSH, and 50 μl of extract. The reaction was initiated by adding freshly prepared DHA (final concentration of 0.8 mM). Monodehydroascorbate reductase (MHAR, EC 1.6.5.4) activity was measured in 50 mM HEPES buffer (pH 7.6) containing 2.5 mM ascorbate, 0.25 mM NADH, and 50 μl of the extract. The assay was initiated by adding 0.4 U cm-3 of ascorbate oxidase and the reaction rate was monitored at 340 nm ($\varepsilon = 6.22\ mM^{-1}\ cm^{-1}$). Glutathione reductase (GR, EC 1.8.1.7) activity was measured in 50 mM HEPES buffer (pH 8.0) containing 0.5 mM EDTA, 0.25 mM NADPH, and 50 μl extract. The reaction was started by adding GSSG to final concentration of 1 mM. Catalase (EC 1.11.1.6) activity was measured at 240 nm according to Aebi (1974). In the above reactions, the change of absorbance was recorded for 3 min by Biochrom Ultrospec 4300 spectrophotometer (Amersham, UK).

Alcohol dehydrogenase (ADH; EC 1.1.1.1) was measured according Blandino et al. (1997) with ethanol and NAD^+. Determination of lactate dehydrogenase (LDH; EC 1.1.1.27) activity was performed according to Hoffman et al. (1986) with pyruvate and NADH. Both activities were measured spectrophotometrically at 340 nm.

Statistical Analysis

All the experiments were repeated at least three times. The data in the text, tables, and figures are expressed as means ± SD of three replicates. The differences with $P \leq 0.05$ were considered as statistically significant.

RESULTS

Germination of Barley Seeds

The seeds of Harrington cultivar were non-dormant with the germination rate 96%. The seeds of Sundre cultivar kept the moderate level of dormancy with germination resistance 1.2 days. Practically all seeds were viable, the treatment with gibberellic acid resulted in alleviation of dormancy and germination rate of 96% for both cultivars (not shown). Therefore, the non-germinated seeds in figures represented viable seeds that kept dormancy at the particular time of assay. Radicle protrusion took place between 10 and 15 h post imbibition. Figure 1 shows the germination of two cultivars as observed at 24 h after imbibition (after radicle protrusion in non-dormant seeds) and at 48 h (when all non-dormant seeds were germinated).

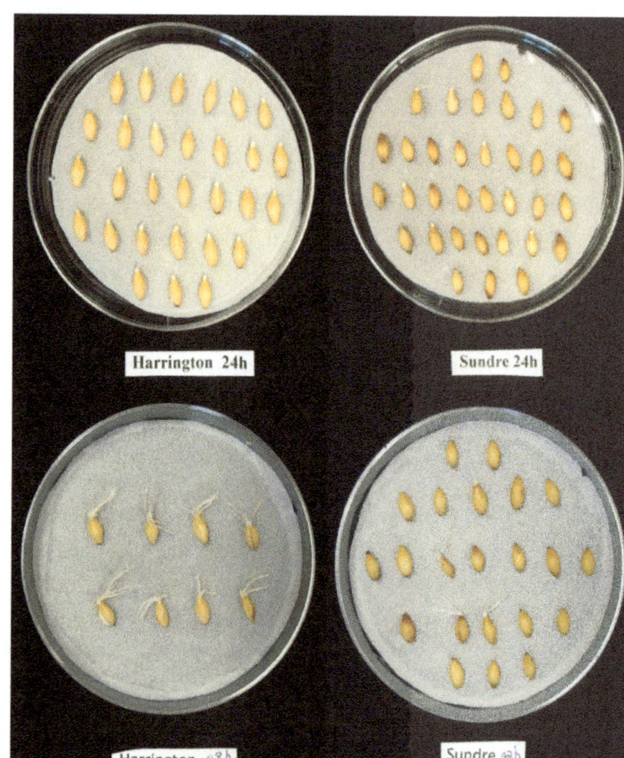

Figure 1: Photos of germinating seeds of Harrington and Sundre cultivars of barley taken at 24 and 48 h post imbibition.

Protein Change, ATP Level and ATP/ADP Ratio

The total protein content in embryo (Figure 2A) started to decrease immediately after imbibition. The protein content decreased almost twofold during the first day and further decreased in next 24 h. The seeds of Sundre that remained non-germinated exhibited stable protein content.

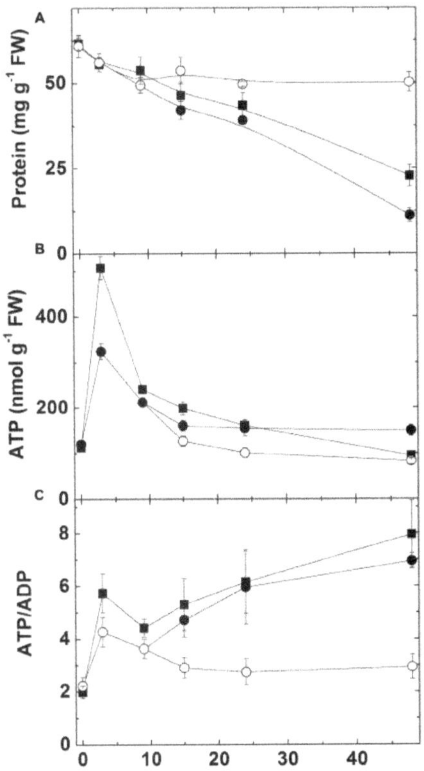

Figure 2: Changes in total protein content (A), ATP (B), and ATP/ADP ratio (C) in embryo of two barley cultivars during germination. Closed squares – Harrington cultivar, closed circles – Sundre cultivar, open circles designate non-germinated seeds ...

The level of ATP in embryo (Figure 2B) sharply increased in the first 3 h after imbibition then declined reaching approximately the same level as in dry seeds after 24 h and declined further on the second day after imbibition. The level of ATP in non-germinated Sundre seeds was constant and similar to the level in dry seeds. The ratio of ATP and ADP (Figure 2C) also increased (from 2 to 6 in Harrington) in first 3 h, then declined to 4 after 10 h post imbibition, and increased again upon radicle protrusion with the same dynamics in both cultivars. The non-germinated seeds of Sundre kept almost the constant ATP/ADP ratio in embryo of approximately 2.5.

NO Metabolism in Embryo of Germinating Seeds

The level of NO, measured with use of hemoglobin, increased (sharply in Harrington) in embryo in the first day starting from imbibition and then the increase became slower (Figure 3A). There was no increase in embryo of seeds of Sundre cultivar that remained non-germinated (even some decrease was observed). The increase of nitrosylation of SH-groups of proteins in embryo (Figure 3B) was continuously increasing after imbibition (only in the seeds remaining non-germinated this increase was slow). The level of free SH-groups in proteins decreased during germination remaining constant in non-germinated seeds (Figure 3C). The lower initial level of NO production in Harrington corresponded to lower rate of nitrosylation in the first hours after germination. The genes encoding the proteins participating in NO metabolism (class 1 phytoglobin and GSNOR) were expressed in germinating seeds (Figures 4A,B). While the expression of GSNOR was practically constant during 48 h post imbibition (slightly decreasing in Sundre), the expression of phytoglobin I gene increased in the first hours after imbibition, and started to decline on the second day post imbibition. There was no significant difference in expression of GSNOR and phytoglobin I genes between germinated and non-germinated seeds of the Sundre cultivar (not shown). The activity of GSNO-reductase increased during germination, more in Sundre than in Harrington, remaining at a lower level in non-germinated seeds (Figure 4C).

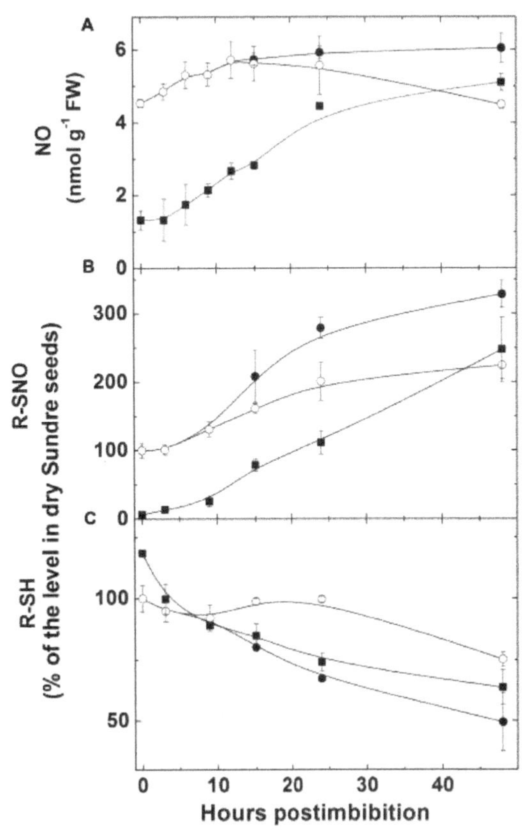

Figure 3: Changes in NO level (A), in the quantity of nitrosylated (-SNO) groups in proteins (B), and the quantity of sulfhydryl groups in proteins (C) in embryo of two barley cultivars during germination. The symbols are the same as in Figure 2...

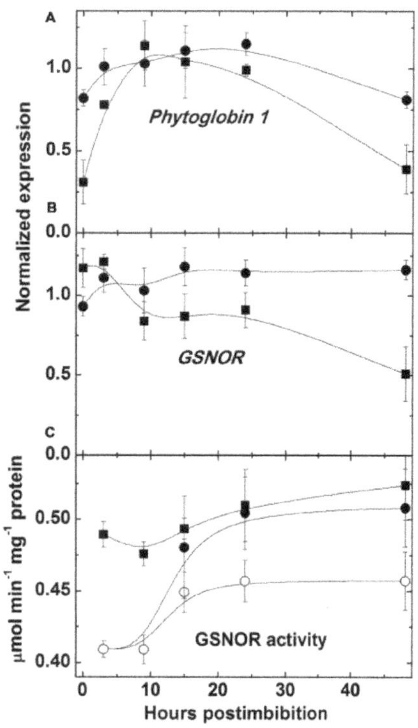

Figure 4: Expression of the genes encoding class 1 phytoglobin (A)and GSNO reductase (B) and GSNO reductase activity (C)in embryo of two barley cultivars during germination. The symbols are the same as in **Figure 2.**

H_2O_2 and O·–2 Levels

The level of superoxide was high in dry seeds of Sundre and low in Harrington. It decreased in Sundre upon imbibition and increased in Harrington (Figure 5A) being significantly higher in embryo after 10 h from imbibition. The content of H_2O_2 (Figure 5B) was higher in the Sundre cultivar remaining constant in non-germinated seeds, and having tendency of increase in germinating seeds in both cultivars, with a sharper increase in Harrington.

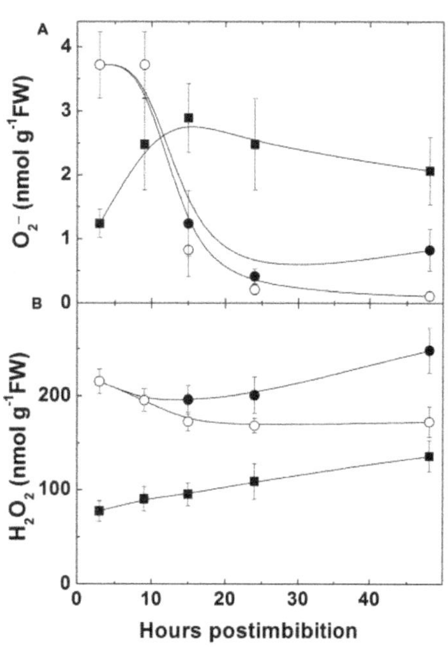

Figure 5: Changes in the level of superoxide anion (A) and of hydrogen peroxide (B) in embryo of two barley cultivars during germination. The symbols are the same as in Figure 2.

Ascorbate and Glutathione Levels and Free SH-Groups

The ratio of ASC to DHA (Figure 6A) was initially higher in Harrington and then decreased, while the level of reduction of ascorbate pool in Sundre was significantly lower than in Harrington. The total ascorbate pool slightly decreased during the first day post imbibition in both cultivars, in dormant seeds it remained constant (Figure 6B). The level of reduction of glutathione decreased in Harrington in the first day, then increased; in Sundre the decrease was observed continuously from the start of imbibition (Figure 6C). These changes took place on the background of decrease in the total pool of glutathione (Figure 6D).

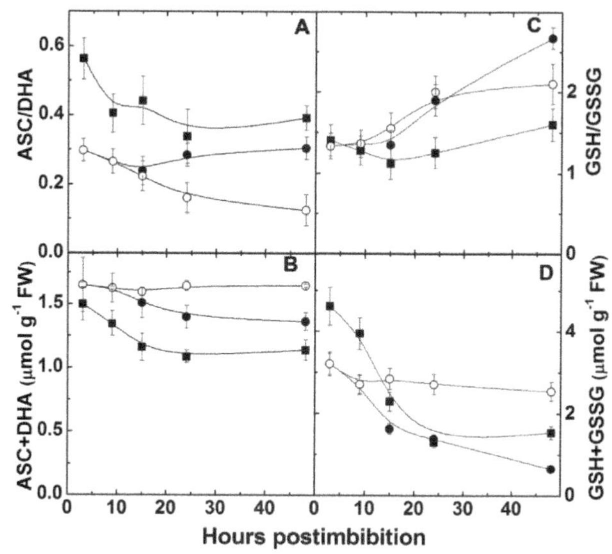

Figure 6: Changes in the ratio ascorbate and dehydroascorbate (A), total ascorbate pool (B), the ratio of reduced and oxidized glutathione (C), and total glutathione pool (D) in embryo of two barley cultivars during germination. The symbols are the same as in .Figure 2.

Activities of Enzymes Involved in Oxidative Metabolism

Catalase activity increased in Harrington in the first hours, then decreased and started to increase on the second day; in Sundre the activity was lower but the increase was continuous (there was no increase in non-germinated seeds; Figure 7A). SOD activity drastically increased after a period of low activity and some fluctuations in first 10–15 h (before radicle protrusion), these changes were similar for Cu/Zn-SOD and Mn-SOD and exhibited the same tendency in both cultivars. It did not show a change in the embryos of the seeds that remained non-germinated (Figure 7A).

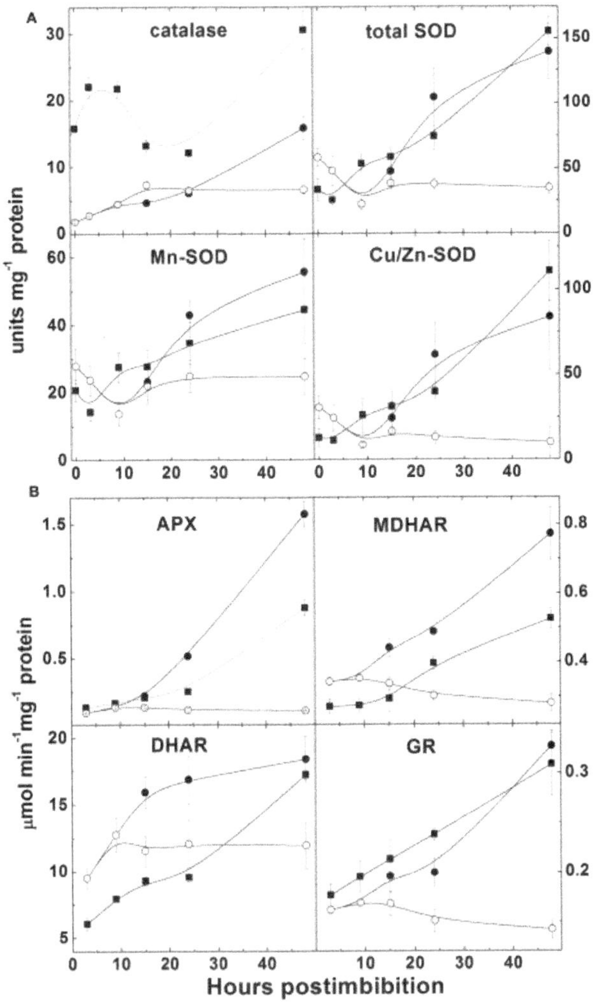

Figure 7: Activities of the enzymes scavenging reactive oxygen species in barley embryos during germination. (A) Activities of catalase and superoxide dismutase (total SOD, Mn-SOD, and Cu/Zn-SOD); (B) Activities of the enzymes of ascorbate-glutathione cycle .

All four enzymes of the ascorbate-glutathione cycle exhibited low initial activity and increased markedly during the first 48 h after imbibition (Figure 7The activities of LDH and ADH were high in the first hours after imbibition and then decreased (Figure 8). They were significantly lower in more dormant Sundre cultivar and did not show change in the seeds that kept dormancy.

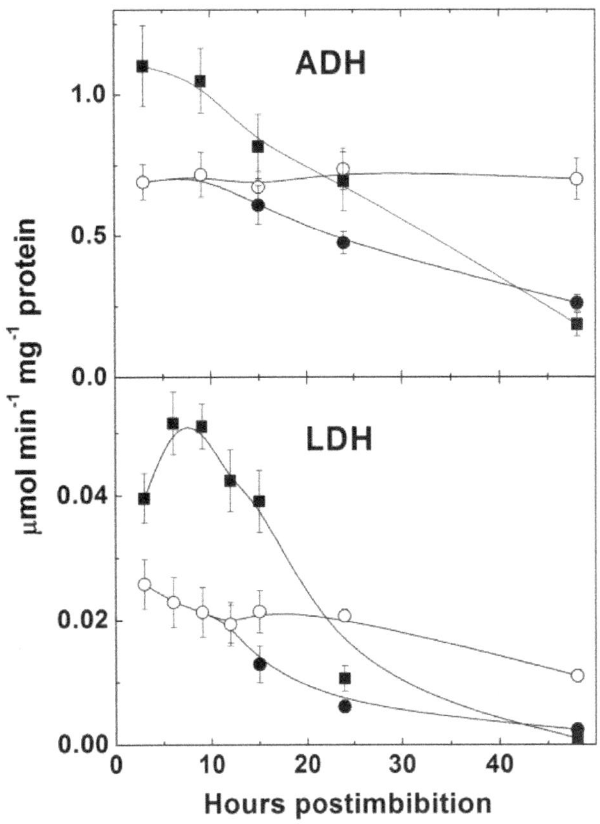

Figure 8: Activities of fermentation enzymes alcohol dehydrogenase (ADH) and lactate dehydrogenase (LDH) in embryo of two barley cultivars during germination. The symbols are the same as in Figure 2.

DISCUSSION

Seed germination is a process that starts from imbibition and results in radicle protrusion initiating seedling development. This process corresponds to the two stages described by Bewley and Black (1994), in the first of which rapid depletion of oxygen takes place and in the second the conditions inside seed are close to anaerobic. After radicle protrusion, the third stage starts in which oxygen concentrations gradually return to aerobic, active mobilization of storage reserves takes place and the seedling begins to develop. In barley seeds, radicle protrusion occurs within the first 24 h after imbibition. In our study the first radicles appeared between 12 and 15 h, and by 24 h all non-dormant seeds developed radicles (Figure 1). The separation between the first and second phases is conventional but we assume that by 3–5 h from imbibition most of oxygen is depleted and then the seeds exist in mostly anaerobic conditions until radicle protrusion.

Germination is characterized by the gradual utilization of storage reserves and a buildup of ATP, which is used for biosynthetic processes (Figure 2). In the seeds remaining dormant the protein level remains without change and the ATP/ADP ratio is constant being near 2.5. The first peak of ATP level is observed at 3 h post imbibition, which probably corresponds to the peak of aerobic metabolism before oxygen becomes depleted. Higher ATP levels in the non-dormant cultivar might reflect higher respiration capacity which in turn could result in higher production of superoxide in the side reactions of mitochondrial metabolism (Møller, 2001). After reaching its peak ATP level sharply declines, reflecting the development of anaerobic conditions, however, the observed decline in the ATP/ADP ratio is not strikingly pronounced and the ATP/ADP ratio, reaching the value 6 in Harrington at 3 h, declines only to the value slightly higher than 4 at 10 h. This indicates that even under highly hypoxic conditions there exist the mechanisms supporting ATP synthesis. Besides fermentation, NO turnover can be a contributor to ATP synthesis in these conditions (Stoimenova et al., 2007). Higher activities of fermentation enzymes (ADH and LDH) were indeed observed in germinating barley seeds in the first hours post imbibition, and the cultivar Harrington having higher ATP/ADP ratio at the first day is characterized by higher levels of

ADH and LDH (Figure 8). It is also characterized by a sharper increase in phytoglobin expression (Figure 4A), which may also contribute to a buildup of ATP via the phytoglobin-nitric oxide cycle (Stoimenova et al., 2007; Gupta and Igamberdiev, 2011). Phytoglobin gene expression precedes the expression of phytoglobin protein which was detected in barley grains in the earlier study (Duff et al., 1998).

Indeed, the production of NO was shown to increase in barley seeds starting immediately from the onset of imbibition (**Figure 5A**). NO levels were initially higher in Sundre (more dormant cultivar) remaining stable in the seeds that kept dormancy, while in Harrington the increase in NO production was faster. The increase was most pronounced at the initial stages of germination, when the seeds become more anaerobic. This corresponds to the mechanism of reductive pathway of NO formation (via nitrite reduction at high redox level) facilitated by transition to anaerobic conditions (Dordas et al., 2003). The oxidative pathway of NO production is considered to be less active in plants and may be absent in germinating seeds (Igamberdiev et al., 2014). It is known that the development of anaerobic conditions in seeds results in a switch from oxygenic respiration to fermentation. An alternative to fermentation pathway could be NO turnover in the phytoglobin-nitric oxide cycle, in which NO is produced by reduction of nitrite and scavenged with formation of nitrate, the reaction catalyzed by a system involving the class 1 phytoglobin (Igamberdiev and Hill, 2004; Gupta and Igamberdiev, 2011). Both pathways operate during seed germination, and the activities of ADH and LDH (Figure 8) and the expression of phytoglobin (Figure 4A) exhibit cultivar-specific patterns.

The production of NO leads to nitrosylation of SH groups in proteins, peptides, free cysteine and its derivatives. The redox and energy level of germinating seeds is controlled via the balance of NO producing and scavenging mechanisms (Bykova et al., 2015). The most important scavenging mechanism for NO is its oxygenation to nitrate with participation of class 1 phytoglobin, which is hypoxically induced (Igamberdiev and Hill, 2004). While NO can be scavenged via phytoglobin-mediated mechanism, the most important RNS species, *S*-nitrosoglutathione is scavenged by GSNOR, which is the class III alcohol dehydrogenase having also the activity of formaldehyde dehydrogenase (Corpas et al., 2013). The activity of this enzyme was high already in dry seeds and it remained higher in the cultivar with high germination rate (Harrington; Figure 4C). The transcript expression of this enzyme (Figure 4B) was relatively constant from the onset of germination; therefore the observed changes in activity could be attributed to the posttranscriptional regulation. On the contrary, phytoglobin expression started to increase from the beginning of germination, remained high between 3 and 24 h and then declined; the changes were more expressed in the cultivar Harrington (Figure 4A). This is in agreement with the reported dependence of phytoglobin expression on the hypoxic conditions (Igamberdiev and Hill, 2004).

While we did not measure the concentration of *S*-nitrosoglutathione in this study, the decline in total glutathione (GSH + GSSG) pool (Figure 6D) can be explained, in particular, by its intensive nitrosylation. Other modifications of thiol groups caused by the increase in ROS formation may be responsible for the observed changes. Nitrosylation and other modifications also explain the decrease of free SH-groups in proteins (Figure 3C), which is supported by the observed increase of nitrosylated (-SNO) groups (Figure 3B), which follows the pattern of NO production with some delay (the increase in nitrosylation becomes more pronounced after 10 h post imbibition). The production of RNS (NO and nitrosylated compounds) during seed germination was suggested to be connected with dormancy alleviation (Bethke et al., 2004) but later it was suggested that inhibition of respiration by NO triggers ROS formation and that results in dormancy alleviation (reviewed inBykova et al., 2015). We observed that in the seeds remaining dormant the increase in NO did not take place and the nitrosylation process did not actively develop (Figure 3).

In the course of germination upon the development of anaerobic conditions we observe oxidation of ascorbate and glutathione pools, while after radicle protrusion (at 15 h post imbibition) the ratios start to increase (Figures 6A,C). The glutathione pool and its reduction is the indicator of the overall cellular environment and important in control of morphogenetic processes (Mitrovic et al., 2012). Schafer and Buettner (2001) suggested that the resulting action of the glutathione redox couple triggers metabolic events and determines morphogenesis. GSH stimulates *Arabidopsis* root growth (Sanchez-Fernandez et al., 1997) favoring cell division and proliferation, while the GSSG content is related to differentiation (Schafer and Buettner, 2001; Stasolla et al., 2004). The changes of glutathione pool and its reduction level observed during germination are much more pronounced than of the ascorbate pool (Figure 6).

The importance of glutathione pool size and its reduction for determination of cell division, growth, and even apoptosis was mentioned in many works (reviewed in Noctor et al., 2012). The quiescent parts of plants such as root quiescent center, and cells in organs such as seeds maintain a highly oxidized intracellular state, in particular reflected in low content of reduction of the glutathione pool (Kranner et al., 2006), and the increase in the total cellular GSH pool is essential for the cells to progress to cell division (Diaz-Vivancos et al., 2010). Glutathione synthesis is required for pollen germination and pollen tube growth (Zechmann et al., 2011). The direct effect of glutathione has been observed in relation to its potential in meristematic cells; however, total tissue concentrations could give indirect indication of the direction of the morphogenetic process. Our study shows that the total pools of glutathione decreased after imbibition while its reduction increased after radicle protrusion. The redox state of glutathione can be tightly controlled by NO production and scavenging via phytoglobin and GSNOR mechanism. Thus NO can be an important regulator of dormancy release and tissue differentiation during seed germination.

The cross-talk between NO and ROS formation is important even during seed germination. The production of ROS occurs actively upon imbibition even in the period of oxygen depletion (Figure5). Antioxidant defense mechanisms protect seeds during the dormancy stage and prevent their germination, while the break of dormancy is achieved in conditions when ROS production breaks these control mechanisms (Bykova et al., 2011a,b; Bykova and Rampitsch, 2013). Some data relate dormancy alleviation with the accumulation of H_2O_2 (Oracz et al., 2007; Bailly et al., 2008), however, the role of superoxide anion was also considered. It has been demonstrated that germination in *Arabidopsis* is mediated by accumulation of O·−2 and H_2O_2 in the radicle (Leymarie et al., 2012). In sunflower (*Helianthus annuus* L.) seeds, germination was shown to be associated with a marked increase in hydrogen peroxide and superoxide anion generation in the embryonic axes resulting from an inhibition of catalase and SOD and from activation of NADPH oxidase (Oracz et al., 2009). The role of NO in germination was also claimed (Bethke et al., 2004), and its interaction with ROS results in the balance of both ROS and RNS compounds that provides the conditions for germination.

The increase of superoxide levels in the non-dormant cultivar Harrington and its decrease in more dormant Sundre was observed during the first day, while the levels of H_2O_2 were lower in the seeds of the non-dormant cultivar (Figure Figure55). This might indicate that the main ROS player in alleviating dormancy in barley is indeed superoxide anion. On the other hand, some role of H_2O_2in promotion of germination follows from the fact that in the seeds remaining non-germinated its level is lower than in the germinating seeds of the Sundre cultivar. The differences between the levels of superoxide and hydrogen peroxide could be related to the activities of corresponding scavenging enzymes (Figure 7A). However, before radicle protrusion the activities of SOD and of the enzymes of ascorbate-glutathione cycle were low and they started to increase drastically mostly on the second day of germination.

The higher activity of catalase in Harrington may explain the lower H_2O_2 level at the onset of germination. Cakmak et al. (1993) also reported that during imbibition of wheat seeds the increase in activities of H_2O_2 scavenging enzymes is much more pronounced than the increase in SOD. In our study the increase in SOD (both the mitochondrial Mn-SOD and cytosolic Cu/Zn-SOD) was observed mainly on the second day of germination, i.e., after radical protrusion and switch to aerobic metabolism. The role of ascorbate-glutathione cycle is important not only for scavenging of H_2O_2 but also for establishing redox potentials of ascorbate and glutathione, which are the best indicators of the overall cellular environment and important in control of morphogenetic processes (Mitrovic et al., 2012; Talukdar, 2012).

CONCLUSION

The production of NO and ROS during germination and their cross-talk are the important events that take place during germination of barley seeds. The class 1 phytoglobin, GSNOR and interaction with ROS play a role in scavenging and turnover of NO and nitrosylated compounds during germination. Although the observed differences between the two cultivars may be not necessarily be attributed to the dormancy levels, it becomes evident that the turnover of NO contributes to the maintenance of redox and energy balance of germinating seeds and may be an important player in alleviation of dormancy.

AUTHOR CONTRIBUTIONS

ZM performed all experiments, participated in discussion of results, contributed to writing the manuscript. FM planned and supervised the experiments on gene expression, analyzed results, contributed to writing the manuscript. NB planned and supervised the experiments on measurement of enzymes and metabolites, analyzed results, contributed to writing the manuscript. AI supervised the design, execution and interpretation of the experiments, prepared the manuscript.

CONFLICT OF INTEREST STATEMENT

The authors declare that the research was conducted in the absence of any commercial or financial relationships that could be construed as a potential conflict of interest.

ACKNOWLEDGMENTS

This work was supported by the grants from the Natural Sciences and Engineering Research Council of Canada (to AI and FM).

REFERENCES

1. Aebi H. (1974). "Catalase," in *Methods of Enzymatic Analysis* Vol. 2 ed. Bergmeyer H. U., editor. (New York, NY: Academic Press;), 673–684.
2. Bailly C., El-Maarouf-Bouteau H., Corbineau F. (2008). From intracellular signaling networks to cell death: the dual role of reactive oxygen species in seed physiology. *C. R. Biol.* 331 806–814. 10.1016/j.crvi.2008.07.022
3. Benamar A., Tallon C., Macherel D. (2003). Membrane integrity and oxidative properties of mitochondria isolated from imbibing pea seeds after priming or accelerated ageing. *Seed Sci. Res.* 13 35–45. 10.1079/SSR2002122
4. Bethke P. C. (2009). Rebirth and death: nitric oxide and reactive oxygen species in seeds.*SEB Exp. Biol. Ser.* 62 17–30.
5. Bethke P. C., Gubler F., Jacobsen J. V., Jones R. L. (2004). Dormancy of *Arabidopsis* seeds barley grains can be broken by nitric oxide. *Planta* 219 847–855. 10.1007/s00425-004-1282-x
6. Bethke P. C., Libourel I. G. L., Jones R. L. (2007). "Nitric oxide in seed dormancy and germination," in *Seed Development, Dormancy and Germination. Annual Plant Reviews*Vol. 27 eds Bradford K., Nonogaki H., editors. (Oxford: Blackwell;), 153–175.
7. Bewley J. D. (1997). Seed germination and dormancy. *Plant Cell* 9 1055–1066. 10.1105/tpc.9.7.1055
8. Bewley J. D., Black M. (1994). *Seeds: Physiology of Development and Germination*, 2nd Edn. Berlin: Springer.
9. Blandino A., Caro I., Cantero D. (1997). Comparative study of alcohol dehydrogenase activity in flor yeast extracts. *Biotechnol. Lett.* 19 651–654. 10.1023/A:1018386731116
10. Bykova N. V., Hoehn B., Rampitsch C., Banks T., Stebbing J. A., Fan T., et al. (2011a).Redox-sensitive proteome and antioxidant strategies in wheat seed dormancy control.*Proteomics* 11 865–882. 10.1002/pmic.200900810
11. Bykova N. V., Hoehn B., Rampitsch C., Hu J., Stebbing J. A., Knox R. (2011b). Thiol redox-sensitive seed proteome in dormant and non-dormant hybrid genotypes of wheat.*Phytochemistry* 72 1162–1172. 10.1016/j.phytochem.2010.12.021

12. Bykova N. V., Hu J., Ma Z., Igamberdiev A. U. (2015). "The role of reactive oxygen and nitrogen species in bioenergetics, metabolism and signaling during seed germination," in*Reactive Oxygen and Reactive Nitrogen Species Signaling and Communication in Plants*, eds Gupta K. J., Igamberdiev A. U., editors. (Berlin: Springer;), 177–195. 10.1007/978-3-319-10079-1_9
13. Bykova N. V., Rampitsch C. (2013). Modulating protein function through reversible oxidation: redox-mediated processes in plants revealed through proteomics. *Proteomics* 13579–596. 10.1002/pmic.201200270
14. Cakmak I., Strbac D., Marscher H. (1993). Activities of hydrogen peroxide-scavenging enzymes in germinating wheat seeds. *J. Exp. Bot.* 44 127–132. 10.1093/jxb/44.1.127
15. Corpas F. J., Alché J. D., Barroso J. B. (2013). Current overview of S-nitrosoglutathione (GSNO) in higher plants. *Front. Plant Sci.* 4:126 10.3389/fpls.2013.00126
16. Dahlgren C., Karlsson A., Bylund J. (2007). Measurement of respiratory burst products generated by professional phagocytes. *Methods Mol. Biol.* 412 349–363. 10.1007/978-1-59745-467-4_23
17. Debska K., Krasuska U., Budnicka K., Bogatek R., Gniazdowska A. (2013). Dormancy removal of apple seeds by cold stratification is associated with fluctuation in H_2O_2, NO production and protein carbonylation level. *J. Plant Physiol.* 170 480–488. 10.1016/j.jplph.2012.11.018
18. Diaz-Vivancos P., Dong Y. P., Ziegler K., Markovic J., Pallardó F., Pellny T. K., et al. (2010). Recruitment of glutathione into the nucleus during cell proliferation adjusts whole cell redox homeostasis in *Arabidopsis thaliana* and lowers the oxidative defence shield.*Plant J.* 64 825–838. 10.1111/j.1365-313X.2010.04371.x
19. Dordas C., Hasinoff B. B., Igamberdiev A. U., Manac'h N., Rivoal J., Hill R. D. (2003).Expression of a stress-induced hemoglobin affects NO levels produced by alfalfa root cultures under hypoxic stress. *Plant J.* 35 763–770. 10.1046/j.1365-313X.2003.01846.x
20. Duff S. M. G., Guy P. A., Nie X., Durnin D. C., Hill R. D. (1998). Hemoglobin expression in germinating barley. *Seed Sci. Res.* 8 431–436. 10.1017/S0960258500004396
21. Gordon A. G. (1971). The germination resistance test – a new test for measuring germination quality of cereals. *Can. J. Plant Sci.* 51 181–183. 10.4141/cjps71-036
22. Gupta A. S., Webb R. P., Holaday A. S., Allen R. D. (1993). Overexpression of superoxide dismutase protects plants from oxidative stress. *Plant Physiol.* 103 1067–1073.
23. Gupta K. J., Igamberdiev A. U. (2011). The anoxic plant mitochondrion as a nitrite: NO reductase. *Mitochondrion* 11 537–543. 10.1016/j.mito.2011.03.005
24. Gupta K. J., Igamberdiev A. U. (2013). Recommendations of using at least two different methods for measuring NO. *Front. Plant Sci.* 4:58 10.3389/fpls.2013.00058
25. Hoffman N. E., Bent A. F., Hanson A. D. (1986). Induction of lactate dehydrogenase isozymes by oxygen deficit in barley root tissue. *Plant Physiol.* 82 658–663. 10.1104/pp.82.3.658
26. Igamberdiev A. U., Bykova N. V., Hill R. D. (2006). Nitric oxide scavenging by barley hemoglobin is facilitated by a monodehydroascorbate reductase-mediated ascorbate reduction of methemoglobin. *Planta* 223 1033–1040. 10.1007/s00425-005-0146-3
27. Igamberdiev A. U., Hill R. D. (2004). Nitrate, NO and haemoglobin in plant adaptation to hypoxia: an alternative to classic fermentation pathways. *J. Exp. Bot.* 55 2473–2482. 10.1093/jxb/erh272
28. Igamberdiev A. U., Ratcliffe R. G., Gupta K. J. (2014). Plant mitochondria: source and target for nitric oxide. *Mitochondrion* 19B, 329–333. 10.1016/j.mito.2014.02.003
29. Jaffrey S. R., Erdjument-Bromage H., Ferris C. D., Tempst P., Snyder S. H. (2001). Protein S-nitrosylation: a physiological signal for neuronal nitric oxide. *Nat. Cell Biol.* 3 193–197. 10.1038/35055104
30. Joshi A. K., Ahmed S., Ames G. F.-L. (1989). Energy coupling in bacterial periplasmic transport systems. *J. Biol. Chem.* 264 2126–2133.
31. Kampfenkel K., Van Montagu M., Inzé D. (1995). Extraction and determination of ascorbate and dehydroascorbate from plant tissue. *Anal. Biochem.* 225 165–167. 10.1006/abio.1995.1127
32. Kranner I., Birtic S., Anderson K. M., Pritchard H. W. (2006). Glutathione half-cell reduction potential: a universal stress marker and modulator of programmed cell death. *Free Radic. Biol. Med.* 40 2155–2165. 10.1016/j.freeradbiomed.2006.02.013
33. Kuthan H., Haussmann H.-J., Werringloer J. (1986). A spectrophotometric assay for superoxide dismutase activities in crude tissue fractions. *Biochem. J.* 237 175–180. 10.1042/bj2370175
34. Leymarie J., Vitkauskaite G., Hoang H. H., Gendreau E., Chazoule V., Meimoun P., et al. (2012). Role of reactive oxygen species in the regulation of *Arabidopsis* seed dormancy.*Plant Cell Physiol.* 53 96–106. 10.1093/pcp/pcr129
35. Logan D. C., Millar A. H., Sweetlove L. J., Hill S. A., Leaver C. J. (2001). Mitochondrial biogenesis during germination in maize embryos. *Plant Physiol.* 125 662–672. 10.1104/pp.125.2.662
36. Lu S., Song J., Campbell-Palmer L. (2009). A modified chemiluminescence method for hydrogen peroxide determination in apple fruit tissues. *Sci. Hortic.* 120 336–341. 10.1016/j.scienta.2008.11.003
37. Mitrovic A., Janosevic D., Budimir S., Pristov J. B. (2012). Changes in antioxidative enzymes activities during *Tacitus bellus* direct shoot organogenesis. *Biol. Plant.* 56 357–361. 10.1007/s10535-012-0098-y
38. Møller I. M. (2001). Plant mitochondria and oxidative stress: electron transport, NADPH turnover and metabolism of reactive oxygen species. *Annu. Rev. Plant Physiol. Plant Mol. Biol.* 52 561–591. 10.1146/annurev.arplant.52.1.561
39. Murphy M. M., Noack E. (1994). Nitric oxide assay using hemoglobin method. *Methods Enzymol.* 233 240–

250. 10.1016/S0076-6879(94)33027-1

40. Murshed R., Lopez-Lauri F., Sallanon H. (2008). Microplate quantification of enzymes of the plant ascorbate-glutathione cycle. *Anal. Biochem.* 383 320–322. 10.1016/j.ab.2008.07.020

41. Nakano Y., Asada K. (1987). Purification of ascorbate peroxidase in spinach-chloroplasts – its inactivation in ascorbate-depleted medium and reactivation by monodehydroascorbate radical. *Plant Cell Physiol.* 28 131–140.

42. Noctor G., Mhamdi A., Chaouch S., Han Y., Neukermans J., Marquez-Garcia B., et al. (2012). Glutathione in plants: an integrated overview. *Plant Cell Environ.* 35 454–484. 10.1111/j.1365-3040.2011.02400.x

43. Oracz K., El-Maarouf Bouteau H., Farrant J. M., Cooper K., Belghazi M., Job C., et al. (2007). ROS production and protein oxidation as a novel mechanism for seed dormancy alleviation. *Plant J.* 50 452–465. 10.1111/j.1365-313X.2007.03063.x

44. Oracz K., El-Maarouf-Bouteau H., Kranner I., Bogatek R., Corbineau F., Bailly C. (2009).The mechanisms involved in seed dormancy alleviation by hydrogen cyanide unravel the role of reactive oxygen species as key factors of cellular signaling during germination. *Plant Physiol.* 150 494–505. 10.1104/pp.109.138107

45. Pérez F. J., Rubio S. (2006). An improved chemiluminescence method for hydrogen peroxide determination in plant tissues. *Plant Growth Regul.* 48 89–95. 10.1007/s10725-005-5089-y

46. Riddles P. W., Blakeley R. L., Zerner B. (1983). Reassessment of Ellman's reagent. *Methods Enzymol.* 91 49–60. 10.1016/S0076-6879(83)91010-8

47. Sanchez-Fernandez R., Fricker M., Corben L., White N. S., Sheard N., Leaver C. J., et al. (1997). Cell proliferation and hair tip growth in the *Arabidopsis* root under mechanistically different forms of redox control. *Proc. Natl. Acad. Sci. U.S.A.* 94 2745–2750. 10.1073/pnas.94.6.2745

48. Schafer F. Q., Buettner G. R. (2001). Redox environment of the cell as viewed through the redox state of the glutathione disulfide/glutathione couple. *Free Radic. Biol. Med.* 30 1191–1212. 10.1016/S0891-5849(01)00480-4

49. Stasolla C., Belmonte M. F., Van Zyl L., Craig D. L., Liu W., Yeung E. C., et al. (2004). The effect of reduced glutathione on morphology and gene expression of white spruce (*Picea glauca*) somatic embryos. *J. Exp. Bot.* 55 695–709. 10.1093/jxb/erh074

50. Stoimenova M., Igamberdiev A. U., Gupta K. J., Hill R. D. (2007). Nitrite-driven anaerobic ATP synthesis in barley and rice root mitochondria. *Planta* 226 465–474. 10.1007/s00425-007-0496-0

51. Sun J., Trumpower B. L. (2003). Superoxide anion generation by the cytochrome bc1 complex. *Arch. Biochem. Biophys.* 419 198–206. 10.1016/j.abb.2003.08.028

52. Talukdar D. (2012). Ascorbate deficient semi-dwarf asfL1 mutant of *Lathyrus sativus*exhibits alterations in antioxidant defense. *Biol. Plant.* 56 675–682. 10.1007/s10535-012-0245-5

53. Taylorson R. B., Hendricks S. B. (1977). Dormancy in seeds. *Annu. Rev. Plant Physiol. Plant Mol. Biol.* 28 331–354. 10.1146/annurev.pp.28.060177.001555

54. Wulff A., Oliveira H. C., Saviani E. E., Salgado I. (2009). Nitrite reduction and superoxide-dependent nitric oxide degradation by *Arabidopsis* mitochondria: influence of external NAD(P)H dehydrogenases and alternative oxidase in the control of nitric oxide levels. *Nitric Oxide* 21 132–139. 10.1016/j.niox.2009.06.003

55. Wünsche H., Baldwin I. T., Wu J. (2011). S-Nitrosoglutathione reductase (GSNOR) mediates the biosynthesis of jasmonic acid and ethylene induced by feeding of the insect herbivore Manduca sexta and is important for jasmonate-elicited responses in *Nicotiana attenuata*. *J. Exp. Bot.* 62 4605–4616. 10.1093/jxb/err171

56. Yuroff A. S., Sabat G., Hickey W. J. (2003). Transporter-mediated uptake of 2-chloro- and 2-hydroxybenzoate by *Pseudomonas* huttiensis strain D1. *Appl. Environ. Microbiol.* 69 7401–7408. 10.1128/AEM.69.12.7401-7408.2003

57. Zaharieva T. B., Abadía J. (2003). Iron deficiency enhances the levels of ascorbate, glutathione, and related enzymes in sugar beet roots. *Protoplasma* 221 269–275. 10.1007/s00709-002-00051-6

58. Zechmann B., Koffler B. E., Russell S. D. (2011). Glutathione synthesis is essential for pollen germination in vitro. *BMC Plant Biol.* 11:54 10.1186/1471-2229-11-54

CHAPTER 7

INCREASING ABSCISIC ACID LEVELS BY IMMUNOMODULATION IN BARLEY GRAINS INDUCES PRECOCIOUS MATURATION WITHOUT CHANGING GRAIN COMPOSITION

Nicole Staroske, Udo Conrad, Jochen Kumlehn, Götz Hensel, Ruslana Radchuk, Alexander Erban, Joachim Kopka, Winfriede Weschke and Hans Weber

Keywords: Abscisic acid, barley, grain composition, grain maturation, immunomodulation, seed development, stress signalling.

INTRODUCTION

In barley, endosperm cellularization is completed at 5–6 days after fertilization (DAF). At 10 DAF, the endosperm begins to accumulate storage products. The pre-storage phase from anthesis until 6 DAF and the storage phase from 10 DAF onwards are separated by a transition stage characterized by transcriptional reprogramming, which promotes the switch of the endosperm into the storage mode. Grain dessication starts at physiological maturity after 20 DAF and grains are fully mature at around 40 DAF (Sreenivasulu *et al.*, 2006).

Abscisic acid (ABA) in plants regulates many developmental processes and responses to environmental stress. In seeds, it is necessary for maturation such as storage product accumulation, desiccation tolerance and dormancy (Finkelstein *et al.*, 2008; Cutler *et al.*, 2010). Levels of ABA are low during early embryogenesis but increase considerably at the onset of maturation, which is generally correlated with the physiological status of seeds. The transition of seeds from pre-storage to maturation is partially controlled by ABA and a network of sugars and ABA (Finkelstein and Gibson, 2002; Weber *et al.*, 2005; Radchuk *et al.*, 2010b). ABA functions in context with other phytohormones. For instance, ABA and gibberellic acids (GAs) are negatively correlated during seed growth and levels fluctuate during seed development, implicating a tightly regulated balance (White *et al.*, 2000; Weier *et al.*, 2014). While GAs stimulate growth by cell elongation (Thiel *et al.*, 2008), ABA functions antagonistically and generally inhibits growth and cell elongation at higher concentrations but is necessary for seed maturation events (Weber *et al.* 2005) including sugar signalling (Finkelstein and Gibson, 2002). A plethora of different genes and second messengers are involved in ABA signalling, such as phospholipases, protein kinases/phosphatases, mitogen-activated kinase, sucrose non-fermenting 1-related kinase 1 (SnRK1), phosphatidic acid, reactive oxygen species and nitric oxide (Hirayama and Shinozaki, 2007). Multiple receptors for ABA have been identified (Ma *et al.*, 2009; Fuchs *et al.*, 2014). While different receptors perceive ABA signals (Boursiac *et al.*, 2013; Nakashima and Yamaguchi-Shinozaki, 2013; Xu *et al.*, 2013), no single receptor acts upstream of the signalling pathway, suggesting that ABA responses are mediated by distinct receptor classes and that signalling components depend on cell type and physiological processes and development (Verslues and Zhu, 2007; Spartz and Gray, 2008).

Plant hormone immunomodulation can alter the hormonal status without directly intervening with its synthesis, degradation or signalling pathways and as such could represent a suitable tool to directly study the functions of ABA during seed maturation (Conrad and Manteuffel, 2001). Correspondingly,

Citation: Nicole Staroske , Udo Conrad, Jochen Kumlehn, Götz Hensel, Ruslana Radchuk, Alexander Erban,Joachim Kopka, Winfriede Weschke and Hans Weber, Increasing abscisic acid levels by immunomodulation in barley grains induces precocious maturation without changing grain composition, doi:10.1093/jxb/erw102.

an ABA-specific single-chain antibody (scFv) was directed into the endoplasmic reticulum (ER) lumen of tobacco seeds, resulting in a developmental switch from maturation to germination, which is characterized by chloroplast formation and reduction of storage oil and protein content (Phillips *et al.*, 1997). The antibody, located in the ER, binds ABA, and as such represents an ABA sink in this compartment (Strauss *et al.*, 2001). Thereby, expression of the ABA-binding single-chain antibody impacts the subcellular distribution of ABA, generating a phenotype similar to the Arabidopsis *aba/abi3* double mutant seeds (Koornneef *et al.*, 1989) and thus delays the concentration of ABA in compartments other than the ER. Immunomodulation in pea embryos by overexpression of an anti-ABA antibody reduces free ABA levels specifically during the transition from pre-storage to storage phase (Radchuk *et al.*, 2010a). This deficiency of free ABA decreases accumulation of seed dry matter, delays the differentiation process and generally down-regulates gene expression related to transcription and translation. These anti-ABA embryos exhibit a widespread repression of sugar uptake and metabolism, diminished starch, amino acid and storage protein biosynthesis. The interaction of ABA with SnRK1 complexes highlights the cooperation between SnRK1, sugar, stress and ABA signalling (Radchuk *et al.*, 2010a).

In this work, we aimed to analyse the effect of ABA immunomodulation on grain development and maturation in barley. Similar to the situation in tobacco, Arabidopsis and pea, ABA immunomodulation drastically increases ABA in barley caryopses from the transition stage onwards. However, it was calculated that the antibody only binds a small fraction of this ABA in the transgenic caryopses, which results in strongly enhanced levels of free ABA. This renders these grains as suitable models to analyse the influence of increased ABA levels on transcriptional and metabolic control of grain development. Metabolite and transcript profiling in these anti-ABA grains expose up-regulation of signal transduction and stress response especially during the transition phase, which is consistent with triggered, enhanced and precocious initiation of maturation processes. The fact that weight and composition of ripe grains remain unchanged indicates that barley caryopses induce specific mechanisms to desensitize ABA signalling. Such mechanisms point to the enormous physiological and metabolic flexibility of barley grains to adjust extreme internal effects in order to ensure and maintain proper grain development.

MATERIALS AND METHODS

Plant material, Transformation and Quantitative PCR

Barley (*Hordeum vulgare* L. cv. Igri) was grown in greenhouses with 16/8h light/dark at 19/14 °C during the generative phase. Stages of grain development were determined as described previously (Weschke *et al.*, 2000). Plant tissue was collected between 10 and 12 am at 3, 7, 10, 14, 20 DAF. Filial grain fractions were manually separated from the pericarp between 7 and 14 DAP, and whole caryopses were sampled at 20 DAF.

Barley plants (*Hordeum vulgare*, cv. Igri) were transformed with an anti-ABA single-chain variable fragment (scFv) gene (Artsaenko *et al.* 1995). Plant transformation was based on infection of embryogenic pollen cultures with *Agrobacterium tumefaciens* (Kumlehn *et al.*, 2006), which can result in instantaneously homozygous plants with regard to the transgene, provided whole genome duplication occurs spontaneously or is chemically induced after transgene integration. In order to achieve grain-specific immunomodulation, the 1Ax1 promoter was used (Halford *et al.*, 1992).

Towards constructing 1Ax1:GFP plants: a *Not*I/*Xba*I 1Ax1::GFP fragment was released from WBVec8/1Ax1::GFP and introduced into the corresponding sites of pNOS-AB-M (DNA Cloning Service, Hamburg, Germany) to generate p1Ax1::GFP-Nos. Subsequently a *Sfi*I fragment harbouring the entire expression cassette was sub-cloned into p6U (DNA Cloning Service, Hamburg, Germany) to generate constructs p6U-1Ax1-GFP. The binary plasmids were introduced into *A. tumefaciens* strain LBA4404pSB1 (Komari *et al.*, 1996). In total eight independent transgenic lines were regenerated with transgene inserts. Three stable homozygous lines – 362, 363 and 364 – were chosen for further analysis.

Copy number of the aABA scFv gene was determined by genomic Southern Blot analysis (Supplementary Fig. S1, available at *JXB* online). 10 µg of genomic DNA from leaves were cut by BamHI and HindIII, separated by electrophoresis and immobilized on nylon filters. Hybridization was done under stringent conditions using a 447bp fragment of the anti-ABA scFv gene labelled with ^{32}P. Copy number was verified by quantitative PCR (Thermocycler 7900HT, Applied Biosystems) using 5ng genomic DNA. First strand synthesis was done using Superscript™III (Invitrogen). Gene-specific primers were: actin 1211rev 5-AGC ACT TCC GGT GGA CAA T-3 , actin 1153 fwd 5-GTG GAT CTC GAA GGG TGA GT-3, aABA-681fwd 5-TGG CAG TGG GTC AGG AAC TA-3, aABA-738rev 5-ATC CTC AGC CTC CAC TCT AC-3. All reactions were performed using Power SYBR Green PCR Master Mix (Applied Biosystems).

Germination Assay

Eight replicates of each 25 grains were tested in a standard germination assay (ISTA 2008 international rules for seed testing, International Seed Testing Association, Bassersdorf, Switzerland). The grains were incubated in trays between two layers of wet filter paper and incubated in a light chamber at a day/night temperature of 20/18 °C (14h light). Regularly, the number of germinated grains was determined by counting. Grains that did not germinate after 12 d were regarded as dormant.

Determination of anti-ABA scFv content, Calculation of free ABA, and Measurement of ABA

Estimation of anti-ABA scFv antibody protein content was performed by western blot analysis as described previously (Fiedler and Conrad, 1995). Determination of the dissociation constant, Kd, of the anti-ABA scFv antibody purified from anti-ABA barley grains with free ABA was done by competition ELISA of affinity-purified scFv protein (Artsaenko *et al.* 1995,Radchuk *et al.* 2010a). Levels of free ABA were calculated using the equation Kd=(scFv)(ABA)/(scFv-ABA complex), (Neri *et al.*, 1996), for different time points and by subtracting the concentration of (scFv–ABA complex) from the total ABA amount. ABA was extracted from plant material and analysed as described previously (Miersch *et al.*, 2008).

RNA Isolation, Labelling, Array Hybridization, and Data Evaluation

Grain material from three biological replicates was harvested at 7, 10, 14 and 20 DAF; and total RNA was extracted from the filial fractions using the Gentra RNA Isolation Kit (Biozyme Scientific, Oldendorf, Germany). RNA integrity was confirmed using the Bioanalyser system (Agilent Technologies). 100ng RNA was used for cRNA synthesis and Cy3-labelling with a Low Input Quick Amp Labelling Kit (Agilent Technologies,www.agilent.com). Labelling efficiency, and amount and quality of cRNA were assured using an ND-1 000 Spectrophotometer (NanoDrop Technologies, Wilmington, USA) and Bioanalyser system. 600ng labelled cRNA was used for fragmentation and array loading (Gene Expression Hybridization Kit, Agilent Technologies, www.agilent.com). Hybridization of the AGILENT Barley Gene Expression Microarray, 4×44K (www.agilent.com) was performed by Atlas Biolabs GmbH, Berlin, Germany. Resulting images were evaluated (determination of spot intensities, background correction) with Feature Extraction V11.5 (Agilent Technologies, www.agilent.com). For further evaluation only spots were considered whose signal intensity was significantly different by at least 2-fold (t<0.05; moderated t-test), (Benjamini and Hochberg, 1995) between wild type and transgenic line. All sequences fulfilling these criteria were subjected to BLAST×2 searches against the NRPEO-data base (ftp://ftp.ncbi.nih.gov/blast/db/FASTA/nr.gz) using the HUSAR software (http://genome.dkfz-heidelberg.de).

Determination of Sucrose, Starch, Globulins/albuMins, free Amino Acids, Total Carbon, and Nitrogen and Metabolite Profiling

After tissue preparation all samples were immediately frozen in liquid nitrogen. Extraction and determination of glucose, fructose, sucrose, starch and globulins/albumins were performed as described (Rolletschek*et al.*, 2002). Free amino acids were performed as described (Thiel *et al.*, 2009). Relative contents of total carbon and nitrogen in dried, powdered samples of grains were measured using an elemental analyser (Vario EL; Elementar Analysensysteme, http://www.elementar.de/cms/en/home/). Statistical analysis was performed using Sigma Stat software (SPSS,http://www.systat.de.).

For GC-MS measurements of polar central metabolites, 30mg of fresh material of filial fractions of grains from line 363 and wild-type Igri, were harvested with six biological replicates at 7, 10, 14 and 20 DAF. Sample preparation, extraction and data evaluation were carried out as described (Erban *et al.*, 2007; Weichert *et al.*, 2010) except that a 0.007mg ml^{-1}stock solution of $^{13}C6$-sorbitol was used as internal standard. The internal standard concentration was reduced to one third, i.e. 500 µl, due to the high volumes of dried polar extract that were analysed in this study. Samples were profiled by GC-MS in splitless injection mode and, for the proofing of highly concentrated metabolites, in 1:30 split mode. Metabolite data were normalized to sample fresh weight and internal standard and are presented as log2-transformed pool size ratios of line 363 over wild type. Calculations, heat map formatting, and two-sample significance testing were performed using Microsoft Excel software assuming heteroscedastic variance of log2-transformed ratios. P-values were corrected for multiple testing (Benjamini and Hochberg, 1995).

RESULTS

Plant transformation and scFv-transgene expression

Transgenic barley expressed the scFv-gene under control of the 1Ax1-promoter, which is strongly active in the endosperm as shown by the expression of a 1Ax1-promoter: green-fluorescent-gene (GFP) fusion, (Fig. 1A). The scFv expression unit includes the LeB4 signal peptide, the endoplasmatic reticulum retention signal (KDEL) and the *c-myc*-tag (Munro and Pelham, 1987) (Fig. 1B). Accumulation of anti-ABA scFv-protein was determined semi-quantitatively in developing and mature caryopses of primary transformands of the anti-ABA plants and in developing caryopses using the *c-myc*-tag and standard proteins (Conrad*et al.*, 2011) (Fig. 2A). In three lines, scFv-protein accumulated from 7 DAF to highest levels at 20 DAF (Fig. 2B, C). These three stable homozygous lines – 362, 363 and 364 – were chosen for further analysis. Transgene copy number of the lines was determined by quantitative PCR and yielded two, four and two copies for lines 362, 363 and 364.

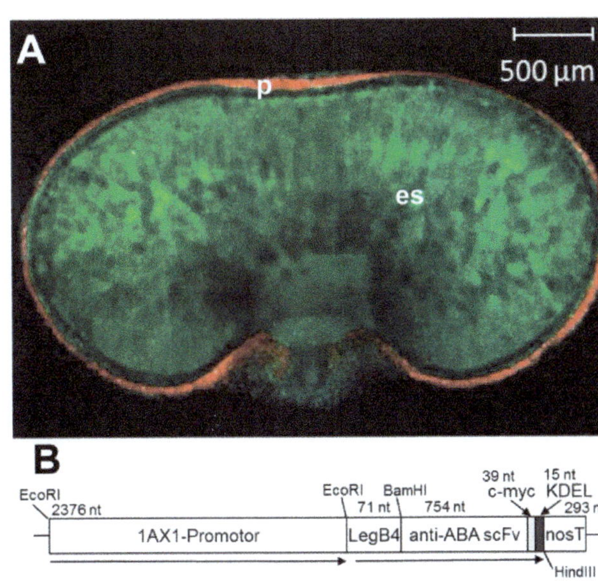

Figure 1: Activity of the 1Ax1-promoter. (A) Green-fluorescence-protein (GFP, green) und chlorophyll-fluorescence (red) in cross-sections of developing barley caryopses, expressing GFP under control of the 1Ax1 promoter from wheat and showing specific activity within the endosperm during the storage phase. es, endosperm; p, pericarp. (B) Schematic view of the anti-ABA scFv-construct. For grain-specific expression the 1Ax1 promoter from wheat was used. The LeguminB4-signal peptide (LeB4) and the ER-retention signal (KDEL) mediate scFv accumulation in the ER. The c-myc tag allows immuno-detection of the anti-ABA scFv.

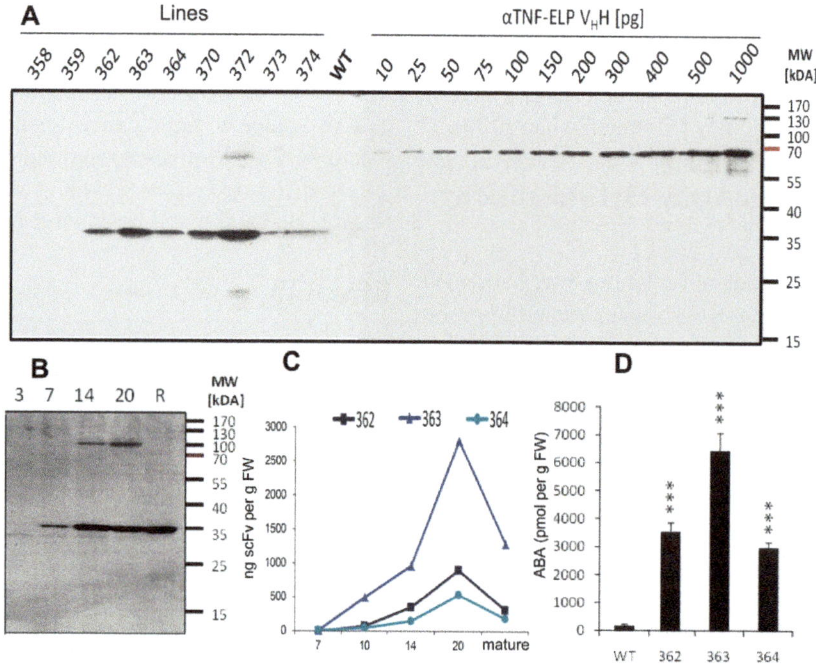

Figure 2: Transgene expression and ABA accumulation. (A) Determination of anti-ABA scFv-protein accumulation in mature caryopses of transformands. 30 μg raw protein and different amounts of αTNF-ELP V_HH-antibody were loaded per lane. After separation (12% SDS-PAA-Gel, reducing conditions) detection was performed by western blot using anti-c-myc antibody followed by anti-mouse IgG peroxidase and ECL-reaction. (B) Amounts of anti-ABA scFv in developing caryopsis at 3, 7, 14 20 DAF and at maturity (R). (C) Semi-quantification of anti-ABA scFv accumulation in lines 362, 363 and 364. (D) ABA accumulation in mature caryopses. ***, significant differences compared to the wild type, $P<0.001$. (This figure is available in colour at *JXB* online.)

Anti-ABA Antibody Expression Interferes with Accumulation of free Unbound ABA in the Grain

Levels of ABA in mature immunomodulated grains increased dramatically as much as 10- to 40-fold, with line 363 showing the largest increase (Fig. 2D). For all three lines, ABA levels in developing grains were not different from the wild type at 3 and 7 DAF but were increasingly higher from 10 DAF until maturation (Fig. 3A). Such high increases of ABA levels have previously been observed in tobacco leaves (Wigger *et al.*, 2002) and seeds (Phillips *et al.* 1997) and in pea seeds (Radchuk *et al.* 2010a) expressing an anti-ABA antibody. The dramatic increase of ABA could be due to (i) increased rate of synthesis, (ii) decreased catabolism or (iii) reduced conjugation within the seeds. Since the anti-ABA antibody sequesters ABA in the ER compartment, it is conceivable that a feedback mechanism regulating ABA synthesis may be activated. However, it can also not be excluded that ABA is masked by the antibody, which prevents catabolism.

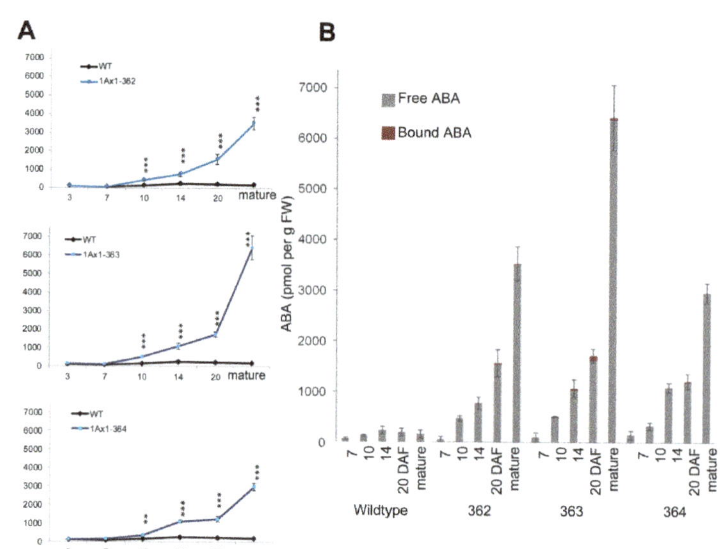

Figure 3: ABA content in developing grains. (A) Means ±SD of ABA-amounts in immunomodulated caryopses of aABA-lines 362, 363 and 364 during development, $n=5$. (B) Amounts of free ABA and ABA complexed with anti-ABA scFv in developing caryopses of immunomodulated lines 362, 363 and 364. Data are means ±SD and were calculated after Neri *et al.* (1996). Significant differences after ANOVA, **, $t<0.01$; ***, $t<0.005$. (This figure is available in colour at *JXB* online.)

Amounts of free versus bound ABA in immunomodulated grains were calculated using concentrations of the scFv:ABA complex, total ABA concentrations, anti-ABA-scFv concentration and the dissociation constant Kd (Phillips *et al.* 1997; Radchuk *et al.* 2010a), (Fig. 3B). Accordingly, almost all ABA was present in the free unbound form. This is due to the fact that the molarity of the ABA in the caryopses is ~300-fold higher compared to the anti-ABA scFv, while binding of antibody and ABA occurs at nearly equimolar concentration to that of the antibody. Therefore, on average at 20 DAF, 99.4% of the antibody but only 5.6% of the ABA is bound (Fig. 3B).

Phenotypical Characterization of anti-ABA Barley Plants

None of the three lines expressing the 1Ax1-promoter-antiABA construct showed any macroscopic, vegetative phenotype different from wild-type Igri after cultivation in greenhouses. Fresh weight and dry weight accumulation of caryopses were not significantly different between the anti-ABA lines and the wild type (Fig. 4) although there was a trend towards lower values for line 363, showing the greatest accumulation of scFv and ABA.

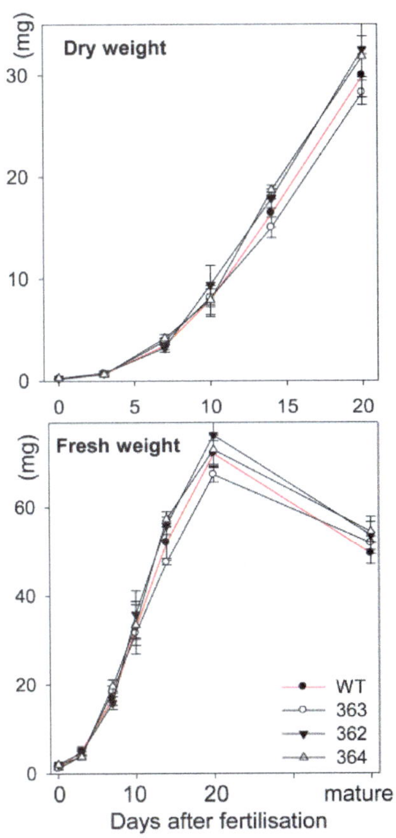

Figure 4: Growth curves of transgenic grains. Fresh weight and dry weight accumulation of developing caryopses of lines 362, 363, 364 and the wild type, means ±SD,n=40. (This figure is available in colour at *JXB*online.).

Surprisingly, a significant portion of the ABA-immunomo dulated grains showed a kink-like phenotype, from ~10 DAF onwards. These grains seemed to be impressed at the dorsal side and sections revealed that this is due to missing or dissolved endosperm cells (Supplementary Fig. S2A–H). The occurrence of this phenotype was observed in three independently grown plant sets and its frequency was from 2% to 10%, with line 363 showing the highest value (Supplementary Fig. S2I).

ABA is well known to inhibit seed germination. Therefore, we tested whether the increased ABA contents in mature grains affects germination. Some 200 grains per line were imbibed and grains that did not germinate after 12 d were regarded as dormant. The proportion of dormant grains in the anti-ABA lines and the wild type was not significantly different. However, the number of grains that had germinated after 2 d was significantly lower for all the three immunomodulated lines compared to the wild type (Fig. 5). This indicates that germination is delayed in the ABA-immunomodulated grains.

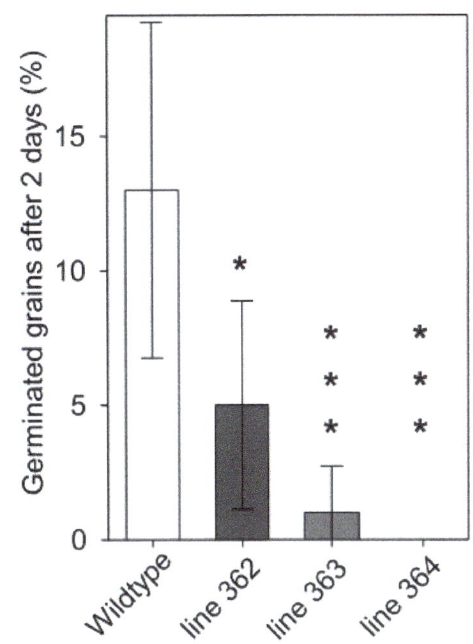

Figure 5: Germination assay. Percentage of germination of immunomodulated and wild-type grains, 2 d after imbibition, ±SD. A standardized germination test was performed starting with eight replicates of 25 grains per line. Significant differences after ANOVA, *,t<0.05; ***, t<0.005; n=25,n=8.

Thousand grain weight of mature seeds, determined in two consecutive years in the greenhouse, was not different between immunomodulated and wild-type plants.

Composition and Metabolite Contents of anti-ABA Grains

Mature grain composition was analysed for all three lines and revealed no significant differences for starch concentration, percentages of total carbon and nitrogen and percentages of prolamins and globulins/albumins. Starch concentration was analysed in grains of line 363 during development and revealed slightly but significantly lower levels in caryopses at only 10 DAF (Supplementary Fig. S3).

Metabolite levels were measured in the filial fractions of immunomodulated grains of line 363 at 7, 10, 14 and 20 DAF using metabolite profiling of a fraction enriched for primary metabolites by gas chromatography-mass spectroscopy (GC-MS), (Erban *et al.*, 2007). In total, 100 assigned metabolites could be determined (Supplementary Table S1). Whereas the levels of the free sugars sucrose, glucose and fructose were clearly lowered compared to the wild type, those of maltose were increased, especially at 20 DAF (Fig. 6A, Supplementary Fig. S3). Levels of sugars involved in cell wall biosynthesis, arabinose, xylose, fucose and myo-inositol were commonly lower in anti-ABA grains, primarily at 10 DAF (Fig. 6B, Supplementary Fig. S3). Free sugars were also measured in whole caryopses of lines 362, 363 and 364 at 14 DAF by an enzymatic assay. All three aABA lines display decreased hexose levels during early maturation (14 DAF) and also increased sucrose to hexose ratios (Supplementary Fig. S4A). Since hexose levels decrease and sucrose to hexose ratios increase during pre-storage to storage phase (Weschke *et al.*, 2000), this indicates an advanced developmental stage compared to the wild type.

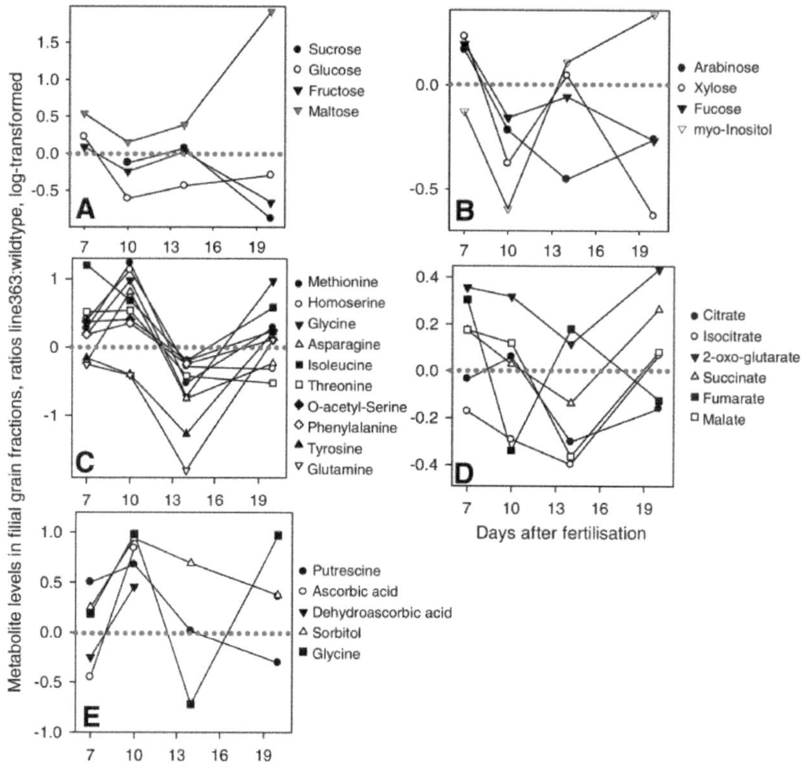

Figure 6: Metabolite profiling. Relative changes in metabolite levels in filial grain fractions of line 363 and wild type at 7, 10, 14 and 20 DAF measured by GC-MS. Data are presented as means (n=6) of log-transformed ratios, significant changes (P<0.05) of at least at one stage are given in bold face (seeSupplementary Fig. S3), (A) soluble sugars, (B) cell wall sugars, (C) amino acids, (D) organic acids, (E) stress-related compounds. (This figure is available in colour at *JXB* online.)

Amounts of free amino acids were somewhat disparate between anti-ABA and wild-type grains. However, a larger set of ten amino acids, among them the most abundant forms Asn and Gln, revealed commonly higher values at 10 DAF and lower levels at 14 DAF (Fig. 6C). Free amino acids in the filial fraction of caryopses of all three lines have been measured by ultra performance liquid chromatography (UPLC). The profiles of the major free amino acids during development compared to the wild type were very similar between the three lines (Supplementary Fig. S4B). Moreover the UPLC-generated values fit quite well to these generated by GC-MS for line 363 (Fig. 6C).

Levels of glycolytic intermediates between anti-ABA and wild-type grains were inconsistent, with either unchanged levels of hexose-phosphates but lower amounts of phosphoenolpyruvate (14 DAF) and higher levels of 3-phospho-glycerate (Supplementary Fig. S3). Whereas amounts of citrate and iso-citrate were lower in anti-ABA grains at 14 DAF, that of 2-oxo-glutarate were higher at 10 and 20 DAF (Fig. 6D, Supplementary Fig. S3). Noticeably, amounts of putrescine, ascorbate, sorbitol and glycine, which are related to stress response and tolerance (Kempa *et al.*, 2008;Maruyama *et al.*, 2014), exhibited higher levels in anti-ABA grains at certain stages (Fig. 6E). Also the amounts of stress-inducible gamma-aminobutyric acid were elevated in the filial caryopses of all three lines (Supplementary Fig. S5).

Transcript Profiling of Developing anti-ABA filial Grain Fractions

To analyse global changes of gene expression due to ABA immunomodulation, comparative transcript profiling was performed for filial grain fractions of line 363 and wild type at 7, 10, 14 and 20 DAF using the 4×44K Barley Gene Microarray (www.agilent.com). Genes were selected whose expression was significantly up- or down-regulated by a factor of at least two and in at least one stage (P<0.05, three biological replicates). Since transcript abundances do not necessarily reflect transcriptional activity, protein content or enzyme activity, all statements on gene identity and function have to be considered as 'putative'. For reasons of simplicity, higher or lower transcript levels were subsequently referred to as down- or up-regulated. In total, 682 genes were de-regulated in anti-ABA grains (Supplementary Table S2). The distribution across the developmental stages revealed that the number of up-regulated genes is very low at 7 DAF but increased by 18-fold from 7 to 10 DAF and decreased again by 2-fold at 14 and 20 DAF (Fig. 7A). Therefore, ABA immunomodulation obviously generates a general up-regulation of gene expression for >97% of all de-regulated genes, especially at 10 DAF. The categorization of up-regulated genes revealed three major categories: signal transduction (24.3%), stress response and tolerance (22%) and maturation and storage (21.5%). Other up-regulated categories accounted for only 3.5–5.3%, such as genes related to transport, primary metabolism, cell proliferation and expansion, transcription and translation (Fig. 7B).

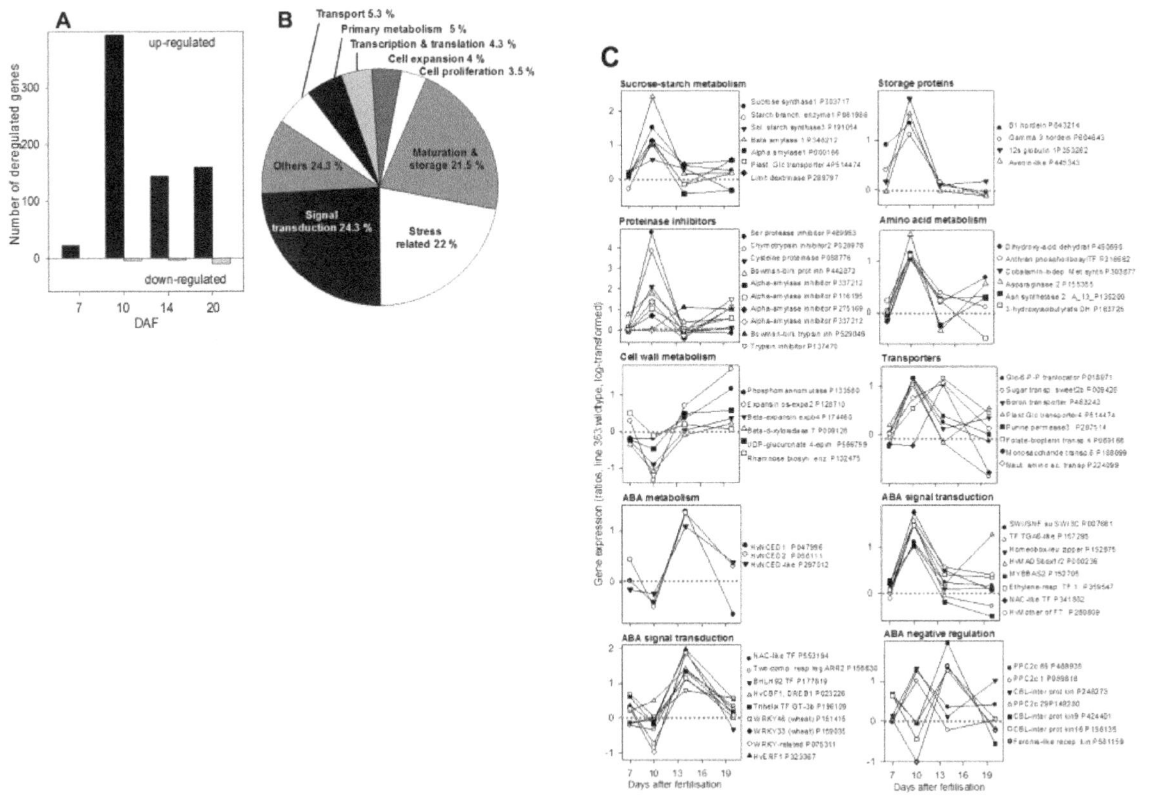

Figure 7: The effect of ABA immunomodulation on gene expression in filial fraction of caryopses of line 363. (A) Numbers of genes up- or down-regulated in filial fractions of caryopses of line 363 at 7, 10, 14, and 20 days after fertilisation (DAF) (B) Categorization of up-regulated genes in the filial grain fraction of line 363 compared to the wildtype IGRI (C) Expression pattern of genes related to specific pathways and physiological states in filial grain, given as ratios. Line 363 to wild type, filial fractions and log-transformed data are extracted from Supplementary Table S2. Sequence IDs (see Supplementary Table S2) are given as P[xxxxx]. (This figure is available in colour at *JXB* online.)

Deregulation of Storage-Associated Gene Expression in anti-ABA Grains

The transition stage of barley grain development is characterized by increased metabolic activity related to starch and storage protein synthesis while growth events such as cell proliferation decreased. At 10 DAF, immunomodulated grains displayed transient transcriptional up-regulation of enzymes of sucrose to starch metabolism such as sucrose synthase, starch synthases and limit dextrinase. In parallel, β-amylase-1, α-amylase-1 and a plastidic glucose transporter were also more highly expressed in anti-ABA grains at 10 DAF. This suggests considerable starch turnover in the anti-ABA grains. However, at later stages, levels of these transcripts remained unchanged (Fig. 7C).

Similar to genes associated with starch metabolism, storage protein genes, such as hordeins, globulins and avenins were also transiently up-regulated at 10 DAF together with a set of maturation-associated protease-inhibitor genes such as α-amylase- and trypsin-inhibitors. The majority of enzymes of amino acid metabolism was also up-regulated at 10 DAF, among them dihydroxy-acid dehydratase, anthranilate phosphoribosyltransferase, methionine synthase and asparagine synthetase-2. The results suggest a general stimulation of storage-related gene transcription in the ABA-immunomodulated caryopses, however only transiently at 10 DAF. At the same time, at 10 DAF, a set of six enzymes involved in cell expansion and the biosynthesis of cell wall-related sugars was transiently down-regulated in anti-ABA grains, among them two expansins, phospho-mannomutase, UDP-glucuronate-4-epimerase and rhamnose-biosynthesis enzyme. Several genes related to sugar and metabolite transport were up-regulated at either 10 DAF, 14 DAF (Fig. 7C) or at 20 DAF (Supplementary Table S2).

Deregulated gene Expression Related to Hormone Functions and signal Transduction

ABA-immunomodulated caryopses accumulate high amounts of ABA (Fig. 3). Three genes, upregulated at 14 DAF in caryopses of line 363, encode 9 cis-epoxy-carotenoid-dioxygenases, HvNCED1, HvNCED2 and HvNCED-like (Fig. 7C). NCEDs were shown to be rate-limiting for ABA biosynthesis (Thompson *et al.*, 2000; Qin and Zeevaart, 2002). Obviously, the up-regulated HvNCEDs were involved in the over-shooting biosynthesis of ABA observed in immunomodulated caryopses.

ABA integrates grain development with stress response and adaption and functions through various positive and negative signalling elements. In anti-ABA grains eight related genes are upregulated at 10 DAF (Fig. 7C). The SWI3 subunit of the SWI/SNF chromatin-remodelling complex might be connected with ABA signalling pathways, suggesting that chromatin-remodelling is involved in ABA response to abiotic stress (Saez *et al.*, 2008). The TGA6-like transcription factor could be involved in pathogenesis-related gene expression and pathogen resistance as seen by knock-out analysis in Arabidopsis (Zhang *et al.*, 2003). The homeobox-leucine zipper protein might be involved in regulating development in response to environmental conditions. In sunflower, such a member, Hahb-4, is rapidly and strongly induced by drought and ABA (Gago *et al.*, 2002). The HvMADS box1/2 (HvOS1/2) protein is related to type I MADS-box genes. HvOS1/2, exhibiting differential expression during barley seed development, is induced by ABA and the protein is detected in specific endosperm subcompartments. DNA methylation differences associated with gene expression suggests epigenetic regulation according to Arabidopsis type I MADS box protein (Kapazoglou *et al.*, 2012). The Myb-BAS2 protein potentially functions as transcriptional activator in ABA-inducible gene expression under drought stress (Abe *et al.*, 2003). The ethylene-response factor-1-protein potentially regulates gene expression under abiotic/biotic stress. In wheat, TaERF1 is induced by drought, salinity, low-temperature, ABA, ethylene and salicylic acid. When overexpressed, stress-related genes are activated leading to improved pathogen and abiotic stress tolerance (Xu *et al.*, 2007). The NAC-like transcription factor might be involved in plant development and stress responses. In rice, ONAC045 was induced by drought, high salt, and low temperature stresses. Arabidopsis overexpressing ONAC045 showed enhanced tolerance to drought and salt (Zheng *et al.*, 2009). The HvMFT, MOTHER OF FT AND TFL1 (MFT) belongs to phosphatidylethanolamine-binding proteins and is known as key regulators of flowering. In wheat and soybean, an MFT homologue negatively regulates germination and is transcriptionally induced and repressed by ABA and gibberellic acid, respectively (Nakamura *et al.*, 2011; Li *et al.*, 2014).

Another set of ten genes, potentially involved in ABA signalling, is up-regulated at 14 DAF. Three are WRKY-like transcription factors, which are known as key components of ABA signalling. Its target genes are involved in ABA-effects and drought stress response (Rushton *et al.*, 2012). A homologue of the bHLH92 transcription factor, in Arabidopsis is strongly salt- and drought-inducible and dependent on ABA (Jiang *et al.*, 2009). A two-component-response regulator, ARR2, could act as transcriptional activator and is up-regulated at 14 DAF. In Arabidopsis, subunits of two-component signalling system are involved in cold stress response (Jeon *et al.*, 2010). Transcriptionally stimulated HvCBF1, which is a member of CBF/DREB1 family genes, could be involved in signalling low temperature, drought and salinity tolerance in plants including barley (Wu *et al.*, 2011). The trihelix-GT3b factor belongs to founding members of the trihelix transcription factor family. Expression of some of these genes is induced by ABA, salt, drought and cold, and overexpression in Arabidopsis increases tolerance to salt, drought and freezing (Kaplan-Levy *et al.*, 2012).

Seven genes, which encode potential negative regulators of ABA signal transduction, are up-regulated at 10 and 14 DAF in anti-ABA grains. Three encode protein phosphatases 2c (PP2C), which interact with kinases, phosphatases, transcription factors and metabolic enzymes thereby inhibiting ABA signals (Yoshida *et al.*, 2015). Three other candidates are related to CBL-interacting protein kinases (CIPKs). This signalling pathway is Ca^{++}-related and responds strongly to abiotic and biotic environmental stimuli. Several CIPK-members have been described as negative regulators of ABA functions (Kim *et al.*, 2003; Pandey *et al.*, 2004). Finally, the FERONIA-like receptor kinase, which was described as positive regulator of auxin-promoted growth, could repress ABA response through activation of ABI2, a PP2C member suppressing ABA responses (Yu *et al.*, 2012).

DISCUSSION

Abscisic acid generally accumulates in response to different biotic and abiotic stresses, delays growth and development and induces stress-tolerance-related gene expression. In seeds, ABA accumulates during the transition to the filling phase and triggers seed maturation, storage activity and the biosynthesis of stress signalling- and tolerance-related proteins (Finkelstein *et al.*, 2002). During late maturation and desiccation, ABA promotes the synthesis of late embryogenesis abundant (LEA) proteins, prevents precocious germination and confers desiccation tolerance and dormancy. ABA functions via a balanced network of biosynthesis and degradation, interaction with other hormones and by integration within a system of negative and positive transcriptional regulators (Raghavendra *et al.*, 2010). Immunomodulation is a suitable tool to alter the hormonal status in seeds, without directly intervening with biosynthesis, degradation or signalling pathways (Conrad and Manteuffel, 2001). In such a way, an anti-ABA antibody gene was directed into barley, whereupon the 1Ax1 promoter mediates strong expression in the endosperm while the anti-ABA antibody is retained within the ER and sequesters ABA.

Anti-ABA Caryopses Accumulate an excess of free ABA

The caryopses of transgenic plants respond to anti-ABA antibody gene expression with increased production of ABA, evidently, by up-regulation of different NCEDs (Fig. 7C). While there is no indication for activated ABA degradation or inactivation, the exact localization of this surplus ABA is still unclear. One possibility is vacuolar sequestration, since transcriptional up-regulation of vacuolar ATPase and H^+-PPiase (Supplementary Table S2), which are known targets for ABA regulation (Fukuda and Tanaka, 2006), improves energized vacuolar transport.

The calculation of free versus antibody-bound ABA reveals a large excess of the free hormone, whose levels increase greatly in the caryopses from 10 DAF onwards. Thus, the situation in anti-ABA barley clearly differs from those of other models, where immunomodulation has been used to alter the ABA status. Whereas in Arabidopsis (Phillips *et al.* 1997), tobacco (Artsaenko *et al.*, 1995) and pea (Radchuk *et al.* 2010a), anti-ABA antibody gene expression decreased free ABA, its levels are increased in the anti-ABA caryopses described in the current study. This is due to the fact that aABA grains produce high ABA levels from which only a small fraction is bound by the antibody. However, since gene expression and protein accumulation of the anti-ABA antibody starts at 7 DAF (Fig. 2B) when endogenous ABA is still very low (Weier *et al.* 2014), transient ABA limitation at that early stage cannot be excluded, which might induce ABA over-production.

Anti-ABA Caryopses Reveal broad Stimulation of Storage and Maturation-related Activities

At the stage when free ABA starts to accumulate (10 DAF), early cell divisions in the caryopses have already terminated and grains are in the transition stage, which is characterized by transcriptional and metabolic phase changes (Kohl *et al.*, 2015). ABA is obviously involved in this switch and decreasing its levels by immunomodulation during this critical period attenuates phase transitions, delays maturation and seed development in tobacco (Phillips *et al.* 1997), bean (Radchuk *et al.*, 2006) and pea (Radchuk *et al.* 2010a). Metabolite and transcript profiling in anti-ABA grains reveal specific changes, which conform to triggered and enhanced ABA functions. This is evident on different levels, such as transcriptional up-regulation of sucrose-to-starch metabolism, storage protein synthesis and ABA-related signal transduction, and leads to gene-induction associated to grain maturation, including proteinase inhibitors, LEA- and numerous stress-associated genes. (Fig. 7C, Supplementary Table S2). ABA is a well known activator of sucrose to starch conversion in rice (Wang*et al.*, 2015) and barley (Sreenivasulu *et al.*, 2006). LEA and proteinase inhibitor genes are responsive to ABA and stress (Skriver and Mundy, 1990). Several up-regulated genes, which are targets for ABA regulation, are involved in the energization of transport, including V-ATPase and H^{+}-PPiase, (Fukuda and Tanaka, 2006) and in mitochondrial energy production such as the mitochondrial ADP:ATP carrier, ATP-synthase (Supplementary Table S2). In Arabidopsis, V-ATPase is required for efficient nutrient acquisition (Krebs *et al.*, 2010). Obviously, ABA promotes nutrient transport during grain maturation via stimulation of membrane energization.

ABA-stimulated storage activity at 10 and 14 DAF is also reflected in metabolite levels. The decrease in free sugars, especially that of glucose, may indicate enhanced demand for starch biosynthesis. Lower levels of citrate and iso-citrate and higher levels of 2-oxo-glutarate could reflect an increased usage of organic acid, particularly for transamination reactions. Levels of many free amino acids are increased at 10 DAF but decreased at 14 DAF (Fig. 7C, Supplementary Fig. S5). This could point to increased biosynthesis at 10 DAF, generated by increased ABA. Subsequently, the lower levels at 14 DAF could indicate limitations due to stimulated storage protein synthesis. In various plant species, ABA and/or drought stress effect accumulation of free amino acids (Urano *et al.*, 2009;Bowne *et al.*, 2012). In accordance, enzymes of amino acid metabolism are up-regulated in anti-ABA caryopses at 10 DAF (Fig. 7C, Supplementary Table S2), such as dihydroxyacid dehydratase involved in the biosynthesis of branched-chain amino acids, whose levels increased under drought stress in wheat (Bowne *et al.* 2012). Anthranilate phosphoribosyl-transferase, asparagine synthetase and methionine synthase are responsive to wounding, salt and cold/drought/salt/ABA, respectively (Pratelli and Pilot, 2014). Grains of three aABA lines display decreased hexose levels and increased sucrose to hexose ratios during early maturation at 14 DAF (Supplementary Fig. S3). High amounts of free hexoses in barley as well as in other seeds indicate a young stage of development, whereas the ratio of sucrose to hexose increases during transition to maturation (Weschke *et al.*, 2000; Weber *et al.*, 2005). The fact that sucrose to hexose ratios are significantly increased in the aABA caryopses at 14 DAF indicate advanced developmental and precocious maturation compared to the wildtype.

ABA Accumulation Induces Precocious Maturation Integrated with Stress Response

The data show that in anti-ABA caryopses levels of free ABA are increased from 10 DAF, which leads to a broad stimulation of storage and maturation-related gene expression.

Unexpectedly, along with starch biosynthesis, several α- and β-amylases are also up-regulated in anti-ABA caryopses, accompanied with lower starch contents at 10 DAF (Supplementary Fig. S3). This could be due to enhanced stress signalling. Similarly, in rice, starch degradation is transcriptionally up-regulated in response to cold or dehydration (Maruyama *et al.* 2014) pointing to an interplay of ABA and stress signalling. Anti-ABA grains display broad constitutive gene induction related to different biotic and abiotic stresses mainly at 10 and 14 DAF (Fig. 7C, Supplementary Table S2). Most of the related genes are ABA- and/or stress-inducible (Ingram and Bartels, 1996) and include alcohol and aldehyde dehydrogenases, peroxidases, chaperones and glutathione-S-transferase, involved in chemical detoxification. The fact that anti-ABA caryopses do not suffer from stress implicates that the increased ABA mimics such an effect and causes gene expression related to stress response and tolerance. It further shows that in anti-ABA grains the precocious accumulation of ABA generates an integrated response of stress and maturation.

In parallel with untimely induction of storage activities on the levels of transcripts and metabolites, cell wall metabolism is down-regulated, which is reflected by lower levels of cell wall sugars (Supplementary Fig. S3) at 10 DAF and transcriptional down-regulation of expansins and several enzymes related to cell wall metabolism (Fig. 7C). Such negative effects on cell wall metabolism could be explained by the general growth-inhibiting effect of ABA. Similarly, in squash hypocotyls ABA inhibits the synthesis of cell wall polysaccharides (Wakabayashi *et al.*, 1991). Also in maize leaves, ABA suppresses leaf elongation rate (Cramer and Quarrie, 2002). These deficiencies in cell wall biosynthesis can also explain the observed kink-like phenotype of a considerable fraction of anti-ABA grains, which may be derived from failure to synthesize intact cell walls. It is tempting to speculate that in the anti-ABA caryopses the advanced increase of ABA during the transition phase negatively interferes with cell expansion and development and induces precocious and untimely maturation.

ABA Sensing and Signalling in anti-ABA grains Integrates Positive and Negative Regulation

Despite the stimulating effect of elevated ABA on storage

activities and grain maturation, grain weight and composition of mature grains are unchanged in anti-ABA plants, although germination is somewhat delayed. This implicates that the advanced maturation triggered by increased ABA is partially balanced at later development, so that maturating grains are not largely affected. In this respect it is noticeable that the stimulation of respective gene expression occurs only transiently at 10 and 14 DAF, whereas at later stages (20 DAF) there are just minor differences (Fig. 7C). It is known that ABA induces specific mechanisms to desensitize its signalling efficiency (Phillips *et al.*, 2007), which will result in decreased perception of ABA levels (Reyes and Chua, 2007). Transcript profiling in the anti-ABA caryopses reveals the activation of a network of positive and negative regulators involved in ABA-related signalling (Fig. 7C,Supplementary Table S2). Several of these elements are possibly involved in triggering transcriptional changes, which are required to switch development during the transition stage towards activating maturation. Such factors include subunits of the SWI/SNF chromatin remodelling complex, homeobox-leucine-zipper proteins and HvOS1/2, implicating that ABA can alter development on the levels of chromatin and DNA methylation and/or epigenetics (Saez *et al.* 2008; Kapazoglu *et al.* 2012). A broad range of elements, up-regulated in the anti-ABA grains, involves stress signalling such as TAG6, Myb BAS2, bHLH92, HvCBF1, Trihelix GT3b, HvERF1, NAC- and WRKY-like transcription factors (Fig. 7C). However, at least seven up-regulated signalling factors in anti-ABA grains potentially might antagonize ABA functions, including PP2C, CBL-CIPK and the FERONIA-like receptor (Pandey *et al.* 2004; Yu *et al.* 2012; Yoshida *et al.* 2015) Also WRKY-like transcription factors can repress ABA signalling (Zhang *et al.*, 2015).

In summary, there is evidence that especially during the later development of grain filling the repression of ABA-signalling could antagonize the effect caused by its over-shooting production. This ultimately leads to normalization of grain development and grain filling and generates mature grains with nearly unchanged dry weight and composition. Such compensation highlights the enormous physiological and metabolic flexibility of barley grain development to adjust the effects of high ABA levels in the anti-ABA grains in order to ensure and maintain proper grain development.

ACKNOWLEDGEMENTS

We thank Swetlana Friedl for support in all questions of bioinformatics and evaluation of microarray data. The excellent technical assistance of Katrin Blaschek, Uta Siebert, Angela Stegmann, Andrea Müller and Ingrid Otto is gratefully acknowledged. We are also grateful for funding by the Bundesministerium für Bildung und Forschung, Germany, in frame of the GABI-grain project (FKZ 0315041A).

REFERENCES

1. Abe H Urao T Ito T Seki M Shinozaki K Yamaguchi-Shinozaki K . 2003.Arabidopsis AtMYC2 (bHLH) and AtMYB2 (MYB) function as transcriptional activators in abscisic acid signaling. The Plant Cell 15, 63–78.
2. Artsaenko O Peisker M zur Nieden U Fiedler U Weiler EW Muntz K Conrad U .1995. Expression of a single-chain Fv antibody against abscisic acid creates a wilty phenotype in transgenic tobacco. The Plant Journal 8, 745–750.
3. Benjamini Y Hochberg Y . 1995. Controlling the false discovery rate – a practical and powerful approach to multiple testing. Journal of the Royal Statistical Society Series B-Methodological 57, 289–300.
4. CrossRef
5. Boursiac Y Leran S Corratge-Faillie C Gojon A Krouk G Lacombe B . 2013. ABA transport and transporters. Trends in Plant Science 18, 325–333. ↵
6. Bowne JB Erwin TA Juttner J Schnurbusch T Langridge P Bacic A Roessner U . 2012.Drought responses of leaf tissues from wheat cultivars of differing drought tolerance at the metabolite level. Molecular Plant 5, 418–429.
7. Conrad U Manteuffel R . 2001. Immunomodulation of phytohormones and functional proteins in plant cells. Trends in Plant Sciences 6, 399–402. CrossRef ↵
8. Conrad U Plagmann I Malchow S Sack M Floss DM Kruglov AA Nedospasov SA Rose-John S Scheller J . 2011. ELPylated anti-human TNF therapeutic single-domain antibodies for prevention of lethal septic shock. Plant Biotechnology Journal 9,22–31.
9. Cramer GR Quarrie SA . 2002. Abscisic acid is correlated with the leaf growth inhibition of four genotypes of maize differing in their response to salinity.Functional Plant Biology 29, 111–115. CrossRef
10. Cutler SR Rodriguez PL Finkelstein RR Abrams SR . 2010. Abscisic acid: emergence of a core signaling network. Annual Review of Plant Biology 61, 651–679.
11. Erban A Schauer N Fernie AR Kopka J . 2007. Nonsupervised construction and application of mass spectral and retention time index libraries from time-of-flight gas chromatography-mass spectrometry metabolite profiles. Methods in Molecular Biology 358, 19–38.
12. Fiedler U Conrad U . 1995. High-level production and long-term storage of engineered antibodies in transgenic tobacco seeds. Biotechnology 13, 1090–1093. ↵
13. Finkelstein R Reeves W Ariizumi T Steber C . 2008. Molecular aspects of seed dormancy. Annual Review of Plant Biology 59, 387–415.
14. Finkelstein RR Gampala SS Rock CD . 2002. Abscisic acid signaling in seeds and seedlings. Plant Cell 14 Suppl , S15–45.
15. Finkelstein RR Gibson SI . 2002. ABA and sugar interactions regulating development: cross-talk or voices in a crowd? Current Opinion in Plant Biology 5,26–32.
16. Fuchs S Tischer SV Wunschel C Christmann A Grill E . 2014. Abscisic acid sensor RCAR7/PYL13, specific regulator of protein phosphatase coreceptors. Proceedings of the National Academy of Sciences, USA 111, 5741–5746.
17. Gago GM, Almoguera C, Jordano J, Gonzalez DH, Chan RL. 2002. Hahb-4, a homeobox-leucine zipper gene potentially involved in abscisicacid-dependent responses to water stress in sunflower. Plant, Cell & Environment 25, 633–640.

18. Halford NG, Field JM, Blair H, Urwin P, Moore K, Robert L, Thompson R, Flavell RB, Tatham AS, Shewry PR. 1992. Analysis of HMW glutenin subunits encoded by chromosome 1A of bread wheat (Triticum aestivum L.) indicates quantitative effects on grain quality. TAG Theoretical and Applied Genetics 83, 373–378.

19. Hirayama T, Shinozaki K. 2007. Perception and transduction of abscisicacid signals: keys to the function of the versatile plant hormone ABA. Trends in Plant Sciences 12, 343–351.

20. Ingram J, Bartels D. 1996. The molecular basis of dehydration tolerancein plants. Annal Review of Plant Biology 47, 377–403.

21. Jeon J, Kim NY, Kim S, et al.2010. A subset of cytokinin two-componentsignaling system plays a role in cold temperature stress response inArabidopsis. Journal of Biological Chemistry 285, 23371–23386.

22. Jiang Y, Yang B, Deyholos MK. 2009. Functional characterization ofthe Arabidopsis bHLH92 transcription factor in abiotic stress. MolecularGenetics and Genomics 282, 503–516.

23. Kapazoglou A, Engineer C, Drosou V, Kalloniati C, Tani E, TsaballaA, Kouri ED, Ganopoulos I, Flemetakis E, Tsaftaris AS. 2012. Thestudy of two barley type I-like MADS-box genes as potential targets ofepigenetic regulation during seed development. BMC Plant Biology 12,166.

24. Kaplan-Levy RN, Brewer PB, Quon T, Smyth DR. 2012. The trihelixfamily of transcription factors – light, stress and development. Trends inPlant Science 17, 163–171.

25. Kempa S, Krasensky J, Dal Santo S, Kopka J, Jonak C. 2008. Acentral role of abscisic acid in stress-regulated carbohydrate metabolism.PLoS One 3, e3935.

26. Kim K-N, Cheong YH, Grant JJ, Pandey GK, Luan S. 2003. CIPK3, acalcium sensor-associated protein kinase that regulates abscisic acid and cold signal transduction in Arabidopsis. The Plant Cell 15, 411–423.

27. Kohl S, Hollmann J, Erban A, Kopka J, Riewe D, Weschke W, Weber H. 2015. Metabolic and transcriptional transitions in barley glumes reveal a role as transitory resource buffers during endosperm filling. Journal of Experimental Botany 66, 1397–1411.

28. Komari T, Hiei Y, Saito Y, Murai N, Kumashiro T. 1996. Vectors carrying two separate T-DNAs for co-transformation of higher plants mediated by Agrobacterium tumefaciens and segregation of transformants free from selection markers. The Plant Journal 10, 165–174.

29. Koornneef M, Hanhart CJ, Hilhorst HWM, Karssen CM. 1989. In vivo inhibition of seed development and reserve protein accumulation in recombinants of abscisic acid biosynthesis and responsiveness mutants inArabidopsis thaliana. Pl ant Physiology 90, 463–469.

30. Krebs M, Beyhl D, Gorlich E, Al-Rasheid KA, Marten I, Stierhof YD, Hedrich R, Schumacher K. 2010. Arabidopsis V-ATPase activity at the tonoplast is required for efficient nutrient storage but not for sodium accumulation. Proceedings of the National Academy of Sciences, USA 107, 3251–3256.

31. Kumlehn J, Serazetdinova L, Hensel G, Becker D, Loerz H. 2006. Genetic transformation of barley (Hordeum vulgare L.) via infection of androgenetic pollen cultures with Agrobacterium tumefaciens. Plant Biotechnology Journal 4, 251–261.

32. Li Q, Fan C, Zhang X, Wang X, Wu F, Hu R, Fu Y. 2014. Identification of a soybean MOTHER OF FT AND TFL1 homolog involved in regulation of seed germination. PLoS One 9, e99642.

33. Ma Y, Szostkiewicz I, Korte A, Moes D, Yang Y, Christmann A, Grill E. 2009. Regulators of PP2C phosphatase activity function as abscisic acid sensors. Science 324, 1064–1068.

34. Maruyama K, Urano K, Yoshiwara K, et al. 2014. Integrated analysis of the effects of cold and dehydration on rice metabolites, phytohormones, and gene transcripts. Plant Physiology 164, 1759–1771.

35. Miersch O, Neumerkel J, Dippe M, Stenzel I, Wasternack C. 2008. Hydroxylated jasmonates are commonly occurring metabolites of jasmonic acid and contribute to a partial switch-off in jasmonate signaling. New Phytologist 177, 114–127.

36. Munro S, Pelham HR. 1987. A C-terminal signal prevents secretion of luminal ER proteins. Cell 48, 899–907.

37. Nakamura S, Abe F, Kawahigashi H, et al. 2011. A wheat homolog of MOTHER OF FT AND TFL1 acts in the regulation of germination. The Plant Cell 23, 3215–3229.

38. Nakashima K, Yamaguchi-Shinozaki K. 2013. ABA signaling in stressresponse and seed development. Plant Cell Reports 32, 959–970.

39. Neri D, Montigiani S, Kirkham PM. 1996. Biophysical methods for the determination of antibody-antigen affinities. Trends in Biotechnology 14, 465–470.

40. Pandey GK, Cheong YH, Kim K-N, Grant JJ, Li L, Hung W, D'Angelo C, Weinl S, Kudla J, Luan S. 2004. The calcium sensor calcineurin b-like 9 modulates abscisic acid sensitivity and biosynthesis in Arabidopsis. The Plant Cell 16, 1912–1924.

41. Phillips J, Artsaenko O, Fiedler U, Horstmann C, Mock HP, Muntz K, Conrad U. 1997. Seed-specific immunomodulation of abscisic acid activity induces a developmental switch. EMBO Journal 16, 4489–4496.

42. Phillips JR, Dalmay T, Bartels D. 2007. The role of small RNAs in abiotic stress. FEBS Letters 581, 3592–3597.

43. Pratelli R, Pilot G. 2014. Regulation of amino acid metabolic enzymes and transporters in plants. Journal of Experimental Botany 65, 5535–5556.

44. Qin X, Zeevaart JA. 2002. Overexpression of a 9-cis-epoxycarotenoid dioxygenase gene in Nicotiana plumbaginifolia increases abscisic acid and phaseic acid levels and enhances drought tolerance. Plant Physiology 128, 544–551.

45. Radchuk R, Conrad U, Saalbach I, Giersberg M, Emery RJ,Kuster H, Nunes-Nesi A, Fernie AR, Weschke W, Weber H. 2010a. Abscisic acid deficiency of developing pea embryos achieved byimmunomodulation attenuates developmental phase transition and storage metabolism. The Plant Journal 64, 715–730.

46. Radchuk R, Emery RJ, Weier D, Vigeolas H, Geigenberger P, Lunn JE, Feil R, Weschke W, Weber H. 2010b. Sucrose non-fermenting kinase 1 (SnRK1) coordinates metabolic and hormonal signals during pea cotyledon growth and differentiation. The Plant Journal 61, 324–338.

47. Radchuk R, Radchuk V, Weschke W, Borisjuk L, Weber H. 2006.Repressing the expression of the SUCROSE NONFERMENTING-1-RELATED PROTEIN KINASE gene in pea embryo causes pleiotropic defects of maturation similar to an abscisic acid-insensitive phenotype.Plant Physiology 140, 263–278.

48. Raghavendra AS, Gonugunta VK, Christmann A, Grill E. 2010. ABA perception and signalling. Trends in Plant Science 15, 395–401.

49. Reyes JL, Chua NH. 2007. ABA induction of miR159 controls transcriptlevels of two MYB factors during Arabidopsis seed germination. The Plant Journal 49, 592–606.

50. Rolletschek H, Hajirezaei MR, Wobus U, Weber H. 2002. Antisenseinhibition of ADP-glucose pyrophosphorylase in Vicia narbonensis seeds increases soluble sugars and leads to higher water and nitrogen uptake. Planta 214, 954–964.

51. Rushton DL, Tripathi P, Rabara RC, et al. 2012. WRKY transcription factors: key components in abscisic acid signalling. Plant Biotechnology Journal 10, 2–11.

52. Saez A, Rodrigues A, Santiago J, Rubio S, Rodriguez PL. 2008. HAB1–SWI3B interaction reveals a link between abscisic acid signaling and putative SWI/SNF chromatin-remodeling complexes in Arabidopsis. The Plant Cell 20, 2972–2988.

53. Skriver K, Mundy J. 1990. Gene expression in response to abscisic acid and osmotic stress. The Plant Cell 2, 503–512.

54. Spartz AK, Gray WM. 2008. Plant hormone receptors: new perceptions. Genes and Development 22, 2139–2148.

55. Sreenivasulu N, Radchuk V, Strickert M, Miersch O, Weschke W, Wobus U. 2006. Gene expression patterns reveal tissue-specific signaling networks controlling programmed cell death and ABA-regulated maturation in developing barley seeds. The Plant Journal 47, 310–327.

56. Strauss M, Kauder F, Peisker M, Sonnewald U, Conrad U, Heineke D. 2001. Expression of an abscisic acid-binding single-chain antibody influences the subcellular distribution of abscisic acid and leads to developmental changes in transgenic potato plants. Planta 213, 361–369.

57. Thiel J, Muller M, Weschke W, Weber H. 2009. Amino acid metabolism at the maternal-filial boundary of young barley seeds: a microdissectionbased study. Planta 230, 205–213.

58. Thiel J, Weier D, Sreenivasulu N, Strickert M, Weichert N, Melzer M, Czauderna T, Wobus U, Weber H, Weschke W. 2008. Different hormonal regulation of cellular differentiation and function in nucellar projection and endosperm transfer cells: a microdissection-based transcriptome study of young barley grains. Plant Physiology 148, 1436–1452.

59. Thompson AJ, Jackson AC, Symonds RC, Mulholland BJ, Dadswell AR, Blake PS, Burbidge A, Taylor IB. 2000. Ectopic expression of a tomato 9-cis-epoxycarotenoid dioxygenase gene causes over-production of abscisic acid. The Plant Journal 23, 363–374.

60. Urano K, Maruyama K, Ogata Y, et al. 2009. Characterization of the ABA-regulated global responses to dehydration in Arabidopsis by metabolomics. The Plant Journal 57, 1065–1078.

61. Verslues PE, Zhu JK. 2007. New developments in abscisic acidperception and metabolism. Current Opinion in Plant Biology 10,447–452.

62. Wakabayashi K, Sakurai N, Kuraishi S. 1991. Effects of abscisic acid on the synthesis of cell-wall polysaccharides in segments of etiolated squash hypocotyl II. Levels of UDP-neutral sugars. Plant and Cell Physiology 32, 427–432.

63. Wang Z, Xu Y, Chen T, Zhang H, Yang J, Zhang J. 2015. Abscisic acid and the key enzymes and genes in sucrose-to-starch conversion in rice spikelets in response to soil drying during grain filling. Planta 241, 1091–1107.

64. Weber H, Borisjuk L, Wobus U. 2005. Molecular physiology of legume seed development. Annual Reviews of Plant Biology 56, 253–279.

65. Weichert N, Saalbach I, Weichert H, et al. 2010. Increasing sucrose uptake capacity of wheat grains stimulates storage protein synthesis. Plant Physiology 152, 698–710.

66. Weier D, Thiel J, Kohl S, Tarkowska D, Strnad M, Schaarschmidt S, Weschke W, Weber H, Hause B. 2014. Gibberellin-to-abscisic acid balances govern development and differentiation of the nucellar projectionof barley grains. Journal of E xperimental Botany 65, 5291–5304.

67. Weschke W, Panitz R, Sauer N, Wang Q, Neubohn B, Weber H, Wobus U. 2000. Sucrose transport into barley seeds: molecular characterization of two transporters and implications for seed development and starch accumulation. The Plant Journal 21, 455–467.

68. White CN, Proebsting WM, Hedden P, Rivin CJ. 2000. Gibberellins and seed development in maize. I. Evidence that gibberellin/abscisic acid balance governs germination versus maturation pathways. Plant Physiology 122, 1081–1088.

69. Wigger J, Phillips J, Peisker M, Hartung W, zur Nieden U, Artsaenko O, Fiedler U, Conrad U. 2002. Prevention of stomatal closure by immunomodulation of endogenous abscisic acid and its reversion by abscisic acid treatment: physiological behaviour and morphological features of tobacco stomata. Planta 215, 413–423.

70. Wu D, Qiu L, Xu L, et al.2011. Genetic variation of HvCBF genes and their association with salinity tolerance in Tibetan annual wild barley. PLoS One 6, e22938.

71. Xu Z-S, Xia L-Q, Chen M, et al. 2007. Isolation and molecular characterization of the Triticum aestivum L. ethylene-responsive factor 1 (TaERF1) that increases

multiple stress tolerance. Plant Molecular Biology 65, 719–732.

72. Xu ZY, Kim DH, Hwang I. 2013. ABA homeostasis and signaling involving multiple subcellular compartments and multiple receptors. Plant Cell Reports 32, 807–813.

73. Yoshida T, Mogami J, Yamaguchi-Shinozaki K. 2015. Omics approaches toward defining the comprehensive abscisic acid signaling network in plants. Plant and Cell Physiology .

74. Yu F, Qian L, Nibau C, et al. 2012. FERONIA receptor kinase pathway suppresses abscisic acid signaling in Arabidopsis by activating ABI2 phosphatase. Proceedings of the National Academy of Sciences, USA 109, 14693–14698.

75. Zhang L, Gu L, Ringler P, Smith S, Rushton PJ, Shen QJ. 2015. Three WRKY transcription factors additively repress abscisic acid and gibberellin signaling in aleurone cells. Plant Sciences 236, 214–222.

76. Zhang Y, Tessaro MJ, Lassner M, Li X. 2003. Knockout analysis of Arabidopsis transcription factors TGA2, TGA5, and TGA6 reveals their redundant and essential roles in systemic acquired resistance. The Plant Cell 15, 2647–2653.

77. Zheng X, Chen B, Lu G, Han B. 2009. Overexpression of a NAC transcription factor enhances rice drought and salt tolerance. Biochimical Biophysical Research Communications 379, 985–989.

CHAPTER 8

EXPRESSION PROFILING OF GENES INVOLVED IN DROUGHT STRESS AND LEAF SENESCENCE IN JUVENILE BARLEY

Gwendolin Wehner, Christiane Balko, Klaus Humbeck, Eva Zyprian and Frank Ordon

Keywords: Barley, Leaf senescence, Drought stress, High-throughput qPCR, Gene expression, eQTL

BACKGROUND

In order to analyse genetic networks and stress response, real time polymerase chain reaction (PCR) is an important tool [1]. For several years high-throughput instruments e.g. the BioMark System from Fluidigm have enabled large scale quantitative PCR studies [2]. Because of this and the possibility to analyse a large number of genotypes easily on expression chips [2] a range of genome wide association studies (GWAS) using expression data were conducted in the last years [3, 4, 5]. Expression quantitative trait loci (eQTL) were detected first in medicinal studies in humans and later also in plants [6, 7, 8, 9, 10]. In plants most eQTL studies were performed for complex pathways and aimed at a better understanding of the molecular networks [11]. Whereas in biotic stress the resistance is often controlled by a single gene, responses to abiotic stresses such as drought stress are controlled by many genes [12, 13, 14] and so these processes are particularly suitable for high throughput expression analyses and genetical genomics approaches [15]. Even in early developmental stages drought stress and drought stress induced premature leaf senescence have major influences on yield formation [16]. Therefore, it is of prime importance to understand regulatory processes of drought stress [17] and leaf senescence [18].

Citation: Gwendolin Wehner, Christiane Balko, Klaus Humbeck, Eva Zyprian and Frank Ordon, "Expression profiling of genes involved in drought stress and leaf senescence in juvenile barley," BMC Plant Biology 201616:3. DOI: 10.1186/s12870-015-0701-4.

In plants drought stress is initiated by water deficit in soil resulting in osmotic and oxidative stress and cellular damage [19]. This leads to defined drought stress responses for instance regarding the maintenance of turgor by an increase of osmoprotective molecules as soluble sugars [20, 21, 22], as well as measurable lower water content and decreased growth in the stressed plants compared to a control [23, 24]. Stress perception is assigned by special receptors, such as abscisic acid (ABA) receptors, hexokinases, or ion channel linked receptors [25]. The stress signal is then transducted for example via serine-threonine kinases, serin-threonine phosphatases, calcium dependent protein kinases, or phospholipases [25]. Finally, the gene expression is regulated by effector genes coding for late embryo abundant (LEA) proteins, dehydrin, or reactive oxygen species (ROS) and transcription factors, such as MYB, WRKY, NAC, AP2/ERF, DREB2, or bZIP to activate stress responsive mechanisms, re-establish homeostasis and protect and repair damaged proteins and membranes [13, 19, 25, 26]. Besides the above mentioned genes, drought stress associated metabolites such as osmoprotectants, polyamines and proteins involved in carbon metabolism and apoptosis are part of drought stress tolerance [12, 27]. Disturbing the regulatory processes in drought stress response results in irreversible changes of cellular homeostasis and the destruction of functional and structural proteins and membranes, leading to cell death [19] and decreased yield formation [28]. A huge transcriptome analysis for drought stress associated genes was done for example in barley [29] and wheat [30] showing differential response of genes involved in drought stress tolerance.

Initiated by external signals e.g. various stresses such as drought, as well as by internal factors for example phytohormones leaf senescence often occurs as a natural degradation process at the final stage of plant development [31]. Drought stress induced leaf senescence proceeds in three steps. Perception of drought stress is the initiation phase in which senescence signals are transferred via senescence associated genes (SAG) [32]. These are regulatory genes which often encode transcription factors regulating gene expression by binding to distinct *cis*-elements of target genes [33]. In the following reorganisation phase resources are transported from source (e.g. roots, leaves) to sink (e.g. fruits, seed) organs being important for yield formation [34]. With this translocation chlorophyll, proteins, lipids and other macromolecules are degraded and the content of antioxidants, ABA and ROS increases induced by a change in gene expression [35, 36]. Differentially expressed genes and their regulation during leaf senescence were identified by transcriptome analysis using microarrays in *Arabidopsis thaliana* [37, 38]. While the genes for photosynthesis and chloroplast development are down-regulated, the genes for the degradation of macromolecules and recycling of resources are up-regulated [39]. For example, expressed genes for chlorophyll degradation are *PA42,Lhcb4* and *psbA* [40] and genes for N mobilization and transport are transcription factors WRKY [41] and NAC [42] as well as glutamine synthetase [38]. Genes differentially expressed can be grouped to those accelerating leaf senescence and genes delaying leaf senescence [43]. The latter possibly resulting in a "stay green" effect and improved drought tolerance [34, 44]. The reorganisation phase is the crucial step for reversibility, after which senescence is irreversible and leads to the final step where leaves and cells often die [45].

In barley (*Hordeum vulgare* L.), a crop plant of worldwide importance, most mechanisms for leaf senescence are still not well understood [18, 34]. The response to drought in juvenile stages is less well documented, as only few studies are focused on early developmental stages [20, 24, 46, 47] whereas a lot of studies were conducted for drought stress in the generative stage [48]. Nevertheless, barley is to some extent a model organism for research at a genome wide level. The barley gene space has been published [49] and with this information gene positions can be compared to these data. Comparing the position of the analysed genes in the Morex genome with positions of the detected eQTL, resulted in the co-localization of eQTL and genes involved in drought stress [11, 50]. Therefore, the present study aimed at the identification of eQTL in barley for genes involved in drought stress in the juvenile phase and early leaf senescence (Table 1) based on a genome wide association study.

Table 1: Primer pairs for the selected genes and the reference gene

	Gene	Functional annotation	Acc. No.	Primer (FOR and REV)	Ampl.
Drought stress genes	A1	ABA inducible gene	GenBank: X78205.1	ACACGGCGCAGTACACCAAGGAGT CCCACCACGGCGTTCACCAC	100 bp
	Dhn1	Dehydrin 1	GenBank: AF181451	GCAACAGATCAGCACACTTCCA GCTGACCCTGGTACTCCATTGT	141 bp
	GAD3	Glutamate decarboxylase 3	GenBank: AY187941	ATGGAGAACTGCCACGAGAA GGAGATCTCGAACTCGTCGT	147 bp
	NADP_ME	NADP-dependent malic enzyme-like	GenBank: XM_003569737	ATGGCGGGAAGATCAGGG ATCCCTCAGCAGGGAATGC	165 bp
	P5CS2	Delta 1-pyrroline-5-carboxylate synthase 2	GenBank: AK249154.1	GTATACATGCACGTGGACCC CAGAGGGTTTTCGCCGAATC	164 bp
Leaf senescence genes	Contig7437	SAG senescence associated gene	GenBank: KF190467.1	GCTGAACGGCTGCCACTCCC GAAACCATCGCGCCTGTGGTG	78 bp
	GSII	Glutamine synthetase 2	GenBank: X53580.1	ACGAGCGGAGGTTGACAG CGCCCCACACGAATAGAG	94 bp
	hv_36467	SAG senescence associated gene	GenBank: AK367894.1	CAGTCCTTTTGCGCAGTTTTC CCAAGCGAGAATGCCTTGTAA	152 bp
	LHC1b20	Light-harvesting complex I	GenBank: S68729.1	CTGACCAAGGCGGGGCTGATGAAC TCGTGGGGCGGGAGGCTGTAG	200 bp
	pHvNF-Y5α	SAG senescence associated gene	GenBank: AK370570	CATGAAGCGAGCTCGTGGAACA GGTGCGAAGGTGGGACTACTCTGA	126 bp

Genes out of GWAS[a]	AVP1	Vacuolar proton-inorganic pyrophosphatase	GenBank: AY255181.1	GACCCTCTCAAGGACACCTC TCCCAACCGGCAAAACTAGA	160 bp
	ETFQO	Electron transfer flavoprotein-ubiquinone oxidoreductase	GenBank: BT000373.1	CCACAACCCTTTCTTGAATCCG GATCTAAGGGCGTGGTGAATTT	160 bp
	SAPK9	Serine/threonine protein	GenBank: AB125310.1	TCATGCAAGACTGTTTCTTGGG TTTCTTCTTGGCACAAAGCATATT	149 bp
	TRIUR3	Protein kinase	GenBank: M94726	ACATTGACGTTGAGAGCAGC GCTACAGAGAATTTGTGACCCA	151 bp
	HvGAPDH	Glyceraldehyde-3-phosphate dehydrogenase	GenBank: DQ196027.1	CAATGCTAGCTGCACCACCAACTG CTAGCAGCCCTTCCACCTCTCCA	165 bp

[a]Genes coding for proteins identified by BlastX of significant marker sequences out of a previous genome wide association study (GWAS) by Wehner et al. [20]

RESULTS

Leaf Senescence

Leaf colour (SPAD, soil plant analysis development) measured at 20 days after drought stress induction (BBCH 25, according to Stauss [51]) being indicative for leaf senescence revealed significant differences between treatments and genotypes but no significant interaction of genotype and treatment was observed at this stage (Fig. 1 and Table 2) giving hint to physiological changes and changes in gene expression.

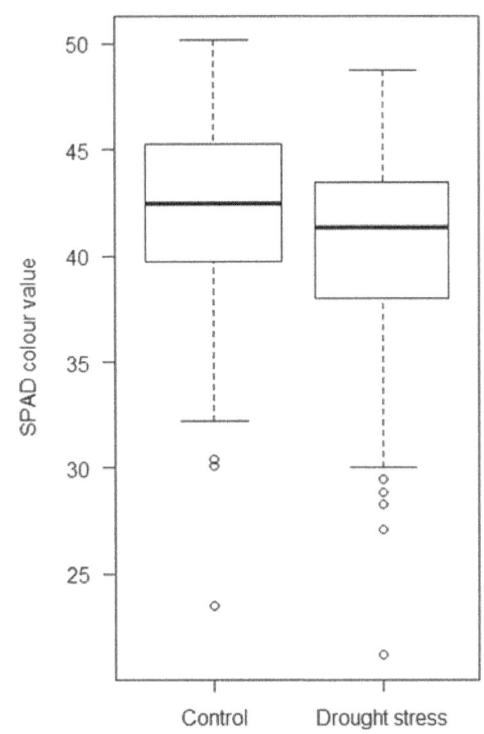

Figure 1: Box whisker plots for status of leaf senescence. Leaf colour (SPAD) for control and drought stress treatment at 27 days after sowing (das) including all 156 analysed barley genotypes.

Table 2: Analysis of variance for leaf colour (SPAD) and the expression of the selected genes

	Trait/Gene	Effect of treatment		Effect of genotype	
		***F* value**	***p* value**	***F* value**	***p* value**
	SPAD	11.2	0.0009	6.6	<2E-16

Drought stress genes	A1	50.1	4.88E-12	8.8	<2E-16
	Dhn1	138.4	<2E-16	23.5	<2E-16
	GAD3	81.8	<2E-16	96.7	<2E-16
	NADP_ME	315.5	<2E-16	4.1	4.63E-09
	P5CS2	229.6	<2E-16	335.4	<2E-16
Leaf senescence genes	Contig7437	0.9	0.342	128.7	<2E-16
	GSII	175.4	<2E-16	65.1	<2E-16
	hv_36467	160.2	<2E-16	46.9	<2E-16
	LHC1b20	102.4	<2E-16	156.7	<2E-16
	pHvNF-Y5α	76.5	<2E-16	196.4	<2E-16
Genes out of GWAS[a]	AVP1	51.4	2.06E-12	37.9	<2E-16
	ETFQO	16.3	5.98E-05	41.3	<2E-16
	SAPK9	9.0	0.00312	5.8	2.88E-07
	TRIUR3	96.5	<2E-16	38.1	<2E-16

[a]Genes coding for proteins identified by BlastX of significant marker sequences out of a previous genome wide association study (GWAS) by Wehner et al. [20]

Relative Expression of Candidate Genes

At the same developmental stage (BBCH 25) expression analyses were conducted for the whole set of 156 genotypes analysing 14 genes (Table 1). The relative expression (-ΔΔCt) ranges from −8.5 to 14.9 (Fig. 2, Additional file 1). In most genotypes all five drought stress related genes (A1, Dhn1, GAD3, NADP_ME and P5CS2) showed a higher expression under stress treatment relative to the control whereas for genes involved in leaf senescence opposite effects were detected for all genes (GSII, hv_36467, LHC1b20 and pHvNF-Y5α) except Contig7437. The genes out of the GWAS [20], i.e. AVP1 and TRIUR3 which are drought stress related genes, were up-regulated, whereas SAPK9 and ETFQO showed a lower expression relative to the control. In total, eight genes were up and six genes were down-regulated relative to the control but not all genotypes responded in the same way.

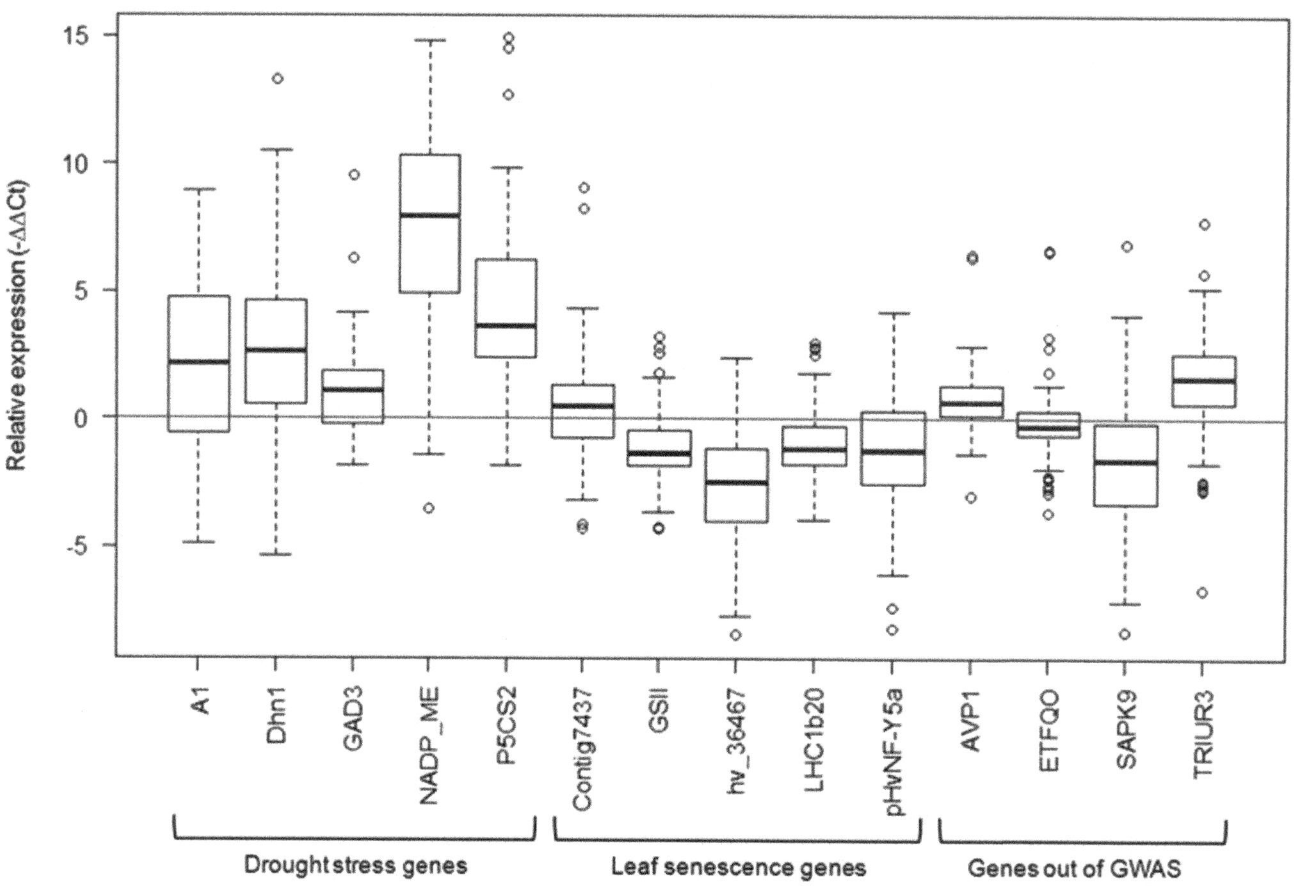

Figure 2: Expression profile for drought stress and leaf senescence genes. Relative Expression (-ΔΔCt) for the selected genes at 26 days after sowing (das) shown in box whisker plots including all 156 analysed barley genotypes.

The mean quality score for all amplifications was 0.954. Because ΔCt and ΔΔCt values were not normally distributed (data not shown) further statistical analysis was done with logarithmic values (log_2). Analysis of variance (ANOVA) revealed significant (p <0.001) effects for genotype and treatment for the 14 genes except Contig7437 (Table 2).

Highest significant correlations for differences in gene expression were identified within groups, i.e. within the group of drought stress genes, leaf senescence genes and genes out of GWAS (Table 3). The highest correlation was observed for the group of drought stress genes between relative expression of GAD3 and P5CS2 (r=0.84), for the group of leaf senescence genes for GSII and pHvNF-Y5a (r=0.64), and for the genes out of GWAS between AVP1 and TRIUR3 (r=0.54). For no gene the differential expression was significantly correlated to the expression differences of all other genes, but ETFQO was correlated to all except Dhn1, and GAD3 and Contig7437 were correlated to all except GSII and AVP1, and SAPK9 and NADP_ME, respectively. Significant correlations were also detected between the relative SPAD values for change in leaf colour and all leaf senescence genes except hv_36467 with the highest coefficients of correlation for GSII (r=0.24) and pHvNF-Y5a (r=0.34). Moreover, significant correlations were observed for relative SPAD values to two genes out of GWAS (r=0.16 for AVP1 and r=0.15 for TRIUR3).

Table 3: Coefficients of correlation for relative expression of the selected genes and the relative SPAD values

		A1	Dhn1	GAD3	NADP_ME	P5CS2	Contig7437	GSII	hv_36467	LHC1b20	pHvNF-Y5α	AVP1	ETFQO	SAPK9	TRIUR3
	Rel. SPAD	0.09	0.02	−0.10	0.01	0	−0.16*	0.24**	−0.13	0.19*	0.34***	0.16*	0.09	−0.15	0.15*
Drought stress genes	A1		0.68***	0.68***	0.44***	0.76***	0.38***	0.15	0.10	−0.16	−0.12	0.14	0.18*	0.37**	−0.11
	Dhn1			0.73***	0.35**	0.72***	0.64***	0.08	0.26**	−0.17*	−0.11	0.12	0.15	0.30*	−0.18*
	GAD3				0.43***	0.84***	0.65***	0	0.17*	−0.31***	−0.28***	0.09	0.20*	0.34**	−0.34***
	NADP_ME					0.49***	0.15	0.29*	0.15	−0.01	0.10	0.27*	0.24*	0.22	0.25*
	P5CS2						0.50***	0.17*	0.13	−0.19*	−0.09	0.10	0.18*	0.40**	−0.18*
Leaf senescence genes	Contig7437							−0.17*	0.45***	−0.24**	−0.35***	0.18*	0.16*	0.21	−0.25**
	GSII								0.09	0.55***	0.64***	0.47***	0.53***	0.18	0.44***
	hv_36467									0.19*	−0.09	0.15	0.30***	0.03	0.01
	LHC1b20										0.49***	0.38***	0.39***	0.10	0.39*
	pHvNF-Y5α											0.42***	0.28***	−0.26*	0.41***
Genes out of GWAS[a]	AVP1												0.46***	0.22	0.54***
	ETFQO													0.17*	0.35*
	SAPK9														0.06

r is significant with *p <0.05, **p <0.01 and ***p <0.001

[a]Genes coding for proteins identified by BlastX of significant marker sequences out of a previous genome wide association study (GWAS) by Wehner et al. [20]

Genome wide Association Study

Significant (p <0.001) marker gene expression associations were detected on all barley chromosomes except 4H with the highest number on chromosome 5H (8 single nucleotide polymorphisms, SNP) (Table 4). The largest transcriptional variance was explained by the marker SCRI_RS_181376 associated to the expression of ETFQO (R^2=11.55 %) and the highest likelihood of odds (LOD) was observed for the marker SCRI_RS_161614 associated to the expression of TRIUR3 (LOD=3.82) on barley chromosome 5H. Five SNP were significantly associated to the relative expression of the genes for drought stress, six to those for leaf senescence and seven to the genes out of the previous GWAS. Within the group of drought stress genes, expression differences of three genes (A1, GAD3 and P5CS2) and within the group of leaf senescence genes expression differences of four genes (Contig7437, GSII, hv_36467 and pHvNF-Y5α) were associated to markers. Out of these, three were located on chromosome 3H at 142.1 cM. This eQTL was detected for the relative expression of two drought stress genes (GAD3 and P5CS2) and one leaf senescence gene (Contig7437) which were also highly and significantly correlated (Table 3). Furthermore, an eQTL was observed for the relative expression of A1 on chromosome 5H at 149.9 cM associated to two markers. Associations for the relative expression of three genes (AVP1, ETFQO and TRIUR3) out of the four GWAS genes were detected on barley chromosomes 3H and 5H. For the expression of TRIUR3 three markers were found on 5H at 44.5 cM, and the expression of AVP1 was associated to a marker on chromosome 5H at 62.5 cM.

Table 4: Significant marker gene expression associations (p <0.001) with positions of eQTL

	Gene (log ΔΔCt)	Marker[b]	Chr.[b]	Pos. in cM[b]	*F*value	*p* value	-log p (LOD)	R^2 in %
Drought stress genes	A1	SCRI_RS_134358	5H	149.9	7.45	8.86E-04	3.05	9.5
	A1	SCRI_RS_165400	5H	150.1	7.45	8.86E-04	3.05	9.5
	GAD3	BOPA2_12_31177	1H	38.0	7.81	6.03E-04	3.22	8.9
	GAD3	BOPA1_4403-885	3H	142.1	12.09	6.67E-04	3.18	6.9
	P5CS2	BOPA1_4403-885	3H	142.1	11.31	9.84E-04	3.40	7.5

Leaf senescence genes	Contig7437	BOPA1_4403-885	3H	142.1	7.36	9.05E-04	3.01	7.1
	GSII	BOPA2_12_30065	7H	40.4	11.36	9.60E-04	3.04	9.5
	hv_36467	BOPA1_6547-1363	1H	111.8	8.11	4.58E-04	3.02	7.9
	hv_36467	BOPA2_12_31461	2H	131.9	13.14	4.00E-04	3.34	11.2
	pHvNF-Y5a	SCRI_RS_152393	6H	64.4	11.48	9.09E-04	3.04	7.8
	pHvNF-Y5a	SCRI_RS_194841	7H	81.5	12.91	4.49E-04	3.35	8.7
Genes out of GWAS[a]	AVP1	SCRI_RS_140294	5H	62.5	13.46	3.42E-04	3.47	9.1
	ETFQO	BOPA1_10126-999	3H	53.3	7.44	8.37E-04	3.08	10.1
	ETFQO	SCRI_RS_181376	5H	143.1	8.34	3.86E-04	3.41	11.5
	TRIUR3	BOPA1_4392-450	5H	44.5	7.64	7.07E-04	3.15	9.9
	TRIUR3	BOPA2_12_30717	5H	44.5	7.64	7.07E-04	3.15	9.9
	TRIUR3	SCRI_RS_41519	5H	44.5	7.64	7.07E-04	3.15	9.9
	TRIUR3	SCRI_RS_161614	5H	139.7	15.17	1.51E-04	3.82	9.8

[a]Genes coding for proteins identified by BlastX of significant marker sequences out of a previous genome wide association study (GWAS) by Wehner et al. [20]

[b]Marker positions are based on Comadran et al. [101]

The five SNP significantly associated to the relative expression of drought stress genes and the seven markers associated to genes out of GWAS all marked *cis* eQTL, while two *trans* eQTL were detected for P5CS2 and AVP1 (Table 5). In contrast, for the six markers significantly associated to leaf senescence genes only one *cis* eQTL was observed for pHvNF-Y5α. In summary, seven *trans* eQTL were detected and eight *cis* eQTL for which the Morex contigs showed a high identity to the gene analysed. Furthermore, *cis* eQTL explained a higher transcriptional variance (R^2) than those in *trans* (Table 4 and Table 5).

Table 5: Positions of the selected genes based on the barley Morex-contigs and their mode of action

	Gene	**POPSEQ[b,c]**	**Chr.[b]**	**cM[b]**	**Identity in %[c]**	**eQTL[d]**
Drought stress genes	A1	morex_contig_38178	5H	156.9	76	*cis*
	GAD3	morex_contig_790741	1H	42.0	81	*cis*
	GAD3	morex_contig_135241	3H	147.0	75	*cis*
	P5CS2	morex_contig_2549060	3H	30.2	76	*trans*
Leaf senescence genes	Contig7437	morex_contig_47765	4H	54.3	94	*trans*
	GSII	morex_contig_274546	7H	70.8	92	*trans*
	hv_36467	morex_contig_138818	1H	132.4	91	*trans*
	hv_36467	morex_contig_458133	2H	58.0	81	*trans*
	pHvNF-Y5a	morex_contig_244610	6H	76.0	100	*trans*
	pHvNF-Y5a	morex_contig_60611	7H	70.8	95	*cis*
Genes out of GWAS[a]	AVP1	morex_contig_80803	5H	44.1	75	*trans*
	ETFQO	morex_contig_6218	3H	51.8	95	*cis*
	ETFQO	morex_contig_1570014	5H	152.4	100	*cis*
	TRIUR3	morex_contig_81592	5H	42.0	88	*cis*
	TRIUR3	morex_contig_160473	5H	129.9	71	*cis*

[a]Genes coding for proteins identified by BlastX of significant marker sequences out of a previous genome wide association study (GWAS) by Wehner et al. [20]

[b]Gene positions are based on POPSEQ map (ibsc 2012)

[c]Morex contigs and identity comes out Blastn of the gene sequences against the Morex genome (ibsc 2012)

[d] *cis* eQTL coincide with the location of the underlying gene (position <10 cM), whereas *trans* eQTL are located in other regions of the genome Druka et al. [11]

DISCUSSION

Drought stress and leaf Senescence Genes

As shown by the significantly decreased SPAD values at 27 days after sowing (das, BBCH 25), drought stress had an accelerating influence on natural leaf senescence in barley (Fig. 1 and Table 2). Furthermore, the drought stress answer in this juvenile stage was observed by differential expression of 14 genes induced by drought stress or leaf senescence (Table 1, Fig. 2).

A1 is a gene which is induced by ABA or abiotic stresses like drought, cold and heat [19, 52, 53]. In the present study expression under drought stress was higher than in the well watered treatment (Fig. 2). This was also shown by several studies first in barley [53] and other species including transgenics [54, 55, 56, 57]. Dehydrins (Dhn) are well known to be expressed under dehydration stress [58]. For instance Dhn1 is described to be up-regulated under drought stress in barley [59, 60] which was also found in this study (Fig. 2). The glutamate decarboxylase gene (GAD3) is regulated by calcium and the protein encoded by this gene catalyzes the reaction of glutamate to γ-aminobutyric acid (GABA) [61, 62]. GABA may be involved in drought stress [63] by up-regulation of genes encoding a GABA receptor [29] which was also shown in the present study (Fig. 2). The NADP-dependent malic enzyme-like (NADP_ME) is involved in lignin biosynthesis, and regulates cytosolic pH through balancing the synthesis and degradation of malate [64]. As described in a drought stress study on barley, this effect is used for control of stomatal closure during the day under water-deficit conditions [29]. Comparable to the present study (Fig. 2) the gene for NADP_ME turned out to be higher expressed under drought stress [29]. The delta 1-pyrroline-5-carboxylate synthase 2 gene (P5CS2) is included in proline synthesis [65]. Content of proline is still controversially discussed as an indicator for drought tolerance [66], but it was shown in a previous study that the proline content increased under drought stress [20]. For approving its role, this gene was selected and showed up-regulation under drought stress (Fig. 2). Up-regulation under drought stress was also observed in tobacco [67] and transgenic rice [68].

The Contig7437 is a senescence associated gene (SAG) which is up-regulated under drought stress, as also shown by Guo et al. [29] in barley for drought stress during the reproductive stage. Other analysed SAGs are hv_36467 and pHvNF-Y5α, which were down-regulated in most genotypes under drought stress in our study (Fig. 2) whereas in literature reverse effects are described. The gene hv_36467 is a SAG12 like gene which is a senescence associated cystein protease and turned out to be up-regulated during natural leaf senescence in barley [69] and during dark induced senescence in tobacco [70]. In *Arabidopsis thaliana* the gene NFYA5 similar to pHvNF-Y5α was analysed by microarrays showing that the expression of this gene was induced by drought stress and ABA treatments [71], as well as under nitrogen stress [72]. Our data indicate a specific regulation of these two genes under different conditions. The protein encoded by the glutamine synthetase 2 (GSII) gene was found in photosynthetic tissues where its main role is the re-assimilation of photorespiratory ammonia [73, 74]. During senescence, the activity of GSII decreased representing down-regulation of associated genes in rice [73], barley and wheat [75] which was confirmed in the present study (Fig. 2). With chlorophyll degradation during leaf senescence the light harvesting complexes (LHC) of PSI and PSII remain stable, but synthesis rates of apoproteins of LHC decrease early in senescence [76]. In the present study LHC1b20 was down-regulated for most genotypes during drought stress induced leaf senescence in juvenile barley (Fig. 2) which was also shown in rice [77] and barley [78, 79] for natural leaf senescence in the generative stage.

In this study, all five selected drought stress genes were up-regulated under drought stress (Fig. 2) according to literature which demonstrates a clear drought stress answer and a good experimental setup for detecting and analysing drought stress response. In contrast, four out of the five selected genes for leaf senescence were down-regulated (Fig. 2) because a few of these genes are involved in photosynthesis and chloroplast development. Results for three of these genes (Contig7437, GSII and LHC1b20) were in accordance with results known from literature, while this was not the case for two of them (hv_36467 and pHvNF-Y5α). However, for all of these genes the adverse effect was detected for some genotypes (Fig. 2). Results revealed that drought stress in early developmental stages of barley leads to premature induced leaf senescence as already observed by physiological parameters [20] and by expression analysis of drought stress and leaf senescence related genes in this study.

Expression differences in three genes (GAD3, P5CS2 and Contig7437) were significantly associated to barley chromosome 3H at 142.1 cM (Table 4). At this position also quantitative trait loci (QTL) were found for drought stress [20, 80] as well as for leaf senescence [81]. These facts and the high correlation of these genes (Table 3) make this eQTL very interesting for marker assisted breeding in barley.

Genes out of GWAS

To verify the QTL identified for drought stress and drought stress induced leaf senescence by Wehner et al.[20] an expression profile and eQTL analysis was conducted with genes coding for proteins identified within respective QTL. The genes ETFQO, SAPK9, TRIUR3 and AVP1 were differentially expressed (Fig. 2).

The protein encoded by the electron transfer flavoprotein-ubiquinone oxidoreductase gene (ETFQO) is located in the mitochondria where it accepts electrons from ETF, transfers them to ubiquinone and acts downstream in the degradation of chlorophyll during leaf senescence [82, 83]. Expression studies showed that ETFQO is up-regulated under darkness induced leaf senescence [83, 84] whereas in this study on drought stress induced leaf senescence no clear direction was observed (Fig. 2). A gene coding for a serine/threonine-protein kinase (SAPK9) was analysed which can be activated by hyperosmotic stress and ABA in rice [85]. In the present study SAPK9 was down-regulated in most genotypes (Fig. 2). Furthermore, the abscisic acid-inducible protein kinase gene (TRIUR3) which is also involved in dehydration stress response [86] was differentially expressed. Until now, no relative expression analysis has been conducted for this gene, but a huge amount of ABA inducible genes are up-regulated under drought stress in rice [87]. In the present study TRIUR3 was also up-regulated under drought stress (Fig. 2). The nucleotide pyrophosphatase/phosphodiesterase gene (AVP1) is a gene which is up-regulated under drought stress [88] which was confirmed in the current study (Fig. 2). Expression of this gene was also observed in transgenics showing a higher drought stress tolerance [89, 90,91, 92].

Three of these genes (SAPK9, TRIUR3 and AVP1) were located within the QTL on barley chromosome 5H at 45 cM [20]. Furthermore, expression differences of two of them (TRIUR3 and AVP1) were again associated to markers on chromosome 5H around 45 cM (Table 4) and this position was also validated in the Morex genome (Table 5). A high and significant correlation between the relative expression data of both genes as well as to the relative SPAD values (Table 3) promotes this finding. At the same position on chromosome 5H two markers which turned out to be significantly associated to SPAD and biomass yield under drought stress treatment were identified [20]. So, these results [20] and those of this study give hint that the two SNP markers, i.e. BOPA1_9766-787 and SCRI_RS_102075 may be used in marker based selection procedures in barley breeding programmes aiming at the improvement of drought stress tolerance.

For the understanding of complex mechanisms, such as the process of drought stress tolerance and drought stress induced leaf senescence as a basis for future breeding activities it is of prime importance to understand how and when regulatory genes are activated and where they are located in the barley genome. Results of this study contribute to elucidate the regulation of drought stress induced leaf senescence during early developmental stages in barley. The present genetical genomics approach helps to localize and understand transcriptional regulation and gene interaction, both from *cis*-acting elements and *trans*-acting factors (Table 5). When analysing the expression regulation of the barley genome, *cis* eQTL were found for the genes A1, GAD3, pHvNF-Y5α, ETFQO and TRIUR3. Markers which were significantly associated to *cis* eQTL explained up to 11.55 % of the transcriptional variance (Table 4 and Table 5). Therefore, most of the strongest eQTL acted in *cis* which was also observed in previous eQTL studies [8, 93, 94].

Factors that act in *trans* regulating the expression levels of the genes of interest were mainly found for the group of leaf senescence genes. Some of these genes are described as SAGs (Contig7437, hv_36467 and pHvNF-Y5α), because up to now little is known about their function. Results of the present study give hint that these SAGs are regulated in *trans*.

CONCLUSION

With respect to the expression of genes involved in drought stress response and early leaf senescence genotypic differences exist in barley. Major eQTL for the expression of these genes are located on barley chromosome 3H and 5H. The eQTL on chromosome 5H coincides with the QTL for drought stress induced leaf senescence identified in a previous GWAS [43]. Respective markers, i.e. BOPA1_9766-787 and SCRI_RS_102075 may be used in future barley breeding programmes for improving tolerance to drought stress and early leaf senescence, respectively.

METHODS

Plant material and Phenotypic Characterisation

Phenotyping, genotyping and QTL analysis were conducted as described in Wehner et al. [20] on a set of 156 winter barley genotypes consisting of 113 German winter barley cultivars (49 two-rowed and 64 six-rowed, [95]) and 43 accessions of the spanish barley core collection (SBCC) [96]. The same set of genotypes as well as the same experimental design was used for expression- and eQTL analysis in the present study. In brief, trials were conducted in greenhouses of the Julius Kühn-Institut in Groß Lüsewitz, Germany and drought stress was applied in a split plot design with three replications per genotype and treatment (control, drought stress). In each pot four plants were sown and all leaves were tied up, except the primary leaf per plant. Drought stress was induced by a termination of watering at the primary leaf stage (BBCH 10, according to Stauss [51]) seven days after sowing (das). From this time drought stress developed slowly till 20 das when the final drought stress level was reached. The drought stress variant was kept at 20 % of the maximal soil water capacity and the control variant at 70 % by weighing the pots resulting in a relative water content (36 das) ranging between 88.8 % and 91.5 % in the control variant and 80.9 % and 86.1 % in the drought stress treatment. The experimental setup and growth conditions for these pot experiment are described in detail as design B in Wehner et al. [20].

At 26 das (BBCH 25) leaf material for RNA extraction was sampled by harvesting one primary leaf per pot taking the middle part for further analyses. Mixed samples out of the three leaf pieces (circa 100 mg) per genotype and treatment (312 samples) each were immediately frozen in liquid nitrogen and stored at −80 °C. Furthermore, to get information on the influence of drought stress on leaf senescence leaf colour (SPAD, Konica Minolta Chlorophyll Meter SPAD-502 Plus, Osaka Japan) was measured 27 das on three primary leaves per pot at five positions each.

RNA Isolation and cDNA Synthesis

The frozen primary leaves were homogenized with a tube pestle (Biozym) in liquid nitrogen. Total RNA from the primary leaves was isolated with the InviTrap Spin Plant RNA Mini Kit (STRATEC Molecular), using lysis solution RP and following the manufacturer's instructions. After incubation for 15 min at room temperature, an additional incubation for 3 min at 55 °C was conducted to get a higher RNA yield. Total RNA yield was measured by Qubit fluorometric quantification (Life technologies) and concentration was adjusted to 50 ng. RNA was used for cDNA synthesis with the QuantiTect Reverse Transcription Kit (Qiagen) following the manufacturer's instructions. cDNA was stored at −20 °C.

Expression Analysis using Quantitative real-time PCR (qPCR)

A high throughput system (BioMark) was used for expression analysis in which four Fluidigm chips (96.96) were analysed for the 312 samples. Default space on these chips allows to analyse 48 genes in two technical replications. Out of these 48 analysed genes (23 genes involved in drought stress, 12 leaf senescence genes, 11 genes coding for proteins out of a previous GWAS [20] and two reference genes), 14 differentially expressed genes revealing clear differences between genotypes and showing a low number of missing values were selected for the present study. Five of these genes were involved in leaf senescence, five in drought stress response and four genes coding for proteins related to leaf senescence or drought stress out of the previous genome wide association study [20] were chosen. In addition, as a reference gene GAPDH was included (Table 1). To identify the gene for those proteins identified in the GWAS studies by Wehner et al. [20] the significant associated marker sequences were compared to the plant nucleotide collection by Blastn (Basic Local Alignment Search Tool, ncbi [www.ncbi.nlm.nih.gov] accessed June 2014) and the gene with the best hit was chosen for primer design.

Primers (Eurofins HPSF purified) were constructed using the primer designing tool of NCBI ([www.ncbi.nlm.nih.gov/tools/primer-blast] accessed June 2014) with a length of 20 bp, annealing temperature of 59 °C and product size of 100–200 bp (Table 1).

qPCR was performed using the high throughput platform BioMark HD System and the 96.96 Dynamic Array IFC (Fluidigm) following the manufacturer's instructions. 5 µl Fluidigm sample premix consisted of 1.25 µl pre-amplified cDNA, 0.25 µl of 20x DNA binding dye sample loading reagent (Fluidigm), 2.5 µl of SsoFast EvaGreen Supermix with low ROX (BioRad) and 1 µl of RNase/DNase-free water. Each 5 µl assay premix consisted of 2 µl of 100 µM primers, 2.5 µl assay loading reagent (Fluidigm) and 0.5 µl RNase/DNase-free water. Thermal conditions for qPCR were: 95 °C for 60 s, 30 cycles of 96 °C for 5 s, 60 °C for 20 s plus melting curve analysis. Data were processed using BioMark Real-Time PCR Analysis Software 3.0.2 (Fluidigm). The quality threshold was set at the default setting of 0.65 and linear baseline correction and automatic cycle threshold method were used.

DATA ANALYSIS

The analysis software (Fluidigm Real- Time PCR Analysis Software) gave cycle threshold (Ct) values and calculated ΔCt values, as well as a quality score for each amplification. Out of these ΔCt values calculated out of the Ct value of the gene of interest minus the Ct value of the housekeeping gene (GAPDH) for each genotype, treatment and replication, the relative expression (ΔΔCt) was calculated out of the ΔCt values for stress treatment minus the ΔCt values for control treatment for each genotype and replication [97]. ΔΔCt values without correction of PCR efficiency were used for calculation, because genes were tested and selected by their efficiency in preliminary experiments. A mean PCR efficiency (Quality Score of Fluidigm) was calculated for all amplifications.

Shapiro-Wilk test for normal distribution and analysis of variance (ANOVA) using a linear model were carried out using R 2.15.1 [98] to test effects of genotype (using ΔΔCt values) and treatment (using ΔCt values). Furthermore, coefficients of correlation (Spearman) were calculated in R between relative expression of the genes and the relative SPAD values [20, 99]. Moreover, for the SPAD values an ANOVA mixed linear model (MLM) was calculated (replication as random) in R to test effects of genotype, treatment and interaction of genotype and treatment. For relative expression as well as for the SPAD values box whisker plots were calculated in R.

Expression Quantitative trait loci (eQTL) Analysis

For the 14 selected genes a genome wide association study (GWAS) for eQTL detection was conducted on the 156 genotypes applying a mixed linear model (MLM) using TASSEL 3.0 [100]. For this purpose a genetic map with 3,212 polymorphic SNP markers with minor allele frequencies larger than 5 % [101], a population structure calculated with STRUCTURE 2.3.4 [102] based on 51 simple sequence repeat (SSR) markers covering the whole genome, a kinship calculated with SPAGeDi 1.3d [103] based on 51 SSRs and the relative expression data (means for replications) were used. For comparability the methods were the same as used for GWAS in Wehner et al. [20]. All results with p values <0.001 (likelihood of odds, LOD=3) were considered as significant marker gene expression associations.

To compare genomic positions of the eQTL with those of the analysed genes, sequences of the genes were compared against high confidential genes (CDS sequences) of the barley Morex genome by Blastn (Basic Local Alignment Search Tool, IPK Barley Blast server [http://webblast.ipk-gatersleben.de/barley/viroblast.php] accessed May 2015) and the Morex contig with the highest identity on the associated linkage group (chromosome) was chosen. With this information eQTL were divided in *cis* and *trans* eQTL. *cis* eQTL coincide with the location of the underlying gene (position <10 cM), whereas *trans* eQTL are located in other regions of the genome [11].

ACKNOWLEDGEMENTS

The authors thank Dr. Brigitte Ruge-Wehling for the lab facilities for RNA isolation, Dr. Ernesto Igartua CSIC, Spain for providing seeds of the SBCC, the Interdisciplinary Center for Crop Plant Research (IZN) of the Martin-Luther-University of Halle-Wittenberg for funding this project and Prof. Dr. Klaus Pillen for close collaboration.

AUTHORS' CONTRIBUTIONS

GW conducted all experiments, including expression, statistical and bioinformatics analyses and mainly wrote the manuscript. EZ provided the Fluidigm BioMark System and supervised the gene expression experiments. CB, KH and FO designed the research, supervised the experimental design and participated in writing the manuscript. All authors approved the final manuscript.

REFERENCES

1. Korenková V, Scott J, Novosadová V, Jindřichová M, Langerová L, Švec D, et al. Pre-amplification in the context of high-throughput qPCR gene expression experiment. BMC Mol Biol. 2015;16(1):5.
2. Spurgeon SL, Jones RC, Ramakrishnan R. High throughput gene expression measurement with real time PCR in a microfluidic dynamic array. PLoS One. 2008;3(2), e1662.
3. Gilad Y, Rifkin SA, Pritchard JK. Revealing the

architecture of gene regulation: the promise of eQTL studies. Trends Genet. 2008;24(8):408–15.

4. Hansen BG, Halkier BA, Kliebenstein DJ. Identifying the molecular basis of QTLs: eQTLs add a new dimension. Trends Plant Sci. 2008;13(2):72–7.

5. Westra H-J, Franke L. From genome to function by studying eQTLs. Biochim Biophys Acta. 2014;1842(10):1896–902.

6. Gibson G, Weir B. The quantitative genetics of transcription. Trends Genet. 2005;21(11):616–23.

7. Potokina E, Prasad M, Malysheva L, Röder M, Graner A. Expression genetics and haplotype analysis reveal cis regulation of serine carboxypeptidase I (Cxp1), a candidate gene for malting quality in barley (Hordeum vulgare L.). Funct Integr Genomics. 2006;6(1):25–35.

8. West MA, Kim K, Kliebenstein DJ, van Leeuwen H, Michelmore RW, Doerge R, et al. Global eQTL mapping reveals the complex genetic architecture of transcript-level variation in Arabidopsis. Genetics. 2007;175(3):1441–50.

9. Drost DR, Puranik S, Novaes E, Novaes CR, Dervinis C, Gailing O, et al. Genetical genomics of Populus leaf shape variation. BMC Plant Biol. 2015;15(1):166.

10. Liu P, Wang CM, Li L, Sun F, Yue GH. Mapping QTLs for oil traits and eQTLs for oleosin genes in jatropha. BMC Plant Biol. 2011;11(1):132.

11. Druka A, Potokina E, Luo Z, Jiang N, Chen X, Kearsey M, et al. Expression quantitative trait loci analysis in plants. Plant Biotechnol J. 2010;8(1):10–27.

12. Vinocur B, Altman A. Recent advances in engineering plant tolerance to abiotic stress: achievements and limitations. Curr Opin Biotechnol. 2005;16(2):123–32.

13. Beck EH, Fettig S, Knake C, Hartig K, Bhattarai T. Specific and unspecific responses of plants to cold and drought stress. J Biosci. 2007;32(3):501–10.

14. Anjum SA, Xie X-y, Wang L-c, Saleem MF, Man C, Lei W. Morphological, physiological and biochemical responses of plants to drought stress. African J Agri Res. 2011;6(9):2026–32.

15. Sreenivasulu N, Sopory S, Kishor PK. Deciphering the regulatory mechanisms of abiotic stress tolerance in plants by genomic approaches. Gene. 2007;388(1):1–13.

16. El Hafid R, Smith DH, Karrou M, Samir K. Physiological responses of spring durum wheat cultivars to early-season drought in a Mediterranean environment. Ann Bot-London. 1998;81(2):363–70.

17. Cattivelli L, Rizza F, Badeck F-W, Mazzucotelli E, Mastrangelo AM, Francia E, et al. Drought tolerance improvement in crop plants: An integrated view from breeding to genomics. Field Crop Res. 2008;105(1-2):1–14.

18. Buchanan-Wollaston V, Earl HJ, Harrison E, Mathas E, Navabpour S, Page T, et al. The molecular analysis of leaf senescence – a genomics approach. Plant Biotechnol J. 2003;1:3–22.

19. Wang W, Vinocur B, Altman A. Plant responses to drought, salinity and extreme temperatures: towards genetic engineering for stress tolerance. Planta. 2003;218(1):1–14.

20. Wehner G, Balko C, Enders M, Humbeck K, Ordon F. Identification of genomic regions involved in tolerance to drought stress and drought stress induced leaf senescence in juvenile barley. BMC Plant Biol. 2015;15(1):125.

21. Blum A. Osmotic adjustment and growth of barley genotypes under drought stress. Crop Sci. 1989;29:230–3.

22. Teulat B, Borries C, This D. New QTLs identified for plant water status, water-soluble carbohydrate and osmotic adjustment in a barley population grown in a growth-chamber under two water regimes. Theor Appl Genet. 2001;103(1):161–70.

23. Jamieson PD, Martin RJ, Francis GS, Wilson DR. Drought effects on biomass production and radiation-use efficiency in barley. Field Crop Res. 1995;43:77–86.

24. Honsdorf N, March TJ, Hecht A, Eglinton J, Pillen K. Evaluation of juvenile drought stress tolerance and genotyping by sequencing with wild barley introgression lines. Mol Breeding. 2014;34(3):1475–95.

25. Bhargava S, Sawant K. Drought stress adaptation: metabolic adjustment and regulation of gene expression. Plant Breeding. 2013;132(1):21–32.

26. Ahuja I, de Vos RC, Bones AM, Hall RD. Plant molecular stress responses face climate change. Trends Plant Sci. 2010;15(12):664–74.

27. Shinozaki K, Yamaguchi-Shinozaki K. Gene networks involved in drought stress response and tolerance. J Exp Bot. 2007;58(2):221–7.

28. Farooq M, Wahid A, Kobayashi N, Fujita D, Basra SMA. Plant drought stress: effects, mechanisms and management. Agron Sust Dev. 2009;29(1):185–212.

29. Guo P, Baum M, Grando S, Ceccarelli S, Bai G, Li R, et al. Differentially expressed genes between drought-tolerant and drought-sensitive barley genotypes in response to drought stress during the reproductive stage. J Exp Bot. 2009;60(12):3531–44.

30. Aprile A, Mastrangelo AM, De Leonardis AM, Galiba G, Roncaglia E, Ferrari F, et al. Transcriptional profiling in response to terminal drought stress reveals differential responses along the wheat genome. BMC Genomics. 2009;10(1):279.

31. Lim PO, Kim HJ, Nam HG. Leaf senescence. Ann Rev Plant Biol. 2007;58:115–36.

32. Sarwat M, Naqvi AR, Ahmad P, Ashraf M, Akram NA. Phytohormones and microRNAs as sensors and regulators of leaf senescence: assigning macro roles to small molecules. Biotechnol Adv. 2013;31(8):1153–71.

33. Balazadeh S, Riaño-Pachón D, Mueller-oeber B. Transcription factors regulating leaf senescence in Arabidopsis thaliana. Plant biology. 2008;10(s1):63–75.

34. Gregersen PL, Culetic A, Boschian L, Krupinska K. Plant senescence and crop productivity. Plant Mol Biol. 2013;82(6):603–22.

35. Munne-Bosch S, Alegre L. Die and let live: leaf senescence contributes to plant survival under drought stress. Funct Plant Biol. 2004;31:203–16.

36. Penfold CA, Buchanan-Wollaston V. Modelling transcriptional networks in leaf senescence. J Exp Bot. 2014;65(14):3859–73.
37. Breeze E, Harrison E, McHattie S, Hughes L, Hickman R, Hill C, et al. High-resolution temporal profiling of transcripts during Arabidopsis leaf senescence reveals a distinct chronology of processes and regulation. Plant Cell. 2011;23(3):873–94.
38. Buchanan-Wollaston V, Page T, Harrison E, Breeze E, Lim PO, Nam HG, et al. Comparative transcriptome analysis reveals significant differences in gene expression and signalling pathways between developmental and dark/starvation-induced senescence in Arabidopsis. Plant J. 2005;42(4):567–85.
39. Humbeck K. Epigenetic and small RNA regulation of senescence. Plant Mol Biol. 2013;82(6):529–37.
40. Miersch I, Heise J, Zelmer I, Humbeck K. Differential degradation of the photosynthetic apparatus during leaf senescence in barley (Hordeum vulgare L.). Plant Biol. 2000;2:618–23.
41. Ay N, Irmler K, Fischer A, Uhlemann R, Reuter G, Humbeck K. Epigenetic programming via histone methylation at WRKY53 controls leaf senescence in Arabidopsis thaliana. Plant J. 2009;58(2):333–46.
42. Guo Y, Gan S. AtNAP, a NAC family transcription factor, has an important role in leaf senescence. Plant J. 2006;46(4):601–12.
43. Li Z, Peng J, Wen X, Guo H. Gene network analysis and functional studies of senescence-associated genes reveal novel regulators of Arabidopsis leaf senescence. J Integr Plant Biol. 2012;54(8):526–39.
44. Borrell AK, Hammer GL, Henzell RG. Does maintaining green leaf area in sorghum improve yield under drought? II. Dry matter production and yield. Crop Sci. 2000;40(4):1037–48.
45. Fischer AM. The Complex Regulation of Senescence. Crit Rev Plant Sci. 2012;31(2):124–47.
46. Ay N, Clauss K, Barth O, Humbeck K. Identification and characterization of novel senescence-associated genes from barley (Hordeum vulgare) primary leaves. Plant Biol. 2008;10(1):121–35.
47. Kleber-Janke T, Krupinska K. Isolation of cDNA clones for genes showing enhanced expression in barley leaves during dark-induced senescence as well as during senescence under field conditions. Planta. 1997;203(3):332–40.
48. Tyagi K, Park MR, Lee HJ, Lee CA, Rehman S, Steffenson B, et al. Fertile crescent region as source of drought tolerance at early stage of plant growth of wild barley (Hordeum vulgare L. ssp. spontaneum). Pak J Bot. 2011;43(1):475–86.
49. Consortium IBGS, Mayer KF, Waugh R, Brown JW, Schulman A, Langridge P, et al. A physical, genetic and functional sequence assembly of the barley genome. Nature. 2012;491(7426):711–6.
50. Kliebenstein D. Quantitative genomics: analyzing intraspecific variation using global gene expression polymorphisms or eQTLs. Plant Biol. 2009;60(1):93.
51. Stauss R. Compendium of Growth Stage Identification Keys for Mono-and Dicotyledonous Plants: Extended BBCH Scale, Autumn 1994: Ciba; 1994.
52. Marttila S, Tenhola T, Mikkonen A. A barley (Hordeum vulgare L.) LEA3 protein, HVA1, is abundant in protein storage vacuoles. Planta. 1996;199(4):602–11.
53. Straub PF, Shen Q, Ho T-hD. Structure and promoter analysis of an ABA-and stress-regulated barley gene, HVA1. Plant Mol Biol. 1994;26(2):617–30.
54. Checker VG, Chhibbar AK, Khurana P. Stress-inducible expression of barley Hva1 gene in transgenic mulberry displays enhanced tolerance against drought, salinity and cold stress. Transgenic Res. 2012;21(5):939–57.
55. Rohila JS, Jain RK, Wu R. Genetic improvement of Basmati rice for salt and drought tolerance by regulated expression of a barley Hva1 cDNA. Plant Sci. 2002;163(3):525–32.
56. Bray EA. Plant responses to water deficit. Trends Plant Sci. 1997;2(2):48–54.
57. Fu D, Huang B, Xiao Y, Muthukrishnan S, Liang GH. Overexpression of barley hva1 gene in creeping bentgrass for improving drought tolerance. Plant Cell Rep. 2007;26(4):467–77.
58. Marzin S, Mihaly R, Pauk J, Schweizer P. A transient assay system for the assessment of cell-autonomous gene function in dehydration-stressed barley. J Exp Bot. 2008;59(12):3359–69.
59. Suprunova T, Krugman T, Fahima T, Chen G, Shams I, Korol A, et al. Differential expression of dehydrin genes in wild barley, Hordeum spontaneum, associated with resistance to water deficit. Plant Cell Environ. 2004;27(10):1297–308.
60. Tommasini L, Svensson JT, Rodriguez EM, Wahid A, Malatrasi M, Kato K, et al. Dehydrin gene expression provides an indicator of low temperature and drought stress: transcriptome-based analysis of barley (Hordeum vulgare L.). Funct Integr Genomics. 2008;8(4):387–405.
61. Serraj R, Shelp BJ, Sinclair TR. Accumulation of γ-aminobutyric acid in nodulated soybean in response to drought stress. Physiol Plant. 1998;102(1):79–86.
62. Ueno H. Enzymatic and structural aspects on glutamate decarboxylase. J Mol Catal. 2000;10(1):67–79.
63. Rhodes D, Handa S, Bressan RA. Metabolic changes associated with adaptation of plant cells to water stress. Plant Physiol. 1986;82(4):890–903.
64. Wheeler MCG, Tronconi MA, Drincovich MF, Andreo CS, Flügge U-I, Maurino VG. A comprehensive analysis of the NADP-malic enzyme gene family of Arabidopsis. Plant Physiol. 2005;139(1):39–51.
65. Szabados L, Savoure A. Proline: a multifunctional amino acid. Trends Plant Sci. 2010;15(2):89–97.
66. Kishor P, Kavi B, Sreenivasulu N. Is proline accumulation per se correlated with stress tolerance or is proline homeostasis a more critical issue? Plant Cell Environ. 2014;37(2):300–11.
67. Dobrá J, Vanková R, Havlová M, Burman AJ, Libus J, Štorchová H. Tobacco leaves and roots differ in the expression of proline metabolism-related genes in the

course of drought stress and subsequent recovery. J Plant Physiol. 2011;168(13):1588–97.

68. Quan R, Hu S, Zhang Z, Zhang H, Zhang Z, Huang R. Overexpression of an ERF transcription factor TSRF1 improves rice drought tolerance. Plant Biotechnol J. 2010;8(4):476–88.

69. Hollmann J, Gregersen PL, Krupinska K. Identification of predominant genes involved in regulation and execution of senescence-associated nitrogen remobilization in flag leaves of field grown barley. J Exp Bot. 2014;65(14):3963–73.

70. Carrión CA, Costa ML, Martínez DE, Mohr C, Humbeck K, Guiamet JJ. In vivo inhibition of cysteine proteases provides evidence for the involvement of 'senescence-associated vacuoles' in chloroplast protein degradation during dark-induced senescence of tobacco leaves. J Exp Bot. 2013;64(16):4967–80.

71. Li W-X, Oono Y, Zhu J, He X-J, Wu J-M, Iida K, et al. The Arabidopsis NFYA5 transcription factor is regulated transcriptionally and posttranscriptionally to promote drought resistance. Plant Cell. 2008;20(8):2238–51.

72. Zhao M, Ding H, Zhu JK, Zhang F, Li WX. Involvement of miR169 in the nitrogen-starvation responses in Arabidopsis. New Phytol. 2011;190(4):906–15.

73. Kamachi K, Yamaya T, Hayakawa T, Mae T, Ojima K. Changes in cytosolic glutamine synthetase polypeptide and its mRNA in a leaf blade of rice plants during natural senescence. Plant Physiol. 1992;98(4):1323–9.

74. Buchanan-Wollaston V. The molecular biology of leaf senescence. J Exp Bot. 1997;48(307):181–99.

75. Gregersen P, Holm P, Krupinska K. Leaf senescence and nutrient remobilisation in barley and wheat. Plant Biol. 2008;10(1):37–49.

76. Humbeck K, Quast S, Krupinska K. Functional and molecular changes in the photosynthetic apparatus during senescence of flag leaves from field-grown barley plants. Plant Cell Environ. 1996;19(3):337–44.

77. Wang Y, Zhang J, Yu J, Jiang X, Sun L, Wu M, et al. Photosynthetic changes of flag leaves during senescence stage in super high-yield hybrid rice LYPJ grown in field condition. Plant Physiol Biochem. 2014;82:194–201.

78. Krupinska K, Mulisch M, Hollmann J, Tokarz K, Zschiesche W, Kage H, et al. An alternative strategy of dismantling of the chloroplasts during leaf senescence observed in a high-yield variety of barley. Physiol Plant. 2012;144(2):189–200.

79. Humbeck K, Krupinska K. The abundance of minor chlorophyll a/b-binding proteins CP29 and LHCI of barley (Hordeum vulgare L.) during leaf senescence is controlled by light. J Exp Bot. 2003;54(381):375–83.

80. Varshney RK, Paulo MJ, Grando S, van Eeuwijk FA, Keizer LCP, Guo P, et al. Genome wide association analyses for drought tolerance related traits in barley (Hordeum vulgare L.). Field Crop Res. 2012;126:171–80.

81. Li W-T, Liu C, Liu Y-X, Pu Z-E, Dai S-F, Wang J-R, et al. Meta-analysis of QTL associated with tolerance to abiotic stresses in barley. Euphytica. 2013;189(1):31–49.

82. Araújo WL, Ishizaki K, Nunes-Nesi A, Larson TR, Tohge T, Krahnert I, et al. Identification of the 2-hydroxyglutarate and isovaleryl-CoA dehydrogenases as alternative electron donors linking lysine catabolism to the electron transport chain of Arabidopsis mitochondria. Plant Cell. 2010;22(5):1549–63.

83. Ishizaki K, Larson TR, Schauer N, Fernie AR, Graham IA, Leaver CJ. The critical role of Arabidopsis electron-transfer flavoprotein: ubiquinone oxidoreductase during dark-induced starvation. Plant Cell. 2005;17(9):2587–600.

84. Engqvist MK, Kuhn A, Wienstroer J, Weber K, Jansen EE, Jakobs C, et al. Plant D-2-hydroxyglutarate dehydrogenase participates in the catabolism of lysine especially during senescence. J Biol Chem. 2011;286(13):11382–90.

85. Kobayashi Y, Yamamoto S, Minami H, Kagaya Y, Hattori T. Differential activation of the rice sucrose nonfermenting1–related protein kinase2 family by hyperosmotic stress and abscisic acid. Plant Cell. 2004;16(5):1163–77.

86. Anderberg RJ, Walker-Simmons M. Isolation of a wheat cDNA clone for an abscisic acid-inducible transcript with homology to protein kinases. P Natl Acad Sci USA. 1992;89(21):10183–7.

87. Rabbani MA, Maruyama K, Abe H, Khan MA, Katsura K, Ito Y, et al. Monitoring expression profiles of rice genes under cold, drought, and high-salinity stresses and abscisic acid application using cDNA microarray and RNA gel-blot analyses. Plant Physiol. 2003;133(4):1755–67.

88. Gaxiola RA, Li J, Undurraga S, Dang LM, Allen GJ, Alper SL, et al. Drought-and salt-tolerant plants result from overexpression of the AVP1 H+-pump. P Natl Acad Sci USA. 2001;98(20):11444–9.

89. Park S, Li J, Pittman JK, Berkowitz GA, Yang H, Undurraga S, et al. Up-regulation of a H+-pyrophosphatase (H+-PPase) as a strategy to engineer drought-resistant crop plants. P Natl Acad Sci USA. 2005;102(52):18830–5.

90. Pasapula V, Shen G, Kuppu S, Paez-Valencia J, Mendoza M, Hou P, et al. Expression of an Arabidopsis vacuolar H+-pyrophosphatase gene (AVP1) in cotton improves drought-and salt tolerance and increases fibre yield in the field conditions. Plant Biotechnol J. 2011;9(1):88–99.

91. Bao A-K, Wang S-M, Wu G-Q, Xi J-J, Zhang J-L, Wang C-M. Overexpression of the Arabidopsis H+-PPase enhanced resistance to salt and drought stress in transgenic alfalfa (Medicago sativa L.). Plant Sci. 2009;176(2):232–40.

92. Kumar T, Khan MR, Abbas Z, Ali GM. Genetic improvement of sugarcane for drought and salinity stress tolerance using Arabidopsis vacuolar pyrophosphatase (AVP1) gene. Mol Biotechnol. 2014;56(3):199–209.

93. Holloway B, Li B. Expression QTLs: applications for crop improvement. Mol Breeding. 2010;26(3):381–91.

94. Holloway B, Luck S, Beatty M, Rafalski J-A, Li B. Genome-wide expression quantitative trait loci (eQTL) analysis in maize. BMC Genomics. 2011;12(1):336.

95. Rode J, Ahlemeyer J, Friedt W, Ordon F. Identification of marker-trait associations in the German winter barley breeding gene pool (Hordeum vulgare L.). Mol Breeding. 2012;30(2):831–43.

96. Igartua E, Gracia MP, Lasa JM, Medina B, Molina-Cano JL, Montoya JL, et al. The Spanish barley core collection. Genet Resour Crop Ev. 1998;45:475–81.

97. Livak KJ, Schmittgen TD. Analysis of relative gene expression data using real-time quantitative PCR and the 2− ΔΔCT method. Methods. 2001;25(4):402–8.

98. RCore. R: A language and environment for statistical computing. R Foundation for Statistical Computing, Vienna, Austria.http://www.r-project.org; 2012.

99. Fischer RA, Maurer R. Drought Resistance in Spring Wheat Cultivars. I Grain Yield Responses. Austral J Agr Res. 1978;29:897–912.

100. Bradbury PJ, Zhang Z, Kroon DE, Casstevens TM, Ramdoss Y, Buckler ES. TASSEL: software for association mapping of complex traits in diverse samples. Bioinformatics. 2007;23(19):2633–5.

101. Comadran J, Kilian B, Russell J, Ramsay L, Stein N, Ganal M, et al. Natural variation in a homolog of Antirrhinum CENTRORADIALIS contributed to spring growth habit and environmental adaptation in cultivated barley. Nature Genet. 2012;44(12):1388–92.

102. Pritchard JK, Stephens M, Donnelly P. Inference of population structure using multilocus genotype data. Genetics. 2000;155(2):945–59.

103. Hardy OJ, Vekemans X. SPAGeDi: a versatile computer program to analyse spatial genetic structure at the individual or population levels. Mol Ecol Notes. 2002;2(4):618–20.

CHAPTER 9

GLOBAL IDENTIFICATION OF MICRORNAS AND THEIR TARGETS IN BARLEY UNDER SALINITY STRESS

Pingchuan Deng , Le Wang , Licao Cui, Kewei Feng, Fuyan Liu, Xianghong Du , Wei Tong, Xiaojun Nie, Wanquan Ji, Song Weining

INTRODUCTION

Salinity is one of the most severe environment factors limiting crop yield, which affects about 954 million hectares of land worldwide at present [1]. It's reported that more than 50% of all arable lands may be salinized because of unreasonable irrigation and climate change by 2050 [2]. Extensive studies have demonstrated that cultivation of salt-tolerant species or varieties is the most economic, efficient and practical approach to alleviate salinity stress in agricultural production [3]. Thus, uncovering the molecular mechanism of salinity response in crops holds the promise for meeting the challenges of food demand increase and global climate change.

MicroRNA (miRNA) is a class of small, non-coding RNAs derived from the processing of longer primary miRNA transcripts, which served as the key players in the gene regulation networks by transcript degradation or translational repression [4]. Since the discovery of miR398 playing an essential role in salt tolerance in *Arabidopsis* [5], a large number of salt-induced miRNAs have been reported in plants up to now, such as miR159, miR160 and so on [5–10]. While some salt related miRNA were identified to show a different expression profile under salinity stress among different plants, such as miR156, miR169 and miR396 [8, 11, 12], which indicated that there might be some species-specific response or tolerance mechanism in miRNA-meditated gene regulation for various plants under salinity stress [13].

As one of the most salt-tolerant cereal crops, barley (*Hordeum vulgare* L.) is the fourth most cultivated crop worldwide and widely grown in arid and semiarid regions [11]. A large number of studies have been conducted to investigate the molecular mechanism and candidate genes responding to salinity stress in barley [12–14]. Some efforts also have been made in recent years to identify miRNA in barley [15–20]. However, the salinity-responsive miRNA in barley have not been well understood up to now. In this study, a combination of small RNA sequencing and mRNA degradome sequencing were used to identify salinity-responsive miRNAs and their corresponding target genes, which will not only provide vital information for the role of miRNAs in salinity stress response, but also shed light on the molecular mechanism of miRNA-mediated genes regulation network under salinity stress in barley and other cereals.

MATERIALS AND METHODS

Plant materials

The barley cultivar Morex, a relative salt tolerant genotype [21] was used in this study. Plant cultures and salinity stress treatment were carried out as previously described [22]. Seeds of barley cv. Morex were sterilized and germinated in vermiculite at 28°C. Then, germinated seeds were transferred to a foam grid with uniform plant density for hydroponic culture. The foam grids were placed on a plastic tanks (length 40 cm,

Citation: Deng P, Wang L, Cui L, Feng K, Liu F, Du X, et al. (2015) Global Identification of MicroRNAs and Their Targets in Barley under Salinity Stress. PLoS ONE 10(9): e0137990. doi:10.1371/journal.pone.0137990.

width 30 cm and depth 20 cm, volume 24 L) to suspend the seedling over one-half- strength Hoagland solution (20 L), with the roots immersed in a continuously aerated nutrient solution. The deionized water was added daily to replace the water lost because of transpiration. The experiment was carried out in a growth chamber (ZPG-400B, Dong Tuo, Heilongjiang, China) with 30/20°C(day/night), a relative humidity of 55–65% and the photoperiod of 14h light (6 model)/ 10h dark (0 model). At the three-leaf stage, plants were exposed to salinity by adding NaCl to the growth medium with 25 mmol/L increments every day, until reaching the final concentration of 100 mmol/L NaCl (EC in the range of 14.5–15.5 dSm^{-1}). Other plants were grown in culture without NaCl as control. Whole treated plants and counterpart controls were harvested at the 3h, 8h and 27h after reaching the final concentration of 100 mM/L NaCl for downstream experiments. A pool of two replicates was used for small RNA sequencing, whereas three individual plants were used for qRT-PCR validation.

Small RNA Library Construction and Sequencing

Total RNA was isolated from the above harvested plants using the TRIzol reagent (Invitrogen, USA) according to the manufacturer's instructions. The RNA quality was examined using gel electrophoresis (28S:18S>1.5) and Bioanalyzer (Agilent2100, RIN≥8.0). The sRNA libraries were constructed following the standard protocol (Illumina, USA). In brief, small RNAs with 16–30 nt were separated from the total RNA by size fractionation. After size selection and purification through agarose gel electrophoresis, the selected fragments were ligated to Solexa adapotors (Illumina, http://www.illumina.com). Then, the ligated RNAs were reverse transcribed and amplified for Illumina sequencing.

The degradome library was constructed using the method as previously described [23]. After salinity stress for 3h, 8h and 27h as described earlier, total RNA from these plants were isolated and pooled with equal amount at each time point. Then, a 5' RNA adaptor consisting of a MmeI recognition site in 3' end was ligated into these fragments using RNA ligase. The ligated fragments were reverse transcribed, followed second strand synthesis, MmeI digestion, ligation of a 3' dsDNA adaptor. Purification of the adaptor-ligated cDNA by agarose gel and then PCR was done to amplify the recovered cDNA for constructing the sequencing library. The high-throughput sequencing was performed on Illumina HiSeq2000 platform according to manufacturer's instructions at BIOMARKEER Biotech Comp (Beijing, China). Further image analysis and base calling were performed using the Illumina standard pipeline.

Identification of Conserved and Novel miRNAs in Barley

After poor-quality and adaptor reads filtered, the sequences with length of 16-30nt were retained for further analysis. The small RNAs that were not miRNAs were removed before miRNA analysis by matching to a number of other sources such as chloroplast genome, plant tRNAs, plant repeats and Rfam database. In brief, all trimmed sequences were mapped to the complete barley chloroplast genome without mismatches and then separated into chloroplast reads and non-chloroplast reads (nuclear-derived reads). Then, other classes of nuclear-derived sRNAs were detected by mapping all non-chloroplast reads to Rfam [24] and repeat sequences from TIGR Hordeum Repeats [25], TIGR Oryza Repeats [25], Triticeae repeat sequence database (TREP) [25] and repbase17.11 [26]. Other classes of chloroplast sRNA were analysed by mapping all the chloroplast-derived reads to Hordeum chloroplast gene, Hordeum chloroplast tRNA and Hordeum chloroplast rRNA [27]. All the reads mapping to a given library would be removed and the remaining reads were used to map another library. Known miRNAs were detected by mapping all unannotated reads to barley miRNAs stored in miRBase (v20.0, June 2013) [28] without mismatch. Then, all the reads mapping to barley miRNAs were removed. For putative homologous miRNAs detection, the remaining reads were further mapped (less than 2 mismatches) to other plants miRNA stored at miRBase (v20.0, June 2013). Furthermore, all the unannotated sequences were used to predict novel miRNAs by searching against barley genome sequence [29]. The 100 nt of the genomic sequences flanking each side of the genome-matched sequences were extracted and used to identify novel miRNAs by the Mireap software program [30]. The secondary structure of precursors was further checked using Mfold program [31]. Then, the criteria described by Meyers *et al.* [32] were used to identify the candidate novel miRNAs.

Furthermore, the differentially expressed miRNA were also identified. Firstly, sequences were computed dividing the number of reads of each sample by the total number of reads of each library, normalized per million. Then, the differentially expressed sequence counts were analyzed by the online service IDEG6 (http://telethon.bio.unipd.it/bioinfo/IDEG6_form/). Because of the lack of biological replicates and technical duplicates in sequencing, several relatively stringent criterial were applied to identify differentially expressed miRNA under salinity stress. We considered a fold change of at least 2 ($|Log_2$ (treatment/control)$| >= 1$) as an indication of significant change and P value cut-off ($P < 0.00001$ simultaneously in Audic Claverie test, Fisher exact test, and Chi-squared 2×2 test with the Bonferroni correction to adjust for pairwise comparison) considered as salinity responsive miRNAs.

Target Prediction for Barley miRNAs and Degradome Sequencing Analysis

To identify the potential targets, all the miRNAs identified in this study were used as a query to search against barley DFCI Gene index (HvGI) release 12 using the psRNATarget program with default parameters excepting that the value of Length for complementarity scoring (hspsize) was set to 18 bp [33]. Sequences with total score less than 3.0 points were considered to be miRNA targets.

The raw data of the degradome sequencing were preprocessed to remove low quality reads and clip adapter sequences. Subsequently, only 20–21 nt sequences with high quality scores were collected for further analysis. The barley miRNAs stored at miRBase (Release 20.0) together with the newly identified in this study were used to create a barley miRNA databases. The barley_HighConf_genes_MIPS_23Mar12_CDSSeq from MIPS barley genome database and barley unigene library from DFCI Gene Index (HVGI, version 12, released on 2011) were used to create a barley gene model (including 109,140 barley genes). Bowtie aligner was used to map barley degradome reads to the barley gene model and only the reads perfectly matched to barley transcript sequences were retained. Then, the degradome analysis and the classification of target categories were performed using CleaveLand pipeline (Version 4.3: November 7, 2013) with the barley gene models and all barley miRNA databases as input sets [34].

Furthermore, GO enrichment analysis was performed to understand the function of the potential miRNA target genes [35]. And then, the GO functional analysis of the putative targets was performed by WEGO (http://wego.genomics.org.cn/cgi-bin/wego/index.pl). GO terms with P-value < 0.05 were considered to be significantly enriched.

Validate the miRNAs and their Targets by qRT-PCR

In this study, two methods were used to cDNA synthesis for real-time quantification of miRNAs. One method was based on adding a poly-A tail and a universal adaptor at the 3' position of mature miRNA. In this method, the one step primeScript miRNA cDNA Synthesis kit (Takara Bio Inc., Otsu', Japan) was used to produce the cDNAs, and then amplified with mature miRNA sequence in combination with the universal adaptor as the primer for miRNA expression analysis. The other method was based on a stem-loop RT primer to detect the expression level of miRNAs [36]. The stem-loop reverse transcription reaction was performed in a volume of 20 μl containing 1 uL DNase I-treated RNA,0.5μl dNTP mix (10mM), 0.25μl SuperScript III Reverse Transcriptase (Invitrogen, 200 units/μl), 4 μl 5× First-Strand buffer, 2 μl DTT (0.1 M), 0.1 μl RNase inhibitor-HPRI (Takara, Dalian, China, 40 units/μl), 11.15 μl nuclease-free water and 1μl stem-loop RT primer for each miRNAs (1μM). The cDNA synthesis was conducted in a Bio-Rad C1000 Touch™ Thermal Cycler using a pulse reverse transcription program [36]. Real-time PCR analysis was carried out using SYBR Premix Ex Taq (Takara) on an ABI PRISM 7300 Real-time PCR System using the following condition: 3min at 95°C followed by 40 cycles of 5 s at 95°C and 31 s at 60°C. The amplification specificity was monitored by melting curve after PCR. The qRT-PCR was performed in three replications. The $2^{-\Delta\Delta Ct}$ method was used to analyze the relative changes in miRNA expression where $\Delta C_T = (C_{TmiRNA} - C_{T18S\ rRNA})$ and $\Delta\Delta C_T = (\Delta C_{T\ treatment} - \Delta C_{T\ control})$ [37].

RESULTS AND DISCUSSION

Deep sequencing of Small RNAs in Barley under Salinity Stress

As shown by previous studies [38], there are three key phases of plant response to salinity stress, including reducing growth rate phase (3h), growth rate recovery phase (8h) and ion-specific responses (27h). To elucidate the miRNA-mediate mechanisms of salinity stress tolerance in barley, six RNA libraries (salinity treated samples and their corresponding control at 3h, 8h and 27h) were generated and sequenced using Illumina technology. An overview of our analysis was presented in Fig 1. An average of 22.34 million sequence tags were obtained, with the largest number of 27.83 million for 27h treatment and the lowest number of 18.75 million for 27h control (high-throughput sequencing data were deposited into the NCBI database with accession no. SRX1176317). After quality filtering, a range from 8.91 million (27h, control) to 14.33 million (27h, treatment) clean reads were remained, of which approximately 65.2% of these clean reads could be mapped to the genome sequence of barley cv. Morex and 5.8% was Exon/Intron tags [29].

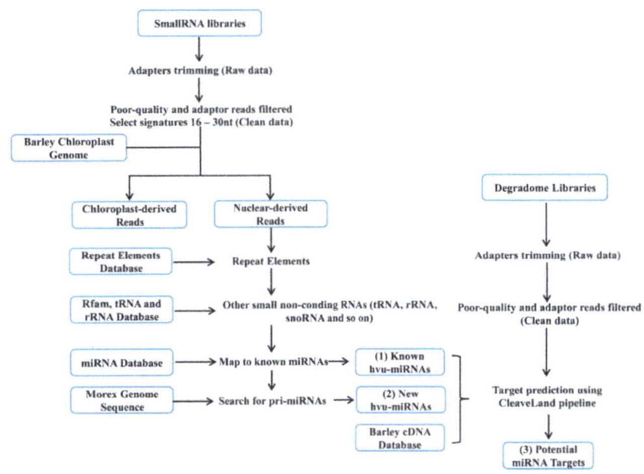

Figure 1: Flow diagram of smallRNA and degradome library analysis.

Note: Six smRNA libraries and one mRNA degradome library were made from whole barley seedling under salinity stress and counterpart control (3h, 8h and 27h). Repeat elements database included repeat sequences from TIGR Hordeum Repeats, TIGR Oryza Repeats, Triticeae repeat sequence database (TREP) and repbase17.11. The miRNAs identified from the analysis of the small libraries (including (1) known miRNA and (2) new miRNAs) and barley miRNAs stored at miRBase (v20.0, June 2013) were used in combination with the degradome libraries to find potential miRNA regulated genes.

Firstly, the chloroplast-derived sRNAs (csRNAs) were identified by mapping all trimmed sequences to the complete barley chloroplast genome (GI: 118430366) without mismatches. Averages of 2,335,565 csRNAs were obtained, of which 82, 322 were unique accounting for 19.94% of the total sRNAs. For total reads, the csRNAs derived from rRNA were the most abundant and these from mRNA were the least prevalent, while for unique reads, csRNAs derived from tRNA were the least prevalent. These results were consistent with that of Chinese cabbage under heat stress [39]. At the same time, 21.49% of the unique csRNA reads and 39.08% of total csRNA reads could be mapped to repeat elements. Among them, 38.99% of the total csRNA were mapped to TIGR Oryza Repeats, 0.01% and 0.08% mapping to TREP and Repbase17.11, respectively.

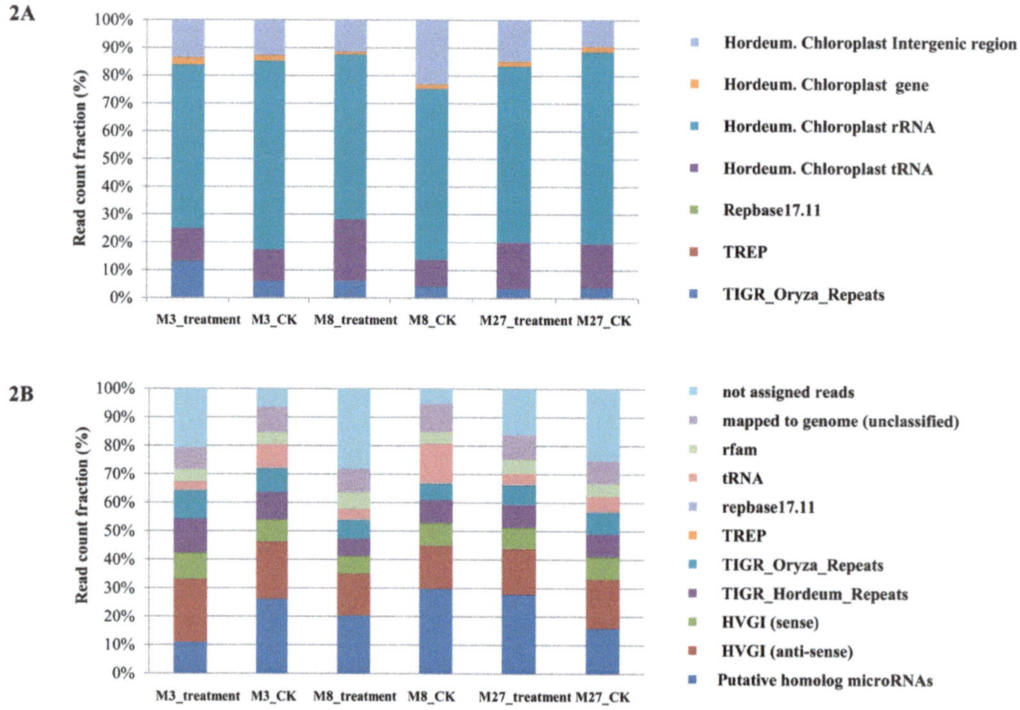

Figure 2. Distribution of different classes of sRNAs derived from barley chloroplast genome (A) and nuclear genome (B).

In contrast to sRNAs derived from the chloroplast genome, nuclear-derived sRNAs were highly redundant under both salinity treatment and control,with an average of 11,715,135 (80.06%) total reads and 2,232,374 (96.31%) unique reads. To classify the function of these small RNA, all the nuclear-derived sRNAs were further grouped by mapped to miRbase (Version 20), TIGR repeat database, TREP repeat database, RepBase, all plants tRNA databases and Rfam. We observed that an average of 82.94% reads could be assigned to the above databases. Among them, 22.09% of them were mapped to plant miRNA, 16.60% to repetitive elements (8.71% for TIGR Hordeum Repeats, 7.53% for TIGR Oryza Repeats, 0.07% for TREP and 0.29% for repbase17.11, respectively), 6.03% to tRNA, 4.70% to ncRNAs from the Rfam database, and 8.49% to un-annotated genomic regions. The antisense strands of barley genes (from the HVGI database), accounting for 17.58%, occupy the highest proportion among all the mapped sRNAs excepting miRNA.

Identification of Conserved and novel miRNAs in Barley

Using the methods as described in the Materials and Methods, a total of 28 barley miRNAs from 20 miRNA families were detected. All the barley miRNAs except Hvu-miR5050 were concurrently detected under salinity stress and normal condition. Hvu-miR5050 was just detected at the 8h and 27h under salinity stress, although with very low reads. In addition, we have identified both mature and star sequences for two barley miRNAs (hvu-miR168 and hvu-miR171). Furthermore, hvu-miR171* showed a higher expression than hvu-miR171 under both salinity stress and normal condition.

After excluding the barley miRNA-mapping reads, the remaining ones were further mapped (less than 2 mismatches) to other plant miRNA in miRBase. A total of 114 putative homologous miRNAs from 99 miRNA families were identified. Among them, 61.97% of miRNAs showed a bias toward 5 uridine or 3 guanosine terminal residues, which was similar to the characteristic of wheat miRNAs [40]. The length of mature miRNAs ranged from 18nt to 24nt. The majority of the small RNA were 21nt in size, occupying for 59.86% of the total conserved miRNAs, followed by 22nt (11.97%), 24nt (10.56%), 20nt (9.86%) and 23nt (3.52%), while the proportion of 19nt and 18nt- sRNA were quite small, only 2.82% and 1.41%, respectively. Furthermore, we found a high abundance for miRNA*sequences of 14 miRNAs [28]. Additionally, the star strand of three miRNAs (bdi-miR159b*, bdi-miR166b*, and ata-miR5168*) showed a higher abundance than their corresponding mature sequence, while other four (ata-miR156d*, osa-miR160f*, ata-miR167b* and ata-miR396e*) showed a less abundance. Two miRNAs showed a similar expression of the mature and star strands (ata-miR172b* and ata-miR396e). The remaining five miRNAs (tae-miR9662a, zma-miR395b, tae-miR1130b, bdi-miR827 and bdi-miR408) were just detected with the star strands.

Furthermore, some novel miRNAs were identified. A total of 10 small RNAs can meet the criteria as describing in Materials and Methods, which were considered as putative novel miRNAs. The length of these novel miRNAs ranged from 18 nt to 24 nt, with 21 nt representing the most abundance class (8/10, 80%). Precursors were also identified among the barley genome sequences. The size of these miRNA precursors ranged from 70 nt to 85 nt with the average of 77 nt. Homologous sequences of miR-n02, miR-n04, miR-n06, miR-n08 and miR-n09 were also found in wheat by aligning the candidate precursor against the survey sequence of *T. aestivum* (https://urgi.versailles.inra.fr/blast/blast.php), while the other five miRNAs were only found in barley. Significantly, the majority of the novel miRNA was detected in all six samples with various expression levels. The obtained sequence count data for these miRNAs in each of the libraries suggested that the depth of sequencing could be sufficient to support estimation of their expression levels. However, some specific-expression of the novel miRNAs was also found, with miR-n07 only detected at 3h salinity treatment, while miR-n09 and miR-n10 in all six libraries except 3h and 8h salinity treatment. These specific-expression miRNA may involve n the regulation network of barley response to salinity stress and play a vital role in regulating gene expression and metabolism process during salinity treatment.

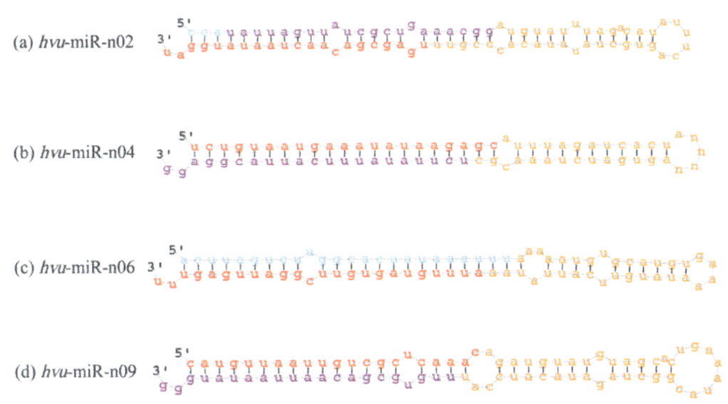

Figure 3. Pre-miRNA stem-loop structures for partial novel miRNAs identified in this study.

Note: Predicted mature miRNA sequences are shown in red.

To validate these newly identified miRNAs, 5 miRNAs were selected randomly to detect expression level in seedling with 3h salinity treatment where the expected product was recovered for all the selected miRNAs by qRT-PCR analysis. Results showed that all the five miRNAs were detected and miR-n02 and miR5179 showed a higher expression, while the miR1862, miR-n01 and miR-n05 showed a lower expression (Fig 4). The results of qRT-PCR analysis also provided the evidence for the actual existence of these novel miRNAs.

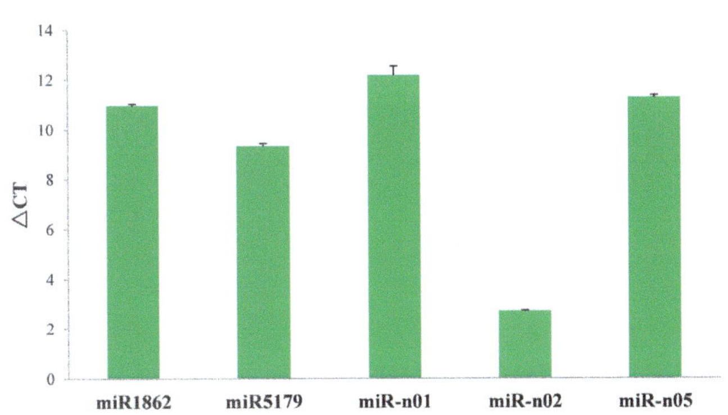

Figure 4: Quantitative real-time PCR analysis of parts miRNA identified in barley.

Note: The amount of expression was normalized to the level of 16s RNA.

Target Genes Characterization for the IDENTIFIED miRNAs by bioinformatics Prediction and Degradome sequencing

To understand the biological function of these identified miRNAs, their targets were first predicted through bioinformatics analysis. Totally, 1589 candidate target genes were identified for 152 miRNA using psRNATarget program (2020 target genes for 142 conserved miRNA and 456 target genes for 10 novel miRNAs, respectively), with the average of 13 target genes per miRNA. In addition, the majority of the target genes (67.92%) were regulated by transcript degradation, while the remaining was regulated by translational repression. A total of 5 miRNAs were found to regulate target genes at two positions. Furthermore, 132 genes targeted by more than one miRNA were also found . These phenomena were also observed in wheat, suggesting that similar miRNA regulation systems and evolutionary events may exist in both wheat and barley [40, 41]. Furthermore, degardome sequencing was used to identify the target genes of these barley miRNAs. It's well known that the majority of the plant miRNAs characterized to date interacts with their target genes by miRNA directed cleavage at the tenth nucleotide of complementarity, resulting in 3' target fragments had a 5' monophosphate and a 3' poly-A tail, which makes it possible to identify target genes by degradome sequencing. We pooled the RNA samples from 3h, 8h and 27h to construct the library for degradome sequencing. Therefore, a total of 5,438,888 raw reads were produced . After quality filtering, 4,928,699 unique reads were obtained and 93.66% of these reads were 20 or 21 nt in size. Among them, 86.68% could be mapped to the barley reference sequence. The computationally predicted gene and cDNA transcripts of barley were also used as the mapping template sequences of the degradome sequencing reads and found that 83.09% of them could be perfectly mapped to the barley gene models, representing 22,972 annotated barley genes. When a mismatches threshold <= 2 was applied, nearly 90% (4,389,591/4,928,699) of unique reads could match to the barley genes and represent 85.24% (93,029/109,140) of the barley gene. Furthermore, using the method of CleaveLand pipeline as described in Material and methods [34], a total of 65 targets were identified for 39 different miRNAs, including 37 conserved miRNAs and 2 novel miRNAs. The cleaved target transcripts were categorized into five classes (categories 0, 1, 2, 3 and 4) based on the abundance of degradome tags indicative of miRNA-mediated cleavage [42]. Among the identified targets, category 2 was the most abundant category occupying for 29.23%, followed by category 3 (27.69%), category 1 (16.92%), category 0 (15.38%) and 4 (10.77%), which was consistent with the results of previous study [42]. The miRNAs were found to be able to target various numbers of genes, ranging from 1 to 6, of which miR951 had the highest number of targets, targeting 6 unique genes. Examples of the 'target plots' (T-plots) of targets for identified barley miRNAs were shown in Fig 5.

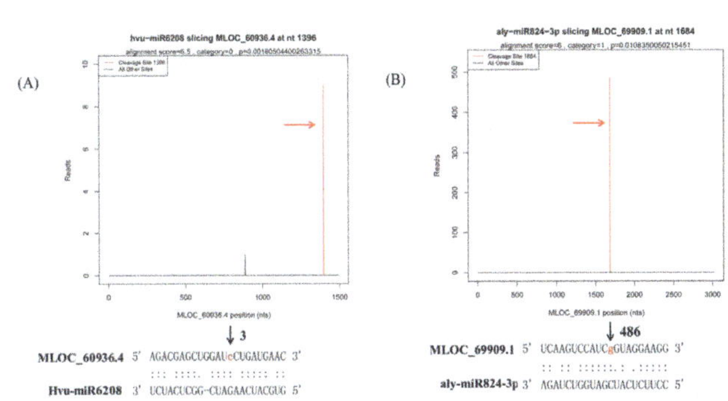

Figure 5. Examples of Target plots (T-plots) of identified miRNA targets using degradome sequencing.

Note: The T-plots show the distribution of the degradome tags along the full length of the target mRNA sequence. Red arrows indicate signatures consistent with miRNA-directed cleavage. mRNA:miRNA alignments along with the detected cleavage frequencies (absolute numbers) are shown above the black arrow. The red colored lowercased nucleotide on the target transcript from 3' end indicates the cleavage site detected in the degradome library.

Expression Analysis of miRNA and Their Targets by qRT-PCR

To validate the targets of the miRNAs, 5 miRNAs and their target genes which were verified by the degradome sequencing, were selected to further confirm by qRT-PCR (Fig 6). The qRT-PCR results showed that four out of these targets have a reverse expression level compared to their counterpart miRNAs. miR164b was down-regulated at 3h and 8h, but up-regulated at 27h, while its expression level showed a gradual increase following the time of salinity stress (Fig 6A). The target gene of miR164b (Oxygen-evolving enhancer protein, MLOC_78630.1) showed a gradual decrease following the time of salinity stress in addition to reverse down or up- regulation patterns of miR164b at the same time points of salinity stress. In contrast to miR164, miR419 showed a gradual decrease whereas its target gene probable cytokinin riboside 5 (MLOC_53792.1) displayed a gradual increase, following the time of salinity stress (Fig 6B). miR169 was up-regulated at the 3h time point, while down-regulated at the 8h and 27 time points. And its target gene polyubiquitin 10 (AK372493) displayed a contrast expression patterns (Fig 6D). miR1507 appeared down-regulated at all the three time points, while its target gene translation initiation factor eIF-2B subunit alpha did not show a strict reverse expression pattern compared with that of miRNA (Fig 6C). This target gene was down-regulated at 3h and 8h, but up-regulated at 27h. It might indicate that there could be some time lag for certain miRNAs to initiate regulation of their target genes. Similar results were also found with miR-n05 and its target gene enolase (GH219301). However, another target gene (B-cell receptor-associated protein, AK355784) of miR-n05 displayed a contrast expression pattern with their miRNAs (Fig 6E). It might indicate that a delayed regulation pattern to initiate regulation of their target genes may be a method for miRNA mediated multiple target genes regulation under salinity stress.

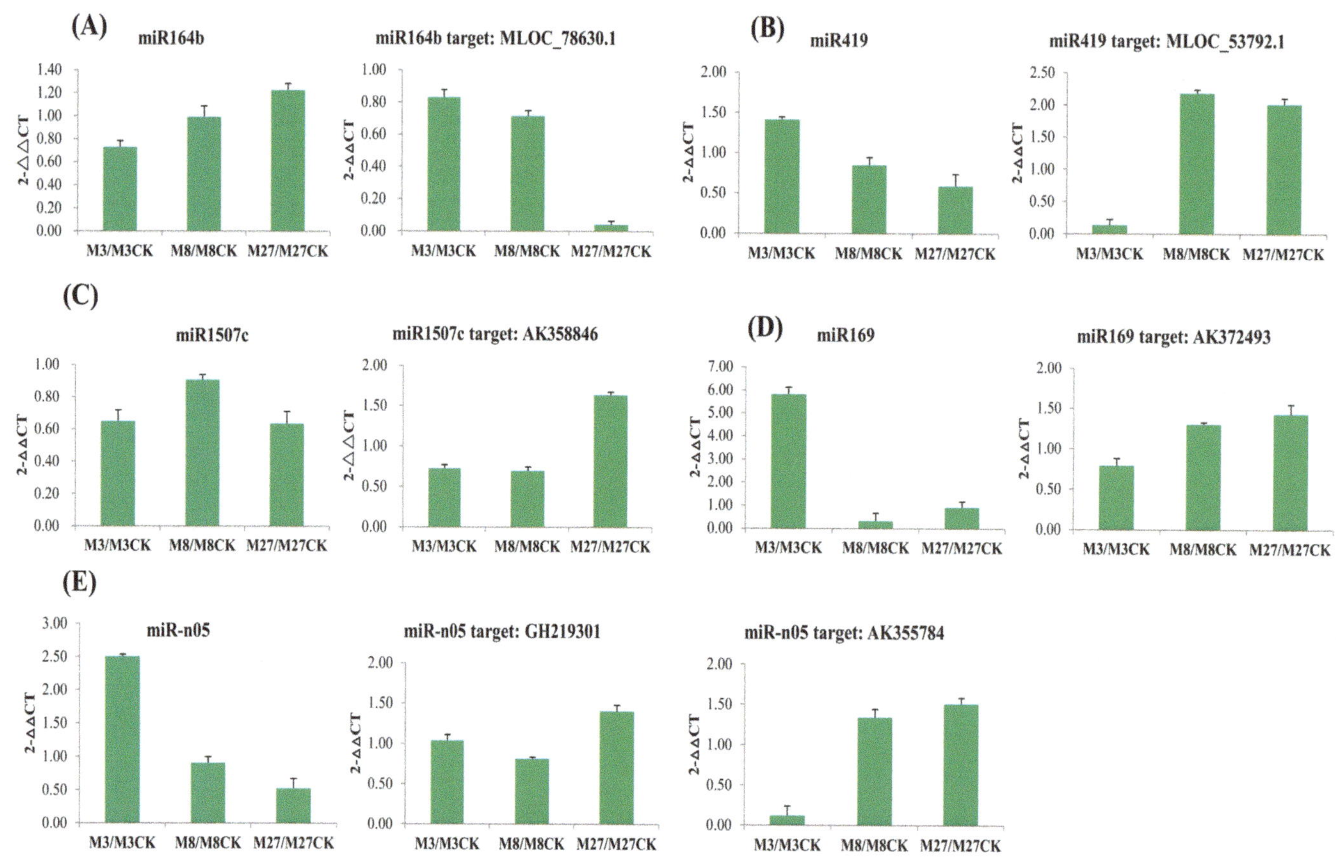

Figure 6: uantitative real-time PCR analysis of miRNAs and their targets under salinity stress and normal condition.

Note: The amount of expression was normalized to the level of 16s RNA. The normalized miRNA levels in control were arbitrarily set to 1.

Analysis of salinity responsive miRNAs in barley

Through comparing the miRNA profiles between salinity treated samples and their control, the salinity responsive miRNAs were further identified as described in Materials and Methods. Among the 28 barley miRNAs, 6 were salinity responsive miRNAs (Table 1). Three miRNAs (hvu-miR168-5p, hvu-miR5048a and hvu-miR444b) were down-regulated while two other miRNAs (hvu-miR171-5p and hvu-miR6213) were up-regulated under salinity stress. The left miRNA (Hvu-miR444a) significantly expressed at all the three time points, which was down-regulated at 8h and 27h but up-regulated at 3h. All of the salinity responsive novel miRNAs except hvu-miR-n01 were up-regulated under salinity stress at the 3h and 8h time points, while all of them except hvu-miR-n05 were down-regulated at the 27h time point. For putative homologous miRNAs, a total of 26 miRNAs and six miRNA* were significant differential regulation at one or more of the three time points.

Table 1: List of differentially expressed miRNAs identified in barley under salinity stress

Proposed miRBase name	miRNA sequence	log2(M3h_treatment/M3h_control)	log2(M8h_treatment/M8h_control)	log2(M27h_treatment/M27h_control)
Barley miRNA				
hvu-miR168-5p	aagccaacugauacuaauugg	-2.27	-1.1	0.47
hvu-miR171-5p	uauuaguuaucgcugaaacgg	-0.63	-0.01	2.79
hvu-miR444a	ucuuauauuucggaauggagg	1.64	-3.32	-1.32
hvu-miR444b	gcuuuguuccuuggauuucca	-1.49	-0.2	-0.6
hvu-miR5048a	auuuguaguguucggauugaguuu	-2.7	-2	0.82
hvu-miR6213	acagauugcuacagacugguc	2.83	0.04	-0.24
Barley novel miRNA				
hvu-miR-n0	aagccaacugauacuaauugg	-1.22	-2.6	-3.25
hvu-miR-n02	gauggcaggugacgguguugacgc	1.89	2.16	-0.45
hvu-miR-n03	agcacuggagguccgaaacca	0.63	2.06	0.35
hvu-miR-n05	gcuuuguuccuuggauuucca	0.98	-1.25	-3.56
hvu-miR-n06	cggccaagagaggaguccucccu	-0.03	1.44	-2.06
Putative homologous miRNAs				
ata-miR167b-3p	aggucaugcuggaguuucauc	-0.99	1.5	0.35
ata-miR167b-5p	ugaagcugccagcaugaucuga	-1.88	3.13	0.76
ata-miR172b-3p	agaaucuugaugaugcugcau	-1.93	-1.93	1.22
ata-miR396e-5p	uuccacagcuuucuugaacug	-1.96	-1.12	0.23
bdi-miR1135	uuucgacaaguaauuccgaccgga	-1.11	-1.81	2.22
bdi-miR159b-3p.1	uuuggauugaagggagcucug	-1.48	-0.42	-2.97
bdi-miR159b-5p.1	gagcuccuaucauuccaauga	1.27	0.55	-1.77
bdi-miR164b	uggagaagcagggcacgugca	-2.11	-0.13	3.16
bdi-miR1878	acuuagucugaacacuauaaaaaa	-3.06	0.1	2.99
bdi-miR319a	ugagggagcuuucuucugucc	-5.51	1.26	0.54
bdi-miR393b	uccaaagggaucgcauugauc	-2.27	5.03	-6.63
bdi-miR408-3p	cugcacugccucuucccuggc	-1.63	0.29	-2.67
bdi-miR5176	uaugccaugucgucacauauc	-6.42	1.64	-0.52
bdi-miR5179	uuuugcucaagaccgcgcaac	-0.97	1.48	0.48
bdi-miR5200	uguagauacucucuaaggcuu	-1.72	-0.02	-0.41
bdi-miR827-3p	uuagaugaccaucagcaaaca	-0.36	-1.11	0.54
gma-miR1507c	gagguguuugggaugagagaa	-1.63	-0.83	-1.34
gma-miR5368	ggacagucucagguagaca	1.16	0.9	-1.49
osa-miR169a	cagccaaggaugacuugccga	-4.27	-5.36	4.96
osa-miR319b	uuggacugaagggugcuccc	-0.9	6.39	4.81
osa-miR419	ugaugaaugcugacgauguug	4.85	3.1	-2.49
osa-miR5071	ucaagcaucauaucguggaca	-1.44	-0.58	-1.37
osa-miR5072	cgauuccccagcggagucgcca	1.71	0.58	-1.57
osa-miR5144	uucuugugcugcugaagagac	-1.47	4.47	-2.21
osa-miR5485	ugacaacugguagcagagcaa	3.59	1.87	-0.73
osa-miR5529	guuucauccauggacaccgca	2.57	1.61	-1.04
osa-miR5538	acugaacucaaucacuugcugc	1.77	1.2	-1.3
osa-miR818c	aaucccuuauauuaugggacgg	3.23	1.93	1.34
ppa-miR894	cguuucacgucggguucacc	1.5	0.59	-0.88
ptr-miR169s	ucagccaaggaugacuugccg	-1.26	3.42	-4.6
tae-miR1122	uagauacauccguaucuaga	0.15	5.2	-0.33
tae-miR9662a-3p	uugaacaucccagagccaccg	-1.33	-0.61	0.22
zma-miR395b-3p	gugaaguguuuggggggaacuc	-3.09	-0.17	-2

Note: The fold change at the three time points sampled is expressed as Log2 (treatment/control). Fold changes more than 1.0-fold are significant changes and marked in bold.

doi:10.1371/journal.pone.0137990.t001

Table 2: The overview of miRNAs highly responsive to salt stress at the three time points

Time after sanility stress	Total miRNAs	Up-regulated miRNAs	Down-regulated miRNAs
3h	32	12 (37.50%)	20 (62.50%)
8h	21	11 (52.38%)	10 (47.62%)
27h	20	6 (30.00%)	14 (70.00%)

doi:10.1371/journal.pone.0137990.t002

Taken together, a total of 44 miRNAs belonging to 39 miRNA families were identified as salinity responsive, of which 6 were barley conserved miRNA and 5 were newly identified in this study. Among them, 29 miRNAs from 27 miRNA families were up-regulated while 27 miRNAs from 25 miRNA families were down-regulated. Previous study has demonstrated that under salinity stress, the majority of the barley functional genes were down-regulated at 8h but up-regulated at the 3h and 27h [22]. In this study, the majority of the salinity responsive miRNAs (57.14%) were up-regulated at the 8h while down-regulated at the 3h and 27h time points (62.50% and 65.00% respectively, Table 2). The contrast expression patterns of the salinity responsive miRNAs and functional genes, suggest that the potential roles of these salt-responsive miRNA playing in regulating the gene expression under salinity stress in barley.

Functional Enrichment for the Identified Salinity Responsive miRNA Target Genes

To gain a better understanding of the functional roles of the targets of salinity responsive miRNA, we further performed the functional enrichment of these targets by Gene Ontology (GO) analysis. The predicted targets were sorted into 616 GO terms. Among them, 396 were Biological Process-related pathways and 155 were Molecular Function-related pathways (Table 3). In Biological Process, we found the frequency of biosynthetic process, metabolic process, defense response and response to stimulus-related (such as response to stimulus, response to stress and response to chemical stimulus) GO terms were increased under salinity stress, whereas the growth and organ growth-related GO terms was decreased. Interesting, nitrogen compound metabolic process GO term was also found to be increased under salinity stress, suggesting that high metabolic process for plants to cope with salinity stress would require re-adjustment of nitrogen (N) homeostasis. In Molecular Function, the frequency of binding, nucleotide binding, catalytic activity and hydrolase activity were increased under salinity stress but protein binding were decreased. The differences of these GO terms for barley under salinity stress seemed consistent with the observations of functional genes reported by Walia *et al.* (2006).

Table 3: Functional Categories of Genes Targeted by salt Responsible miRNAs in Barley

GO ID	Term	Observed	Expected	P value
Biological Process (Total: 15)				
GO:0008152	metabolic process	156 (65.5%)	6050 (32.8%)	0
GO:0006807	nitrogen compound metabolic process	81 (34.0%)	2321 (12.6%)	0
GO:0009058	biosynthetic process	59 (24.8%)	3281 (17.8%)	0.005
GO:0043170	macromolecule metabolic process	95 (39.9%)	3963 (21.5%)	0
GO:0044237	cellular metabolic process	129 (54.2%)	5309 (28.8%)	0
GO:0044238	primary metabolic process	121 (50.8%)	4907 (26.6%)	0
GO:0009987	cellular process	147 (61.8%)	6990 (37.9%)	0
GO:0044237	cellular metabolic process	129 (54.2%)	5309 (28.8%)	0
GO:0003006	reproductive developmental process	15 (6.3%)	707 (3.8%)	0.049
GO:0043473	pigmentation	50 (21.0%)	2530 (13.7%)	0.001
GO:0019222	regulation of metabolic process	36 (15.1%)	1461 (7.9%)	0
GO:0050896	response to stimulus	59 (24.8%)	3146 (17.0%)	0.002
GO:0065007	biological regulation	53 (22.3%)	2776 (15.0%)	0.002
GO:0040007	growth	7 (2.9%)	1578 (8.6%)	0.002
GO:0044085	cellular component biogenesis	8 (3.4%)	2155 (11.7%)	0
Molecular Function (Total: 9)				
GO:0003824	catalytic activity	130 (54.6%)	4672 (25.3%)	0
GO:0016491	oxidoreductase activity	24 (10.1%)	936 (5.1%)	0.001
GO:0016740	transferase activity	51 (21.4%)	1265 (6.9%)	0
GO:0016787	hydrolase activity	43 (18.1%)	1628 (8.8%)	0
GO:0005488	binding	150 (63.0%)	5600 (30.3%)	0
GO:0000166	nucleotide binding	81 (34.0%)	967 (5.2%)	0
GO:0001882	nucleoside binding	66 (27.7%)	612 (3.3%)	0
GO:0043167	ion binding	36 (15.1%)	700 (3.8%)	0
GO:0005515	protein binding	10 (4.2%)	2773 (15.0%)	0

Note: Observed, numbers of genes observed in this study; Expected, numbers of genes in this same category in the GO enrichment analysis program.

doi:10.1371/journal.pone.0137990.t003

Comparative Analysis of the Salinity Induced miRNAs at the three time Points

Comparative analysis of these salinity responsive miRNAs at the three key phases was further performed to identify the main candidate miRNAs in salinity response of barley. Most of the up-regulated or down-regulated miRNAs showed a time specific expression pattern. For up-regulated miRNAs, 10, 8 or 4 of them were found to be expressed specifically at the 3h, 8h and 27h time point, respectively. And 2 of these miRNAs were shared by two time points (1 between the 3h and 8h, 1 between the 8h and 27h), with only one miRNA (osa-miR818c) shared by all the three time points (Fig 7A). Among the down-regulated miRNAs, there were just one miRNAs (hvu-miR-n01) shared by three time points, while 12 miRNAs were shared by two time points (Fig 7B). Finally, 9, 2 and 6 down-regulated miRNAs were specifically expressed at the 3h, 8h and 27h time points, respectively.

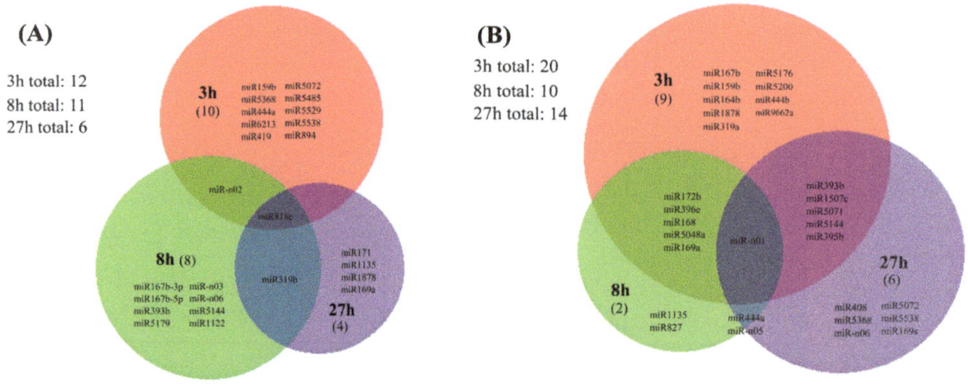

Figure 7: Venn diagram of salt responsive miRNAs at 3h, 8h and 27h.

Note: (A) Up-regulated miRNAs in response to salinity stress. (B) Overlap among the down-regulated miRNAs.

In the present study, a number of miRNAs with targets involved in modulating leaf and root development displayed significant differential expression under salinity stress. Among them, miR164b was down-regulated while miR160f was up-regulated. NAC-domain proteins, the targets of miR164b, have been reported to be salinity induced in many species [7, 43]. Previous studies showed that NAC1 might be an early auxin-responsive gene and positively regulate the lateral root development [44]. The increasing accumulation of NAC1 transcripts resulting from the down-regulated miR164b would promote lateral root development at the 3h time point. At the same time, miR172b was down-regulated while its target gene auxin response factor (ARF) was up-regulated at the 3h and 8h time points. ARF regulates various important processes in plants growth, development and responses to environmental stress [45,46]. The down-regulation of miR172b would increase the expression of ARF, thus promoting auxin response and enhance leaf development. Moreover, it has been reported that a decrease in endogenous cytokinins (CKs) levels could enhance salinity stress tolerance [47]. In this study, a cytokinin biosynthesis related gene (cytokinin riboside 5'-monophosphate phosphoribohydrolase, MLOC_53792.1) was found down-regulated at the 3h time point, while up-regulated at the 8h and 27h as a result of regulation by miR419,indicating that miR419 might play a role in regulating barley growth under salinity stress. Our results indicate that these miRNAs might play a role in modulating barley morphological development to cope with salinity stress.

Salinity damage might be the result of excess apoplastic ion concentrations or ion toxicity effects on metabolic processes in the symplast including inhibited most enzymes activity, damaged membranes and macromolecules [38]. Heat-shock proteins (Hsps), acting as molecular chaperones and maintaining the stabilization of proteins and membranes, were one of the major types of specific stress-induced proteins that accumulated upon water, salinity and extreme temperature stress [38, 48]. miR396e was predicted to target Hsp80, responsible for both heat-shock and developmental regulation [49]. In our study, miR396e was down-regulated at the 3h and 8h time points of salinity stress, likely resulting in the upturn of Hsp80 transcripts and consequently maintaining the stabilization of proteins and membranes. On the other hand, plant also degraded some mis-folded proteins and those proteins no longer needed by the cell to cope with stress. Heat shock protein HslVU was a proteasome-like degradation complex and responsible for 70 to 80% of the protein degradation *in vivo* [50]. miR5175a was down-regulated at 8h time point and predicted to target HslVU, which might assure the expression of HslVU degradation of some mis-folded proteins in response to stress.

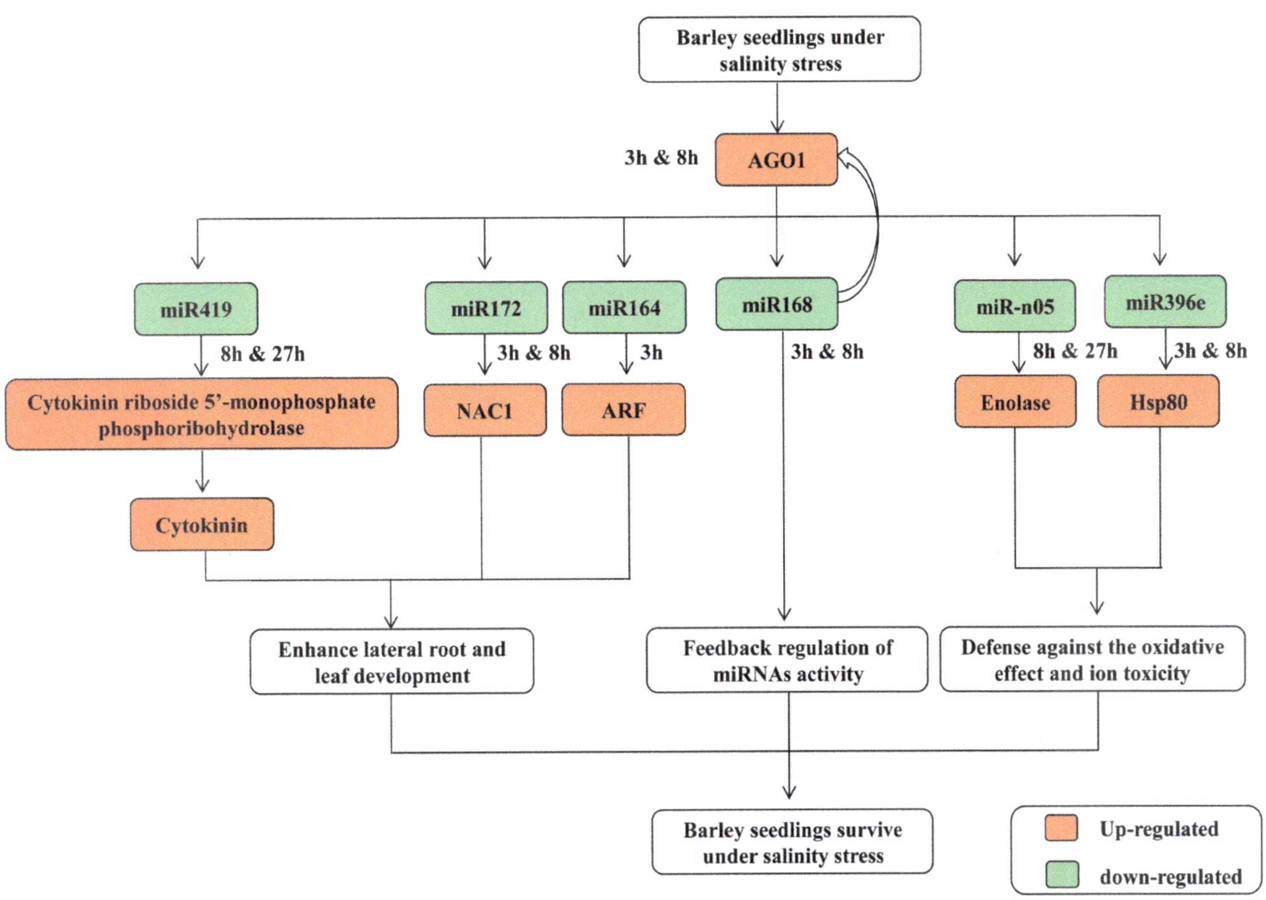

Figure 8: The potential regulatory network of salinity-responsive miRNAs for barley seedling in the early hours of salinity stress.

Note: the data contributing to the network were collected from 3h, 8h and 27h salinity stress (100 mm/L NaCl) in barley seedling, and show the potential roles of the salinity responsive miRNAs inbarley.

Recently, wheat enolase was identified as a non-toxic cryopreservation agent and it could protect cells and tissues against oxidative stress [51]. In our study, miR-n05 was detected to target enolase (GH219301) by degradome sequencing. qRT-PCR also showed that the enolase was up-regulated at the 27h time point resulting from the down-regulation of miR-n05 (Fig 6E), indicating that miR-n05 might be involved in defense against the oxidative effect generation during salinity stress.

There also may be a negative feedback regulation in the miRNA pathway for barley under salinity stress. In plants such as Arabidopsis [5], maize [52] and cotton [53], miR168 was detected to target AGO1 (ARGONAUTE 1)) gene and involved in miRNAs biogenesis and function. AGO1 genes (Hv.19452 and Hv.26206) were also found targeted by miR168 in barley [54]. In this study, miR168 was found to be a salinity responsive miRNA and down-regulated at the 3h and 8h time points. Since AGO1 was a key regulator for miRNAs biogenesis, the down-regulated of AGO1 expression under salinity stress might induce further changes of numerous miRNA activities. This might explain why there were more salinity responsive miRNAs at the 3h and 8h time points than at the 27h (Table 2).

Here a model has been proposed for the role of miRNAs in the salinity responses of barley seedling under a lower salinity stress. Fig 8 showed that the salinity responsive miRNAs were involved at post-transcriptional levels in the regulation of the metabolic and morphological processes of barley seedling under salinity stress.

Comparative Analysis of Salinity Responsive miRNAs in Plants

Currently, a large number of studies have been performed to reveal the biological mechanism and process of miRNA-mediated salinity stress response in plants, meanwhile an increasing number of salinity-responsive miRNA have been identified [5–8, 10, 53, 55]. In this study, 34 miRNA families were found to be salt-responsive in barely. Compared to other plants, 14 of these miRNA families were also found to be salinity responsive in other plant species. Among them, miR159, miR164, miR167 miR168, miR171, miR172 and miR393 have been demonstrated to be involved in signal transduction by targeting MYB transcription factor, F-box and Nodulation signaling pathway 2 proteins. In addition, three miRNAs including miR319, miR393 and miR396 were targeting genes involved in plant morphological modulation such as plants leaf and root development. Morphological modulation and signal transduction were common survival strategies for plants to adjust to stress [43, 56]. Thus, plants might evolve a conserved miRNA-mediated regulatory mechanism in morphological modulation and signal transduction to response to salinity stress.

Although the majority of the salinity responsive miRNA existed in various plants, the expression patterns of some miRNA were different among species. For example, miR396 was up-regulated expressed in *Arabidopsis* [5], but down-regulated in rice and cotton [52, 53]. Similarly, miR167 was down-regulated in maize [52], but up-regulated in *Arabidopsis* in response to salinity stress [5]. The reasons for these conflicting results might be attributed to the differential experimental conditions or the species specific regulatory networks under salinity stress. These conflicting results highlight the need for more in-depth and detailed characterizations of salinity responsive miRNAs in plants.

Furthermore, some specific salinity responsive miRNA families were also identified. Four miRNAs (namely miR818, miR5071, miR5072 and miR5200), were only found in switchgrass (*Panicum virgatum*) and barley, indicating that these miRNA families might be grass-specific salinity responsive miRNAs. Additionally, 11 miRNA families were newly identified salt responsive miRNA in barley in this study, which could be considered as the barely-specific salt-responsive miRNAs. Their corresponding target genes enriched in the functions of transport, cell wall biogenesis, signal transduction, transcription regulation and other abiotic stress. At the same time, some of them have been reported to display differential expression under other stresses. For example, miR444 has been found to be significantly down-regulated under drought stress in barley [18]. These newly identified salt-responsive miRNAs in barley shed light on the molecular mechanism of barley salt resistance, and also provide invaluable resource to uncover the elite miRNAs for salt resistance improvement in barley and other cereals.

CONCLUSIONS

Barley, renowned as the most salt-tolerant cereal, is an important crop worldwide and widely grown in arid and semiarid regions. In this study, a combination of small RNA and mRNA degradome sequencing was performed to identify miRNAs and their targets in barley under salinity stress. A total of 152 miRNAs (142 conserved and 10 novel ones) were identified, of which 44 miRNAs (39 conserved and 5 novel ones) belonging to 39 families were found as salinity responsive. 14 out of the conserved miRNA families were also found to be salinity responsive miRNAs in other plant species, while 4 and 11 salinity responsive miRNA families were only found in grass and barley, respectively. Attenuated plant growth and decreased metabolic rate are common survival strategies employed to divert energy and other resources to cope with stress condition. 86 and 37 target gene families were found to involve in metabolic process and response to stimulus respectively. Overall, there were at least 39 barley miRNA families and 123 corresponding target gene families response to salinity, which may consist to a miRNA regulatory network in barley. These results could facilitate future research to understand the molecular mechanism of salinity response in barley and related cereals to improve their salinity tolerance.

AUTHOR CONTRIBUTIONS

Conceived and designed the experiments: PCD SW XD. Performed the experiments: PCD LW. Analyzed the data: PCD LW LCC KWF. Contributed reagents/materials/analysis tools: LW FYL WT. Wrote the paper: PCD XJN WQJ SW.

REFERENCES

1. Si J, Zhou T, Bo W, Xu F, Wu R. Genome-wide analysis of salt-responsive and novel microRNAs in Populus euphratica by deep sequencing. BMC genetics. 2014;15(Suppl 1):S6. doi: 10. 1186/1471-2156-15-S1-S6 pmid:25079824.

2. Wang W, Vinocur B, Altman A. Plant responses to drought, salinity and extreme temperatures: towards genetic engineering for stress tolerance. Planta. 2003;218(1):1–14. doi: 10. 1007/s00425-003-1105-5 pmid:14513379.
3. Shannon MC. Adaptation of plants to salinity. Advances in agronomy. 1997;60:75–120. doi: 10. 1016/s0065-2113(08)60601-x
4. Bartel DP. MicroRNAs: genomics, biogenesis, mechanism, and function. Cell. 2004;116(2):281–97. pmid:14744438 doi: 10. 1016/s0092-8674(04)00045-5
5. Liu H-H, Tian X, Li Y-J, Wu C-A, Zheng C-C. Microarray-based analysis of stress-regulated microRNAs in Arabidopsis thaliana. Rna. 2008;14(5):836–43. doi: 10. 1261/rna. 895308. pmid:18356539
6. Yin Z, Li Y, Yu J, Liu Y, Li C, Han X, et al. Difference in miRNA expression profiles between two cotton cultivars with distinct salt sensitivity. Molecular biology reports. 2012;39(4):4961–70. doi: 10. 1007/s11033-011-1292-2. pmid:22160515
7. Frazier TP, Sun G, Burklew CE, Zhang B. Salt and drought stresses induce the aberrant expression of microRNA genes in tobacco. Molecular biotechnology. 2011;49(2):159–65. doi: 10. 1007/s12033-011-9387-5. pmid:21359858
8. Lu S, Sun YH, Chiang VL. Stress-responsive microRNAs in Populus. The Plant journal: for cell and molecular biology. 2008;55(1):131–51. doi: 10. 1111/j. 1365-313X. 2008. 03497. x pmid:18363789.
9. Zhang B. MicroRNA: a new target for improving plant tolerance to abiotic stress. J Exp Bot. 2015;66(7):1749–61. doi: 10. 1093/jxb/erv013 pmid:25697792.
10. Eren H, Pekmezci M, Okay S, Turktas M, Inal B, Ilhan E, et al. Hexaploid wheat (Triticum aestivum) root miRNome analysis in response to salt stress. Annals of Applied Biology. 2015. doi: 10. 1111/aab. 12219
11. Greenway H, Munns R. Mechanisms of salt tolerance in nonhalophytes. Annual review of plant physiology. 1980;31(1):149–90. doi: 10. 1146/annurev. pp. 31. 060180. 001053
12. Munns R, Rawson H. Effect of salinity on salt accumulation and reproductive development in the apical meristem of wheat and barley. Functional Plant Biology. 1999;26(5):459–64. doi: 10. 1071/pp99049
13. Ozturk ZN, Talamé V, Deyholos M, Michalowski CB, Galbraith DW, Gozukirmizi N, et al. Monitoring large-scale changes in transcript abundance in drought- and salt-stressed barley. Plant molecular biology. 2002;48(5–6):551–73. pmid:11999834 doi: 10. 1023/a:1014875215580
14. Ueda A, Kathiresan A, Inada M, Narita Y, Nakamura T, Shi W, et al. Osmotic stress in barley regulates expression of a different set of genes than salt stress does. Journal of Experimental Botany. 2004;55(406):2213–8. pmid:15361537 doi: 10. 1093/jxb/erh242
15. Lv S, Nie X, Wang L, Du X, Biradar SS, Jia X, et al. Identification and characterization of MicroRNAs from barley (Hordeum vulgare L.) by high-throughput sequencing. International journal of molecular sciences. 2012;13(3):2973–84. doi: 10. 3390/ijms13032973. pmid:22489137
16. Scoles G, Dryanova A, Zakharov A, Gulick PJ. Data mining for miRNAs and their targets in the Triticeae. Genome / National Research Council Canada = Genome / Conseil national de recherches Canada. 2008;51(6):433–43. doi: 10. 1139/G08-025. pmid:18521122
17. Ozhuner E, Eldem V, Ipek A, Okay S, Sakcali S, Zhang B, et al. Boron Stress Responsive MicroRNAs and Their Targets in Barley. PloS one. 2013;8(3):e59543. doi: 10. 1371/journal. pone. 0059543. pmid:23555702
18. Hackenberg M, Gustafson P, Langridge P, Shi BJ. Differential expression of microRNAs and other small RNAs in barley between water and drought conditions. Plant biotechnology journal. 2014. doi: 10. 1111/pbi. 12220 pmid:24975557.
19. Kantar M, Unver T, Budak H. Regulation of barley miRNAs upon dehydration stress correlated with target gene expression. Functional & integrative genomics. 2010;10(4):493–507. doi: 10. 1007/s10142-010-0181-4
20. Schreiber AW, Shi B-J, Huang C-Y, Langridge P, Baumann U. Discovery of barley miRNAs through deep sequencing of short reads. BMC genomics. 2011;12(1):129. doi: 10. 1186/1471-2164-12-129
21. Witzel K, Weidner A, Surabhi GK, Borner A, Mock HP. Salt stress-induced alterations in the root proteome of barley genotypes with contrasting response towards salinity. J Exp Bot. 2009;60(12):3545–57. doi: 10. 1093/jxb/erp198 pmid:19671579; PubMed Central PMCID: PMC2724703.
22. Walia H, Wilson C, Wahid A, Condamine P, Cui X, Close TJ. Expression analysis of barley (Hordeum vulgare L.) during salinity stress. Functional & integrative genomics. 2006;6(2):143–56. doi: 10. 1007/s10142-005-0013-0
23. German MA, Pillay M, Jeong D-H, Hetawal A, Luo S, Janardhanan P, et al. Global identification of microRNA—target RNA pairs by parallel analysis of RNA ends. Nature biotechnology. 2008;26(8):941–6. doi: 10. 1038/nbt1417. pmid:18542052
24. Burge SW, Daub J, Eberhardt R, Tate J, Barquist L, Nawrocki EP, et al. Rfam 11. 0: 10 years of RNA families. Nucleic acids research. 2013;41(D1):D226–D32. doi: 10. 1093/Nar/Gks1005 WOS:000312893300031.
25. Ouyang S, Buell CR. The TIGR Plant Repeat Databases: a collective resource for the identification of repetitive sequences in plants. Nucleic acids research. 2004;32:D360–D3. doi: 10. 1093/Nar/Gkh099. WOS:000188079000086. pmid:14681434
26. Jurka J, Kapitonov VV, Pavlicek A, Klonowski P, Kohany O, Walichiewicz J. Repbase update, a database of eukaryotic repetitive elements. Cytogenetic and genome research. 2005;110(1–4):462–7. doi: 10. 1159/000084979. WOS:000231064600047. pmid:16093699
27. Cui LY, Veeraraghavan N, Richter A, Wall K, Jansen RK, Leebens-Mack J, et al. ChloroplastDB: the chloroplast genome database. Nucleic acids research. 2006;34:D692–D6. doi: 10. 1093/Nar/Gkj055. WOS:000239307700146.

pmid:16381961

28. Kozomara A, Griffiths-Jones S. miRBase: annotating high confidence microRNAs using deep sequencing data. Nucleic acids research. 2014;42(Database issue):D68–73. doi: 10. 1093/nar/gkt1181 pmid:24275495; PubMed Central PMCID: PMC3965103.

29. Mayer KF, Waugh R, Brown JW, Schulman A, Langridge P, Platzer M, et al. A physical, genetic and functional sequence assembly of the barley genome. Nature. 2012;491(7426):711–6. doi: 10. 1038/nature11543 pmid:23075845.

30. Chen X, Li Q, Wang J, Guo X, Jiang X, Ren Z, et al. Identification and characterization of novel amphioxus microRNAs by Solexa sequencing. Genome biology. 2009;10(7):R78. doi: 10. 1186/gb-2009-10-7-r78 pmid:19615057; PubMed Central PMCID: PMC2728532.

31. Zuker M. Mfold web server for nucleic acid folding and hybridization prediction. Nucleic acids research. 2003;31(13):3406–15. pmid:12824337 doi: 10. 1093/nar/gkg595

32. Meyers BC, Axtell MJ, Bartel B, Bartel DP, Baulcombe D, Bowman JL, et al. Criteria for annotation of plant MicroRNAs. The Plant Cell Online. 2008;20(12):3186–90. doi: 10. 1105/tpc. 108. 064311

33. Dai XB, Zhao PX. psRNATarget: a plant small RNA target analysis server. Nucleic acids research. 2011;39:W155–W9. doi: 10. 1093/Nar/Gkr319. WOS:000292325300026. pmid:21622958

34. Addo-Quaye C, Miller W, Axtell MJ. CleaveLand: a pipeline for using degradome data to find cleaved small RNA targets. Bioinformatics. 2009;25(1):130–1. doi: 10. 1093/bioinformatics/btn604 pmid:19017659; PubMed Central PMCID: PMC3202307.

35. Ashburner M, Ball CA, Blake JA, Botstein D, Butler H, Cherry JM, et al. Gene ontology: tool for the unification of biology. The Gene Ontology Consortium. Nature genetics. 2000;25(1):25–9. doi: 10. 1038/75556 pmid:10802651; PubMed Central PMCID: PMC3037419.

36. Varkonyi-Gasic E, Wu R, Wood M, Walton EF, Hellens RP. Protocol: a highly sensitive RT-PCR method for detection and quantification of microRNAs. Plant methods. 2007;3:12. doi: 10. 1186/1746-4811-3-12 pmid:17931426; PubMed Central PMCID: PMC2225395.

37. Livak KJ, Schmittgen TD. Analysis of relative gene expression data using real-time quantitative PCR and the 2(-Delta Delta C(T)) Method. Methods. 2001;25(4):402–8. doi: 10. 1006/meth. 2001. 1262 pmid:11846609.

38. Munns R. Comparative physiology of salt and water stress. Plant, cell & environment. 2002;25(2):239–50. pmid:11841667. doi: 10. 1046/j. 0016-8025. 2001. 00808. x

39. Wang L, Yu X, Wang H, Lu YZ, de Ruiter M, Prins M, et al. A novel class of heat-responsive small RNAs derived from the chloroplast genome of Chinese cabbage (Brassica rapa). BMC genomics. 2011;12. doi: 10. 1186/1471-2164-12-289 WOS:000292250800001.

40. Deng P, Nie X, Wang L, Cui L, Liu P, Tong W, et al. Computational Identification and Comparative Analysis of miRNAs in Wheat Group 7 Chromosomes. Plant Mol Biol Rep. 2014;32(2):487–500. doi: 10. 1007/s11105-013-0669-x

41. Yao Y, Guo G, Ni Z, Sunkar R, Du J, Zhu JK, et al. Cloning and characterization of microRNAs from wheat (Triticum aestivum L.). Genome biology. 2007;8(6):R96. doi: 10. 1186/gb-2007-8-6-r96 pmid:17543110; PubMed Central PMCID: PMC2394755.

42. Liu H, Qin C, Chen Z, Zuo T, Yang X, Zhou H, et al. Identification of miRNAs and their target genes in developing maize ears by combined small RNA and degradome sequencing. BMC genomics. 2014;15:25. doi: 10. 1186/1471-2164-15-25 pmid:24422852; PubMed Central PMCID: PMC3901417.

43. Ding D, Zhang L, Wang H, Liu Z, Zhang Z, Zheng Y. Differential expression of miRNAs in response to salt stress in maize roots. Annals of Botany. 2009;103(1):29–38. doi: 10. 1093/aob/mcn205. pmid:18952624

44. Guo HS, Xie Q, Fei JF, Chua NH. MicroRNA directs mRNA cleavage of the transcription factor NAC1 to downregulate auxin signals for arabidopsis lateral root development. The Plant cell. 2005;17(5):1376–86. doi: 10. 1105/tpc. 105. 030841 pmid:15829603; PubMed Central PMCID: PMC1091761.

45. Fedoroff NV. Cross-talk in abscisic acid signaling. Science Signaling. 2002;2002(140):re10. doi: 10. 1126/scisignal. 1402002re10

46. Achard P, Cheng H, De Grauwe L, Decat J, Schoutteten H, Moritz T, et al. Integration of plant responses to environmentally activated phytohormonal signals. Science. 2006;311(5757):91–4. pmid:16400150 doi: 10. 1126/science. 1118642

47. Ha S, Vankova R, Yamaguchi-Shinozaki K, Shinozaki K, Tran LS. Cytokinins: metabolism and function in plant adaptation to environmental stresses. Trends Plant Sci. 2012;17(3):172–9. doi: 10. 1016/j. tplants. 2011. 12. 005 pmid:22236698.

48. Vierling E. The roles of heat shock proteins in plants. Annual review of plant biology. 1991;42(1):579–620. doi: 10. 1146/annurev. arplant. 42. 1. 579

49. Koning AJ, Rose R, Comai L. Developmental expression of tomato heat-shock cognate protein 80. Plant physiology. 1992;100(2):801–11. pmid:16653061; PubMed Central PMCID: PMC1075629. doi: 10. 1104/pp. 100. 2. 801

50. Jain R, Chan MK. Support for a potential role of E. coli oligopeptidase A in protein degradation. Biochemical and biophysical research communications. 2007;359(3):486–90. doi: 10. 1016/j. bbrc. 2007. 05. 142 pmid:17553460.

51. Grondin M, Chow-Shi-Yee M, Ouellet F, Averill-Bates DA. Wheat enolase demonstrates potential as a non-toxic cryopreservation agent for liver and pancreatic cells. Biotechnology journal. 2015;10(5):801–10. doi: 10. 1002/biot. 201400562 pmid:25740431.

52. Ding D, Zhang L, Wang H, Liu Z, Zhang Z, Zheng Y. Differential expression of miRNAs in response to salt stress in maize roots. Ann Bot. 2009;103(1):29–38. doi:

10. 1093/aob/mcn205 pmid:18952624; PubMed Central PMCID: PMC2707283.

53. Xie F, Wang Q, Sun R, Zhang B. Deep sequencing reveals important roles of microRNAs in response to drought and salinity stress in cotton. J Exp Bot. 2015;66(3):789–804. doi: 10. 1093/jxb/eru437 pmid:25371507; PubMed Central PMCID: PMC4321542.

54. Dryanova A, Zakharov A, Gulick PJ. Data mining for miRNAs and their targets in the Triticeae. Genome / National Research Council Canada = Genome / Conseil national de recherches Canada. 2008;51(6):433–43. doi: 10. 1139/G08-025 pmid:18521122.

55. Lu S, Sun YH, Chiang VL. Stress-responsive microRNAs in Populus. The Plant Journal. 2008;55(1):131–51. doi: 10. 1111/j. 1365-313X. 2008. 03497. x. pmid:18363789

56. Zhu JK. Salt and drought stress signal transduction in plants. Annual review of plant biology. 2002;53:247–73. doi: 10. 1146/annurev. arplant. 53. 091401. 143329 pmid:12221975; PubMed Central PMCID: PMC3128348.

CHAPTER 10

TRANSCRIPTOME ANALYSIS REVEALED THE DROUGHT-RESPONSIVE GENES IN HULLESS BARLEY

Xingquan Zeng , Lijun Bai, Zexiu Wei, Hongjun Yuan , Yulin Wang, Qijun Xu, Yawei Tang and Tashi Nyima

Keywords: Hordeum vuglare, DGE, Transcriptome sequencing, Drought-responsive genes, ABA, Tetrapyrrole

BACKGROUND

Hulless barley (*Hordeum vuglare* L. var. *nudum* Hook. f.) is one of the most important crops in China, especially in Tibet Plateau, for over half of the total food production. With caryopses that thresh free from the pales, hulless (naked) barley provides an attractive advantage for the human consumption [1]. It is cultivated in the valleys and in the higher land on Tibet. Drought-induced water deficit greatly affects crop growth and development, and endangers crop agronomic yields. Crops have also formed various molecular and physiological changes to prevent water deficit. Drought tolerance crops maintain turgor and continue metabolism in cells even at low water potential, mainly by protoplasmic tolerance, synthesis of osmolytes or/and compatible solutes [2]. Signal transduction molecules play important roles in this process by mediating the transmission of the stress signals via complex signal transduction pathways. Numerous fundamental molecular aspects of tetrapyrroles and abscisic acid (ABA) are available [3, 4, 5]. Tetrapyrroles are the active cores of some compounds with crucial biochemical roles in living systems, such as chlorophyll, heme, siroheme and phytochromobilin [6]. Heme biosynthesis is transcriptionally responsive to reactive oxygen species (ROS)-mediated stress signaling in *Arabidopsis* [7]. ABA is a vital stress-responsive phytohormone sensitive to these cellular changes, particularly to the loss of turgor [8]. The ABA signal transduction pathway comprises the ABA-bound pyrabactin resistance/regulatory component of ABA receptor (*PYR/RCAR*) proteins [9], type 2C protein phosphatases (*PP2C*) [10], and SNF1-related kinases (*SnRK2*) [11], *NCED* [12], *ABF* [13], *MYB/MYC* [14], *AP2/ERF* family [15], *LEA* [16] and *DHN* [17]. In *Arabidopsis*, the molecular drought response mechanism can be divided into ABA-dependent and ABA-independent pathways [18]. In soybean, ABA treatment influences the expression of drought response genes [19].

Breeding for drought tolerance is particularly challenging because of the genetic complexity of this trait. On the one hand, hulless barley has a complex diploid genome (HH), with the genome size of 5000 Mb, larger than that of human [20]. On the other hand, drought tolerance has been well documented to result from cooperative interactions among multiple morphological, physiological, and biochemical characters. Different genotypes may have diverse responses to drought stress [18, 21, 22]. Therefore, efficient improvement requires an indepth understanding of the gene expression regulation mechanisms in response to drought stress.

Many researchers have analyzed Tibetan hulless barley, and many genes associated with drought stress responses in plants are known. Honsdorf et al. [23] has detected the drought tolerance QTL in wild barley, while Chen et al. [24] has analyzed grain development and nutrient storage in Tibetan hulless barley. However, little has been known about display the changes of gene expression in Tibetan hulless barley during the whole drought response process. In the study, we used transcriptome-seq to identify differentially expressed genes in

Citation: Xingquan Zeng Lijun Bai, Zexiu Wei, Hongjun Yuan†, Yulin Wang, Qijun Xu, Yawei Tang and Tashi Nyima, Transcriptome analysis revealed the drought-responsive genes in hulless barley, DOI: 10.1186/s12864-016-2685-3.

response to drought stress in leaf tissue of a Tibetan hulless barley drought-resistibility cultivar, which were grown in different levels of relative soil moisture content (RSMC) and water recovery. This analysis serves as a reference for future studies on Tibetan hulless barley response to various stresses, such as to drought, cold and salt.

RESULTS

Analysis of transcriptome-seq data

In the present study, cDNA libraries were constructed from leaves harvested at eight days after drought stress and during rehydration with three biological replicates, and then sequenced using the Illumina HiSeq™ 2000 platform. After cleaning and checking the read quality, we obtained almost 53.99 billion 200 bp paired-end clean reads. Among the clean reads, 100 % had quality scores at the Cycle Q20 level (a base quality greater than 20 and an error probability of 0.01). The statistics of sequencing sample and data were in Table 1.

Table 1: Statistics of transcriptome sequencing data in different samples

Genotype	Samples	Raw Reads (M)	Clean Reads (M)	Q20 (%)	GC (%)
Himalaya 10	A1	75.61	67.79	96.64	55.83
Himalaya 10	B2	75.11	67.95	96.82	54.84
Himalaya 10	C3	75.84	64.22	95.22	55.57
Himalaya 10	D4	75.15	66.79	96.74	55.45
Himalaya 10	E5	73.23	65.40	96.69	56.21
Himalaya 10	F6	75.41	69.19	97.53	54.47
Himalaya 10	G7	74.46	68.07	97.49	55.49
Himalaya 10	H8	73.41	66.20	97.37	56.19
Himalaya 10	I9	69.88	64.24	97.37	56.63

The clean reads of each sample (A1-I9) were mapped to the full gene set of Tibetan hulless barley. The 9 transcriptome data could map to 28,077 genes, which covered 71.6 % of the whole gene-set.

DEGs Analysis and Validation of Hulless Barley Sequencing Data

Samples A1, D2, C3, D4, E5 and F6 were divided into two groups: one includes A1, B2 and C3, which is under water-sufficient conditions; the other consists of D4, E5 and F6, which is under withholding water conditions. Then, the comparison analysis between the two groups was performed use Noiseq method, and 313 DEGs were finally obtained. The pairwise comparison was also executed among samples F6, G7, H8 and I9. We filtered genes whose RPKM value are below five in all four samples and removed genes that the frequency of occurrence is less than three in pairwise comparison, and then we got 632 DEGs.

Among these DEGs, genes with distinctly changed expression profiling were confirmed using qPCR (Fig. 1). The results of this experiment were basically consistent with RNA-seq data.

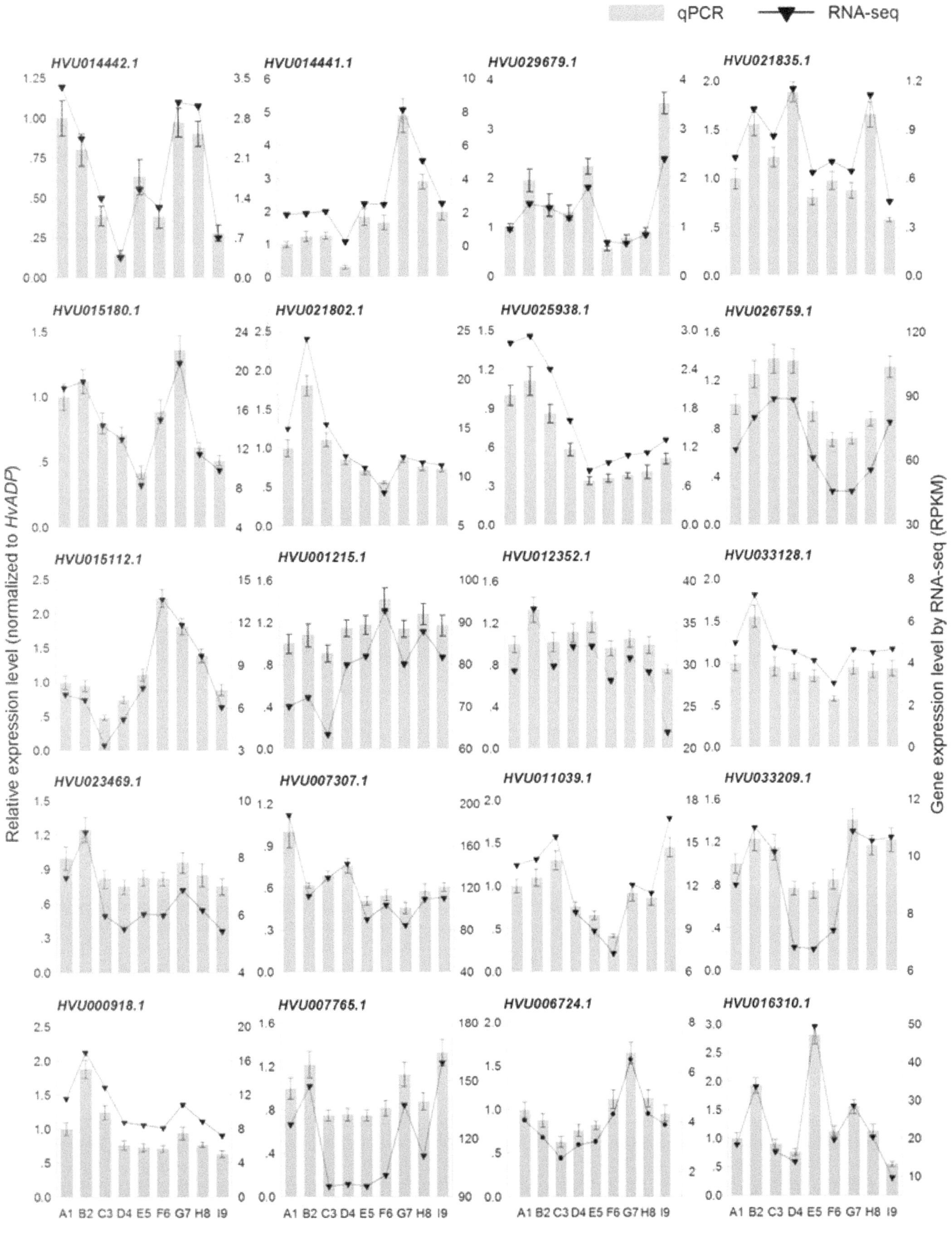

Figure 1: Expression of 20 differentially expressed genes in response to drought stress treatment. Vertical bar charts with simple error bars (left y-axis) represent quantitation of 20 genes transcripts in nine samples, using qPCR. Values are means ± SE (n=3). Line and scatter plot (right y-axis) represents transcript abundance (RPKM) of nine samples for each gene detected by RNA-seq. The A1-F6 indicates the relative soil moisture content (RSMC) of 33.4, 27.5, 21.1, 15.5, 9.8 and 4.8 %, respectively, and G7-I9 indicates the 2 h, 4 h and 8 h after water recovery from 4.8 % to 33.4 %, respectively

Expression Patterns and Cluster Analysis DEGs

The expression patterns and cluster analysis were conducted by *Mev* v4.7.4 software with K-Means clustering method and Pearson correlation as distance calculation method. The number of clusters is set to nine. Gene expression pattern for 853 potential drought-resistant related genes clustering into nine clusters (Fig. 2). 93 Genes in cluster IV were obviously down-regulated in low RSMC level, while 108 genes in cluster VI and 119 genes in cluster VIII were both up-regulated. Interestingly, 98 genes in cluster IX were up-regulated in the early phase of recovery process from sample G7 to H8 (Fig. 3). Furthermore, cluster IX shows up-regulation profile both in high and low RSMC level which indicates stress response to abnormal water content in soil. The other genes expression for the cluster IV, VI, VII, and VIII showed down-regulated pattern under drought stress (Additional file 1: Figure S1).

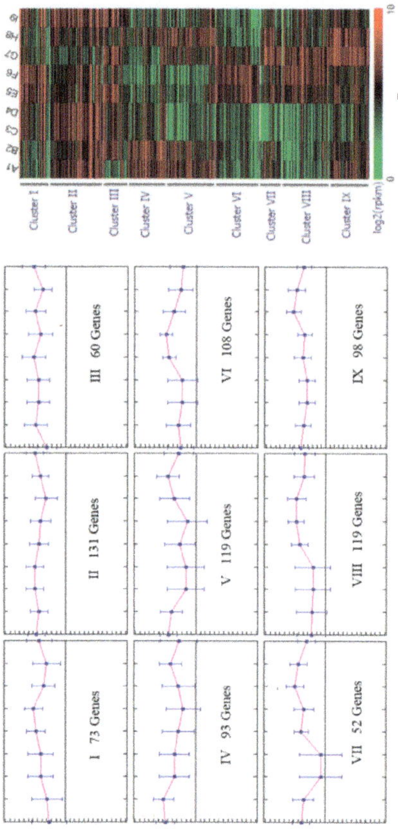

Figure 2: Gene expression pattern for 853 potential drought-resistant related genes clustering into nine clusters. **a** The horizontal axis shows the nine samples A1, B2, C3, D4, E5, F6, G7, H8, I9, while the vertical axis shows the mean value of log-transformed RPKM value for genes in clusters. Error bars were presented for each sample in each cluster. **b** Gene expression pattern for all 853 DEGs. The columns show the nine samples A1, B2, C3, D4, E5, F6, G7, H8, I9, while the rows show the log-transformed RPKM values

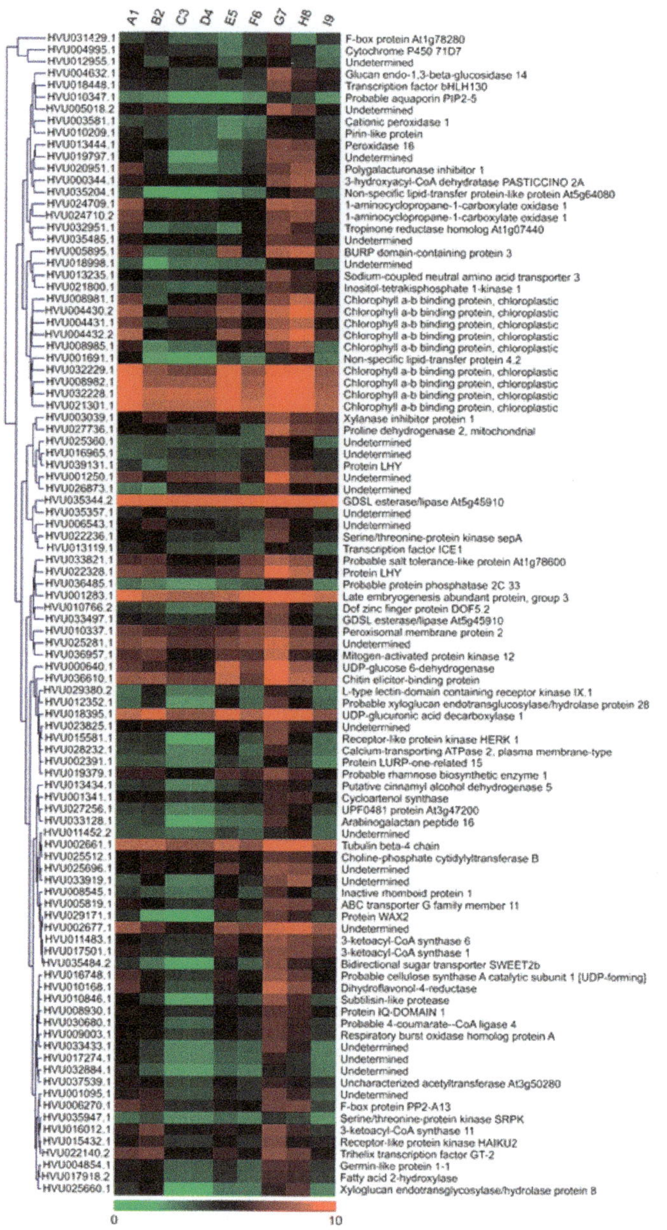

Figure 3: Gene expressions for the cluster IX showing down-regulated pattern under drought stress. The columns show the nine samples A1, B2, C3, D4, E5, F6, G7, H8, I9, while the rows show the log-transformed RPKM values of 98 genes in cluster IX, respectively. Hierarchical clustering of expression pattern for genes in was shown at the left of heat-map figure

GO and KEGG analyze DEGs

The combination of above two strategies results in a final 853 DEGs. We then conducted GO annotation and KEGG pathway analysis of these 853 genes. The assigned functions of these genes covered a broad range of GO categories (Fig. 4). Under the cellular component category, the parts of membrane-bounded organelle, cytoplasmic membrane-bounded vesicle, intrinsic to membrane, cytoplasmic part, intracellular part, plastid part, chloroplast part were prominently represented. Under the category of molecular function, the parts of catalytic activity, transferase activity, transferring phosphorus-containing groups, ion binding, cation binding, oxidoreductase activity represented the majorities of the category. For the biological process category, many genes were classified into the oxidation-reduction process, protein phosphorylation, phosphorus metabolic process, phosphate-containing compound metabolic process, regulation of biological process, biological regulation, transport, and cellular ketone metabolic process. GO enrichment of nine clusters were illustrated in Fig. 5. Of the listed twenty-one GO terms, majority GO terms fell into cluster IX. Tetrapyrrole binding, photosystem and photosynthetic membrane were the most affected in cluster IX. Only one GO term was included in cluster I and cluster VIII, respectively. GO terms focused on cellular component category, whereas DEGs were almost balanced distribution in three categories.

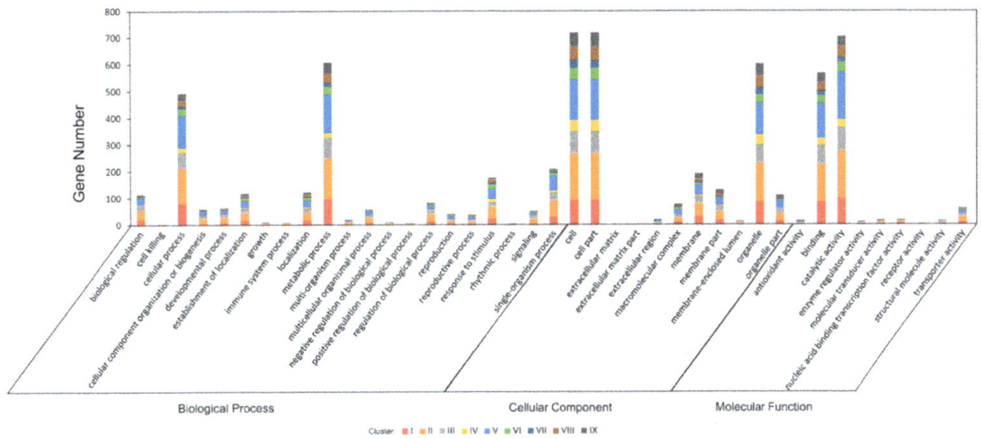

Figure 4: GO annotation of these 853 genes in each category. X-axis indicates GO terms, while the Y-axis indicates the percent of DGEs.

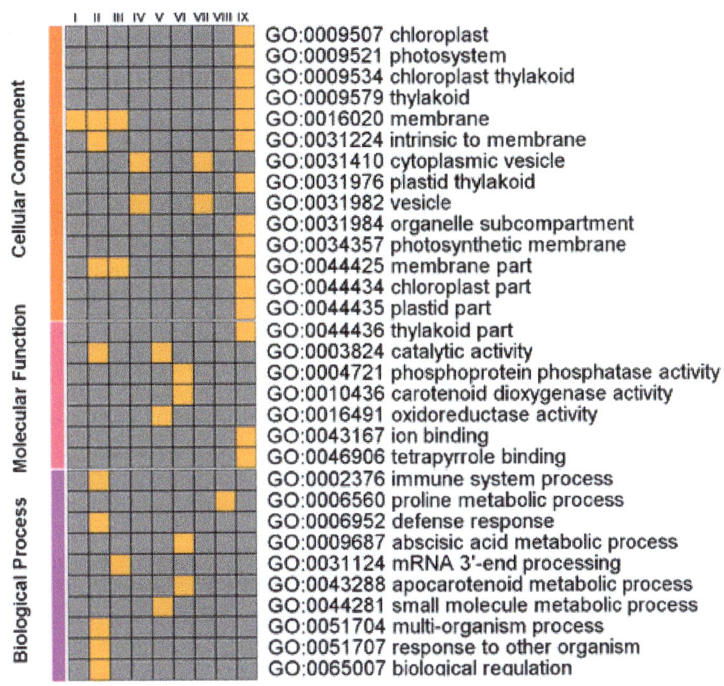

Figure 5: GO terms distribution in nine clusters (I~IX). Each column indicated each cluster, and each row indicated GO degree of enrichment. Red indicated greater enrichment, whereas the grey indicated lower enrichment.

KEGG pathway analysis showed that these genes were mainly located in RNA transport, mRNA surveillance pathway, plant hormone signal transduction, defense-related gene induction, and glycerophospholipid metabolism pathway (Table 2).

Table 2: KEGG pathway of the 853 potential drought resistance related genes

Pathway	Gene Number	Pathway ID
Metabolic pathways	218	ko01100
RNA transport	134	ko03013
Biosynthesis of secondary metabolites	118	ko01110
mRNA surveillance pathway	109	ko03015
Plant hormone signal transduction	56	ko04075
Plant-pathogen interaction	53	ko04626
Glycerophospholipid metabolism	40	ko00564

Endocytosis	37	ko04144
Ether lipid metabolism	35	ko00565
Starch and sucrose metabolism	26	ko00500
Phenylpropanoid biosynthesis	24	ko00940
Cysteine and methionine metabolism	23	ko00270
Pyrimidine metabolism	23	ko00240
Purine metabolism	20	ko00230
Galactose metabolism	20	ko00052
Spliceosome	19	ko03040
RNA polymerase	17	ko03020
Circadian rhythm – plant	17	ko04712
Flavonoid biosynthesis	15	ko00941
Pentose and glucuronate interconversions	15	ko00040
Amino sugar and nucleotide sugar metabolism	14	ko00520
Stilbenoid, diarylheptanoid and gingerol biosynthesis	14	ko00945
Fructose and mannose metabolism	14	ko00051
Glycolysis/Gluconeogenesis	13	ko00010
Photosynthesis - antenna proteins	13	ko00196
Phenylalanine metabolism	12	ko00360
Protein processing in endoplasmic reticulum	11	ko04141
Carotenoid biosynthesis	11	ko00906
Limonene and pinene degradation	10	ko00903
Cyanoamino acid metabolism	10	ko00460

Drought Defense-related Genes and Pathways

We also analyzed the genes related to ABA-dependent and independent signaling pathway of drought stress responses, including *NCED*, *PYR/PYL/RCAR*, *SnRK2*, *ABF*, *MYB/MYC*, and *AP2/ERF* transcription factors. The typical ABA-dependent and independent signaling pathways responsive to drought stress was illustrated in Fig. 6a. The pathways were described by Mustilli et al. [25] and Yoshida et al. [26]. Almost all genes were up-regulated except *PP2C* down-regulated. Dehydration first induced the expression of *NCED* in chloroplast and Ca^{2+} accumulation in nucleus. ABA biosynthesis induction of *NCED* triggered *PYR/PYL/RCAR* and *MYB/MYC*. *PYR/PYL/RCAR* located both in nucleus and the outside blocked the expression of *SnRK2* by inhibiting *PP2C*synthesis. *SnRK2* located in cytoplasm may accelerate the expression of *SLAC1* but repress the expression of*KAT1* located in cell wall, and then result in stomatal closure. This was one response formation in cytoplasm to drought. *AP2/ERF* caused by Ca^{2+} accumulation, *ABF* acted by *SnRK2*, and *MYB/MYC* induced by ABA together promoted the expression of *LEA* and *DHN* in nucleus, and finally mediated drought response. Figure 6bexhibited the expression pattern of representative genes allocated in Fig. 6a mentioned pathways. The expression of *NCED* dramatically decreased during water recovery. *PYR/PYL/RCAR*, *PP2C*, and *MYC* still kept a relative balance expression. The expression of *MYB*, *LEA*, *DHN* rapidly increased when RSMC reached 9.8 %. All genes presented an abundant expression during the RSMC maintained from 9.8 % to 4.8 %.

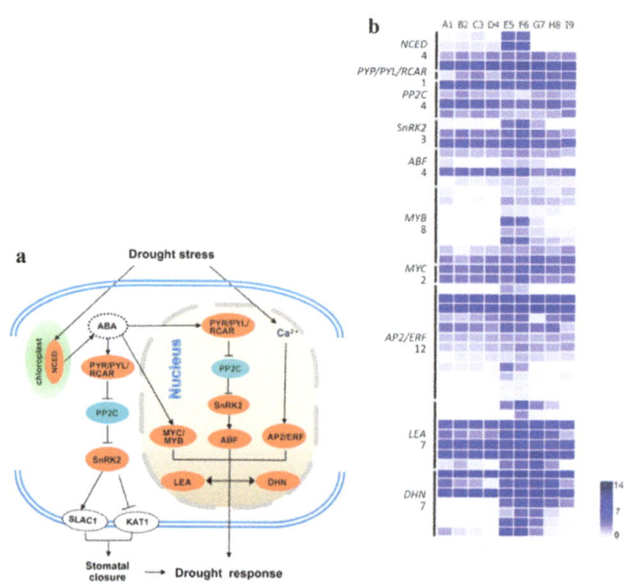

Figure 6: Expression of stress adaptation related genes. **a**, The typical ABA-dependent and independent signaling pathways responsive to drought stress. Red and blue ellipses indicate up-regulated and down-regulated genes identified by transcriptomes under drought stress, respectively. **a** was described by Mustilli et al. [22] and Yoshida et al.

[23]. **b**, The expression pattern of representative genes allocated in above mentioned pathways. The expression levels were assessed by log2-transformed RPKM values. The A1-F6 indicates the relative soil moisture content (RSMC) of 33.4, 27.5, 21.1, 15.5, 9.8 and 4.8 %, respectively, and G7-I9 indicates the 2 h, 4 h and 8 h after water recovery from 4.8 % to 33.4 %, respectively. *NCED*, 9-cis-epoxycarotenoid dioxygenase; *SnRK2*, SNF1-related protein kinase 2; *ABF*,*AREB/ABF*, *MYB/MYC*, *AP2/ERF*, transcription factors, *LEA*, late embryogenesis abundant protein; *DHN*, dehydrin. The number below each gene indicates the member of each gene family.

DISCUSSION

Drought, salinity and cold are the major environmental factors impacting on survival and productivity of hulless barley in Tibetan Plateau of China. Barley is known to be relatively tolerant to abiotic stresses among the major cereal crops and, thus, is often grown in more marginal sites [27]. Tibetan hulless barley cultivar, Himalaya 10, has developed a strong tolerance and adaptation to drought deficit. Using high-throughput RNA sequencing technology, we compared in detail the transcriptional differences and overlap between different levels of RSMC and water recovery, and display the changes of gene expression in Tibetan hulless barley during the whole drought response process.

In this study, the mRNA of Himalaya 10 with good drought tolerance was sequenced using Illumina HiSeq™ 2000 in the manner of PE91, with Sera-mag Magnetic Oligo (dT) Beads. A clear bioinformatic map of mRNA involved in multiple biological processes was produced. As a result, 53.99 G clean reads were collected from nine samples under different RSMC and water recovery, which met the requirements for further analysis. Saturability analysis indicated a qualified coverage of most genes based on our data size. In addition, the clean reads of Q20 occupied over than 95 % of the total, suggesting high quality sequencing. TopHat package was used to blast the transripome data to the Tibetan hulless barley. It has been found that 71.6 % of the reads were mapped to the reference genome. Major mapping reads indicated reliable transcriptome data. These non-mapped tags most likely represent regions where the reference genome is incomplete [28], or there are allelic sequence differences between the reference genome and the cultivar Himalaya 10 used in this study. Another reason may be that RNA-seq data for reference genome annotation should represent all major tissue types, developmental stages and responses to abiotic and biotic stresses [29].

DGE global analysis provided a comprehensive dataset responding to drought stress in leaves of hulless barley seedlings. We identified nine clusters for all DGEs and coarsely assigned them to 31 functional categories ($p < 0.05$) (Fig. 4). Interestingly, there were no other overlapping GO functional enrichments between clusters except catalytic activity, membrane, vesicle, indicating that these genes of different clusters were predicted to be involved in many plant biological processes, including defense [30]. These GO functions were enriched in IX cluster (Fig. 5). Tetrapyrrole binding, photosystem and photosynthetic membrane were the most affected in IX cluster. The former finding implied that tetrapyrroles were the structural backbone of chlorophyll and heme were essential for primary photochemistry, light harvesting, and electron transport [31]. Tetrapyrrole-binding proteins of cHBP1 and cHBP2 have properties suitable for tetrapyrrole carrier proteins [32]. Tetrapyrrole binding protein of genomes uncoupled 4 (GUN4) regulates chlorophyll synthesis and plastid-to-nucleus signal transduction by binding both the product and the substrate of Mg-chelatase, an enzyme that produces magnesium-protoporphyrin IX (Mg-Proto) [33]. Tetrapyrrole biosynthesis had recently been implicated in wilting avoidance [34, 35]. Heme mediated chloroplast-to-nucleus signalling upon drought stress [36, 37]. Enhanced tetrapyrrole biosynthesis was likely to confer drought tolerance via retrograde signaling and induction of drought-responsive gene expression [38].

Pathway enrichment analysis revealed 30 pathways were significantly affected by drought stress (Table 2). Plant hormone signal transduction pathway, for instance, ABA signal transduction pathway was a significant different during the whole drought response process in our study. Transcription factors of *NCED*,*PYR/PYL/RCAR*, *PP2C*, *SnRK2* located in cytoplasm together lead to stomatal response and then transcription factors of *PYR/PYL/RCAR*, *PP2C*, *SnRK2*, *ABF*, *MYB/MYC*, *AP2/ERF*, *LEA*, and *DHN* located in nucleus, together mediated drought response (Fig. 6a). Soil drying first inducted ABA release in roots and then was distributed throughout the plant via the transpiration stream [18]. ABA induced reduction of leaf growth rate and stomatal closure, which triggered stress proteins and various metabolites to protect cells against drought stress [4, 37] when ABA was produced endogenously via water deficit, and then plant tolerance to drought was increased [39, 40, 41]. Drought stress signals can also be generated by osmotic stress-induced Ca^{2+}expansion, which promoted Ca^{2+} channels and induces protein kinases and resulted in drought-responsive gene expression [42, 43, 44]. Du et al. [45] has compared *DHN* between wild barley and Tibetan hulless barley associated with drought stress resistance. Liang et al. [46] indicated that *LEA* genes (*HVA1* and *Dhn6*) might participate in adaptive responses to water deficit in different ways in Tibetan hulless barley. Qian et al. [47] indicate that the differential *HVA1* gene has a functional role in the dehydration tolerance in Tibetan hulless barley.

CONCLUSIONS

In summary, this study provided a comprehensive analysis of drought-responsive genes and transcriptome expression profiles of Tibetan hulless barley leaves by combined DGE, RNA-seq, and computational approaches. Our results revealed 853 potentially drought-responsive genes in Tibetan hulless barley. We analyzed the genes related to ABA-dependent and independent signaling pathway of drought stress responses. Additionally, we also observed the genes related to tetrapyrrole binding of drought stress responses. This result filled up drought-resistant related genes in Tibetan hulless barley in the available literature. Characterizing the components of these pathways will contribute to improve drought tolerance in Tibetan hulless barley.

METHODS

Plant growth under drought condition

An elite hulless barley cultivar Himalaya 10, with good drought tolerance, is used for drought tolerant gene analysis. The RSMC of the original soil sample using for planting were measured and adjusted to 33.4 % by adding proper volume of water. The seedlings of Himalaya 10 were growing under same condition in a greenhouse with a temperature of 23 °C/15 °C (day/night) and a relative humidity of 10–20 %. Three plants from each pot at given condition were considered as biological replicates. Prior to drought stress treatment, these seedlings were well-watered by supplying with proper amount of water every two days to maintain the RSMC at 33.4 %. Different RSMCs at stable status were described by Zhang et al. [48] and Wang et al. [49]. Based on their descriptions, we weighted each pot twice at 8:00 a.m. and 8:00 p.m. Since the soil and plant of each pot was relatively constant, we only add certain water to maintain the RSMC of each pot. Irrigation was removed when the seedlings grew to two-and-a-half leaf stage (18 days after sowing) at six different levels by deficit altering supply water. Eight days after drought stress, leaf samples were harvested from all six groups. The drought stress levels of these groups were evaluated with their score of RSMC (33.4, 27.5, 21.1, 15.5, 9.8 and 4.8 %). After drought treatment, the remaining seedlings of 4.8 % RSMC group were rewatered to restore the RSMC to 33.4 %, leaf samples during rehydration (2 h, 4 h and 8 h after RSMC recovered to 33.4 %) were also collected (Fig. 7). All these leaf samples mentioned above were fast frozen in nitrogen immediately and stored at −80 °C for RNA-seq analysis respectively.

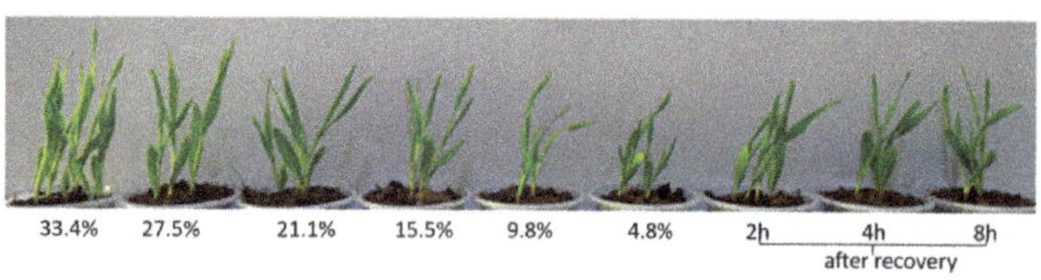

Figure 7: Himalaya 10 underwent a series of RSMCs (33.4, 27.5, 21.1, 15.5, 9.8 and 4.8 %) and rewatered conditions to 33.4 % for 2 h, 4 h and 8 h.

RNA Isolation and Construction of Hulless barley RNA-seq Library

After the total RNA extraction and DNase I treatment, we used magnetic beads with Oligo (dT) to isolate mRNA. The mRNA molecules were fragmented into 200 bp and cDNAs were synthesized taking mRNA fragments as templates. The cDNA fragments were purified and resolved with EB buffer for end repair, single nucleotide A (adenine) addition and connection of adapters. After PCR amplification, Agilent 2100 Bioanaylzer and ABI StepOnePlus Real-Time PCR System were used in quantification and qualification of the sample library. At last, the 200 bp library was sequenced using Illumina HiSeq™ 2000 in the manner of PE91. The clean reads were obtained by filtering out adaptor-only reads, reads containing more than 5 % unknown nucleotides, and low-quality reads which the percentage of low quality bases (base quality ≤ 10) is more than 20 %.

Gene Expression Levels and Identification of Differentially Expressed Genes (DEG)

The gene expression was calculated by RPKM method (reads per kb per million reads) [50]. The RPKM method is able to eliminate the influence of different gene length and sequencing discrepancy on the calculation of gene expression, which could facilitate comparing the difference of gene expression among samples.

DEGs were identified between two samples in the method described below, which was intend to detect the significance of digital gene expression profiles [51]. The clean reads for gene A is x, x yields to the Poisson distribution. The total clean read numbers of the sample 1 and sample 2 are N1 and N2 respectively. Gene A holds x and y reads in sample1 and sample2. The probability of gene A expressed equally between two samples could be calculated by the following formula:

$$2\sum_{i=0}^{y} p(i/x) \quad \left(if \sum_{i=0}^{y} p(i/x) \leq 0.5 \right)$$

Also, the p value was corrected for false positive (type I errors) and false negative (type II errors) using FDR method [52]. We use FDR ≤ 0.001 and the absolute value of Log2Ratio ≥ 1 as the threshold to identify DEGs.

Gene Ontology (GO), pathway annotation and enrichment analyses were based on the Gene Ontology Database (http://www.geneontology.org/) [53] and Kyoto Encyclopedia of Genes and Genomes (KEGG) pathway (http://www.genome.jp/kegg/) [54], respectively. When we investigated pathways in which different genes were involved and enriched, q-value was used to aid identification according to the previous description [55].

DEGs and drought defense-related genes

Samples A1, D2, C3, D4, E5 and F6 were divided into two groups: one includes A1, B2 and C3, which is under water-sufficient conditions; the other consists of D4, E5 and F6, which is under withholding water conditions. Then, the comparison analysis between the two groups was performed use Noiseq method [56], and these DEGs were finally obtained. The pairwise comparison was also executed among samples F6, G7, H8 and I9. We filtered genes whose RPKM value are below five in all four samples and removed genes that the frequency of occurrence is less than three in pairwise comparison, and then we got those DEGs. The combination of above two strategies results in a final DEGs. We then conducted GO annotation and KEGG pathway analysis of these genes. The expression patterns and cluster analysis were conducted by *Mev* v4.7.4 software [57] with K-Means clustering method and Pearson correlation as distance calculation method.

qPCR analysis

To validate the results of the RNA-seq data, expression of the genes the same samples used for Transcriptome-seq analysis was performed by real-time quantitative reverse transcription-polymerase chain reaction (RT-PCR) with the fluorescent intercalating dye SYBRGreen in a detection system (Opticon 2; MJ Research, Waltham, USA), using a hulless barley gene (*HvADP*) as a standard control [58]. A two-step RT-PCR procedure was performed in the experiments. First, 2 μg of purified total RNA was reversely transcribed into cDNAs which were used as templates for PCR reactions using gene-specific primers (Table 3). Second, quantitative PCR was performed using PCR Master Mix (Toyobo, Osaka, Japan) according to the manufacturer's instructions. Relative quantification of gene expression was determined using the comparative Ct method.

To achieve optimal amplification, PCR conditions for each primer combination were optimized for annealing temperature, and PCR products were verified by melting curve analysis and confirmed on an agarose gel. Mean values and standard errors were calculated from three independent experiments with three biological replicates of hulless materials, and the data were normalized with the relative efficiency of each primer pair.

Table 3: Primers of qRT-PCR assay used for eight RNA-seq libraries in this study

Gene_id	Amplicon size (bp)[a]	Forward primer (5'- > 3')	Reverse primer (5'- > 3')
HVU014442.1	142	ATTTCTTCGACTGGGGCCTG	ATGACCTTGCCGTCGATCTC
HVU014441.1	190	CAGGAGCCTGAGTAGATGCG	AATGAGAGGCCGACCACAAG
HVU029679.1	194	AAGAAACTGATCCGAGGCGG	ATTCGTCCGGCCCGTATTTT
HVU021835.1	180	GCGGCTATATCCCACCTTCC	AAGAGCTGAGGTGAAGCGAC
HVU015180.1	208	TGGAGAAACGGATCGAAGCC	TGCAGCCCAGCTAGAAAAGG
HVU021802.1	206	CCTTCACCTCCAGGAACGTC	CACTCCGATTCCACTTGGCT
HVU025938.1	180	TGTCACCGCTGAACCAATCA	TTCACTGGGATTCACGGACG
HVU026759.1	86	GAAGTCGCGCGAATCTGTTC	GAGCAGAGCAGCCAGATCAA
HVU015112.1	215	AAGATCCCAACAAGGCGAGG	CCTGGCTTGCTCCTCTTTGA
HVU001215.1	229	AAGCATGCCGTCTTACACCA	GGCCATCGGAGAGTGCATTA
HVU012352.1	163	TGGCGTGTCAGCTGAGATTT	GGCCGTGAAGGACCAAAAAC
HVU033128.1	214	CCATCAGACTGTGGCGTCTT	GAACCCTCTGCGCATAGACA
HVU023469.1	244	AGCACTACTACGGCACCAAC	CTGAAACCGGCGTACTCCTT
HVU007307.1	216	AGATGGGGTGCGAACTTGAG	CAGGGACAGGACCATCCAAC
HVU011039.1	115	GGTGATCATGCGTGTGTTCG	CTTCTTGGCCGAGTCCTTGT
HVU033209.1	234	AGCCCAAACCTACCAAGCTC	ATCATAGCGCAGGAGCCATC
HVU000918.1	87	GTAGCTATCCACGGTCACGG	TGGAACTGTAAGCGTCCACC
HVU007765.1	208	CGCATGCTGATGGAAAAGGG	ATTGCACCGCACTCAACAAC
HVU006724.1	135	TTTGTGTGGGGAGGTCGATG	CATGCACTCTTCGGTGACCT
HVU016310.1	245	TGAGGATGAAGCGAGTGCAG	TGGGGACTAGCACGCAAAAT

[a]The amplicon lengths for all the genes are within 70–250 bp range to satisfy the requirement for the primer pairs used for real time RT-PCR analysis using SYBR green according to manufacturer's guide

[a]The amplicon lengths for all the genes are within 70–250 bp range to satisfy the requirement for the primer pairs used for real time RT-PCR analysis using SYBR green according to manufacturer's guide

DECLARATIONS

Acknowledgments

This work was supported by grants from the National Program on Key Basic Research Project (2012CB723006), the National Science and Technology Support Program (2012BAD03B01), and the Tibet Autonomous Region Financial Special Fund (2014CZZX001).

AUTHORS' CONTRIBUTIONS

TN, XZ and LB conceived and designed the study. XZ, LB, ZW, HY, YW, QX and YT performed the analyses. XZ and LB wrote the manuscript with the participation of all authors. All authors have read and approved the manuscript.

REFERENCES

1. Baik BK, Ullrich SE. Barley for food: characteristics, improvement, and renewed interest. J Cereal Sci. 2008;48(2):233–42.
2. Nguyen HT, Babu RC, Blum A. Breeding for drought resistance in rice: physiology and molecular genetics considerations. Crop Sci. 1997;37(5):1426–34.
3. Tanaka R, Kobayashi K, Masuda T. Tetrapyrrole metabolism in *Arabidopsis thaliana*. Arabidopsis Book. 2011;9:e0145.
4. Umezawa T, Nakashima K, Miyakawa T, Kuromori T, Tanokura M, Shinozaki K, Yamaguchi-Shinozaki K. Molecular basis of the core regulatory network in ABA responses: sensing, signaling and transport. Plant Cell Physiol. 2010;51(11):1821–39.
5. Lee SC, Luan S. ABA signal transduction at the crossroad of biotic and abiotic stress responses. Plant Cell Environ. 2012;35(1):53–60.
6. Tanaka R, Tanaka A. Tetrapyrrole biosynthesis in higher plants. Annu Rev Plant Biol. 2007;58:321–46.
7. Nagai S, Koide M, Takahashi S, Kikuta A, Aono M, Sasaki-Sekimoto Y, Ohta H, Takamiya K and Masuda T. Induction of isoforms of tetrapyrrole biosynthetic enzymes, AtHEMA2 and AtFC1, under stress conditions and their physiological functions in Arabidopsis. Plant Physiol. 2007;144:1039–51.
8. Schroeder JI, Kwak JM, Allen GJ. Guard cell abscisic acid signalling and engineering drought hardiness in plants. Nature. 2001;410:327–30.
9. Park SY, Fung P, Nishimura N, Jensen DR, Fujii H, Zhao Y, Lumba S, Santiago J, Rodrigues A, Chow TF, Alfred SE, Bonetta D, Finkelstein R, Provart NJ, Desveaux D, Rodriguez PL, McCourt P, Zhu JK, Schroeder JI, Volkman BF, Cutler SR. Abscisic acid inhibits type 2C protein phosphatases via the PYR/PYL family of START proteins. Science. 2009;324(5930):1068–71.
10. Ma Y, Szostkiewucz I, Korte A, Moes D, Yang Y, Christmann A, Grill E. Regulators of PP2C phosphatase activity function as abscisic acid sensors. Science. 2009;324(5930):1064–8.
11. Melcher K, Ng LM, Zhou XE, Soon FF, Xu Y, Suino-

Powell KM, Park SY, Weiner JJ, Fujii H, Chinnusamy V, Kovach A, Li J, Wang YH, Li JY, Peterson FC, Jensen DR, Yong EL, Volkman BF, Cutler SR, Zhu JK, Xu HE. A gate-latch-lock mechanism for hormone signaling by abscisic acid receptors. Nature. 2009;462:602–8.

12. Schwartz SH, Tan BC, Gage DA, Gage DA, Zeevaart JA, McCarty DR. Specific oxidative cleavage of carotenoids by VP14 of maize. Science. 1997;276(5320):1872–4.
13. Choi H, Hong J, Ha J, Knag J, Kim SY. ABFs, a family of ABA-responsive element binding factors. J Biol Chem. 2000;275(3):1723–30.
14. Abe H, Urao T, Ito T, Seki M, Shinozaki K, Yamaguchi-Shinozaki K. *Arabidopsis* AtMYC2 (bHLH) and AtMYB2 (MYB) function as transcriptional activators in abscisic acid signaling. Plant Cell. 2003;15(1):63–78.
15. Zhang G, Chen M, Li L, Xu Z, Chen X, Guo J, Ma Y. Overexpression of the soybean *GmERF3* gene, an AP2/ERF type transcription factor for increased tolerances to salt, drought, and diseases in transgenic tobacco. J Exp Bot. 2009;60(13):3781–96.
16. Hundertmark M, Hincha DK. LEA (Late Embryogenesis Abundant) proteins and their encoding genes in Arabidopsis thaliana. BMC Genomics. 2008;9:118.
17. Robertson M. Increased dehydrin promoter activity caused by HvSPY is independent of the ABA response pathway. Plant J. 2003;34(1):39–46.
18. Shinozaki K, Yamaguchi-Shinozaki K. Gene networks involved in drought stress response and tolerance. J Exp Bot. 2007;58(2):221–7.
19. Marcolino-Gomes J, Rodrigues FA, Fuganti-Pagliarini R, Bendix C, Nakayama TJ, Celaya B, Molinari HB, de Oliveira MC, Harmon FG, Nepomuceno A. Diurnal oscillations of soybean circadian clock and drought responsive genes. PLoS One. 2014;9(1):e86402.
20. The International Barley Genome Sequencing Consortium. A physical, genetic and functional sequence assembly of the barley genome. Nature. 491:711–16. doi:10.1038/nature11543.
21. Neumann PM. Coping mechanisms for crop plants in drought-prone environments. Ann Bot. 2008;101(7):901–7.
22. Deikman J, Petracek M, Heard JE. Drought tolerance through biotechnology: improving translation from the laboratory to farmers' fields. Curr Opin Biotech. 2012;23(2):243–50.
23. Honsdorf N, March TJ, Berger B, Tester M, Pillen K. High-throughput phenotyping to detect drought tolerance QTL in wild barley introgression lines. PLoS One. 2014;9(5):e97047.
24. Chen X, Long H, Gao P, Deng G, Pan Z, Liang J, Tang Y, Tashi N, Yu M. Transcriptome assembly and analysis of Tibetan hulless barley (*Hordeum vulgare* L. var. *nudum*) developing grains, with emphasis on quality properties. PLoS One. 2014;9(5):e98144.
25. Mustilli AC, Merlot S, Vavasseur A, Fenzi F, Giraudat J. Arabidopsis OST1 protein kinase mediates the regulation of stomatal aperture by abscisic acid and acts upstream of reactive oxygen species production. Plant Cell. 2003;14(12):3089–99.
26. Yoshida R, Hobo T, Ichimura K, Mizoguchi T, Takahashi F, Aronso J, Ecker JR, Shinozaki K. ABA-activated SnRK2 protein kinase is required for dehydration stress signaling in arabidopsis. Plant Cell Physiol. 2002;43:1473–83.
27. Jana S, Wilen RW. Breeding for abiotic stress tolerance in barley. In: Ashraf M, Harris PJC, editors. Abiotic stresses: plant resistance through breeding and molecular approaches. 1st ed. New York, USA: Food Products Press; 2005. p. 491–505.
28. Zeng X, Long H, Wang Z, et al. The draft genome of Tibetan hulless barley reveals adaptive patterns to the high stressful Tibetan Plateau. Proc Natl Acad Sci U S A. 2015;112(4):1095–100.
29. Consortium PGS, Xu X, Pan S, Cheng S, Zhang B, Mu D, Ni P, Zhang G, Yang S, Li R, et al. Genome sequence and analysis of the tuber crop potato. Nature. 2011;475:189–95.
30. Shinozaki K, Yamaguchi-Shinozaki K, Seki M. Regulatory network of gene expression in the drought and cold stress responses. Curr Opin Plant Biol. 2003;6:410–7.
31. Papenbrock J, Grimm B. Regulatory network of tetrapyrrole biosynthesis--studies of intracellular signalling involved in metabolic and developmental control of plastids. Planta. 2001;213(5):667–81.
32. Takahashi S, Ogawa T, Inoue K, Masuda T. Characterization of cytosolic tetrapyrrole-binding proteins in *Arabidopsis thaliana*. Photochem Photobiol Sci. 2008;7(10):1216–24.
33. Larkin RM, Alonso JM, Ecker JR, Chory J. GUN4, a regulator of chlorophyll synthesis and intracellular signaling. Science. 2003;299:902–6.
34. Allen SM, Luck S, Mullen J, Sakai H, Sivasanker S, Tingey SV, Williams RW. Drought tolerant plants and related constructs and methods involving genes encoding ferrochelatases. United States Patent US8158859 B2. 2010.
35. Phung TH, Jung HI, Park JH, Kim JG, Back K, Jung S. Porphyrin biosynthesis control under water stress: sustained porphyrin status correlates with drought tolerance in transgenic rice. Plant Physiol. 2011;157:1746–64.
36. Voigt C, Oster U, B€ornke F, Jahns P, Dietz KJ, Leister D, Kleine T. In-depth analysis of the distinctive effects of norflurazon implies that tetrapyrrole biosynthesis, organellar gene expression and ABA cooperate in the GUN-type of plastid signalling. Physiol Plant. 2010;138:503–19.
37. Woodson JD, Perez-Ruiz JM, Schmitz RJ, Ecker JR, Chory J. Sigma factor-mediated plastid retrograde signals control nuclear gene expression. Plant J. 2013;73:1–13.
38. Nagahatenna DSK, Langridge P, Whitford R. Tetrapyrrole-based drought stress signalling. Plant Biotechnol J. 2015;13:447–59.
39. Kuromori T, Mizoi J, Umezawa T, Yamaguchi-Shinozaki K, Shinozaki K. Drought stress signaling network. Mol

Biol. 2014;2:383–409.

40. Atkinson CJ, Davies WJ, Mansfield TA. Changes in stomatal conductance in intact aging wheat leaves in response to abscisic-acid. J Exp Bot. 1989;40(9):1021–8.

41. Munns R, Sharp RE. Involvement of abscisic-acid in controlling plant-growth in soils of low water potential. Austra J Plant Physiol. 1993;20(5):425–37.

42. Schulz P, Herde M, Romeis T. Calcium-dependent protein kinases: hubs in plant stress signaling and development. Plant Physiol. 2013;163:523–30.

43. Zou JJ, Wei FJ, Wang C, Wu JJ, Ratnasekera D, Liu WX, Wu WH. *Arabidopsis* calcium-dependent protein kinase CPK10 functions in abscisic acid-and Ca^{2+}-mediated stomatal regulation in response to drought stress. Plant Physiol. 2010;154:1232–43.

44. Pei ZM, Murata Y, Benning G, Thomine S, Klusener B, Allen GJ, Grill E, Schroeder JI. Calcium channels activated by hydrogen peroxide mediate abscisic acid signalling in guard cells. Nature. 2000;406:731–4.

45. Du JB, Yuan S, Chen YE, Sun X, Zhang ZW, Xu F, Yuan M, Shang J, Lin HH. Comparative expression analysis of dehydrins between two barley varieties, wild barley and Tibetan hulless barley associated with different stress resistance. Acta Physiol Plant. 2011;33(2):567–74.

46. Liang JJ, Deng GB, Long H, Pan ZF, Wang CP, Cai P, Xu D, Nima ZX, Yu MQ. Virus-induced silencing of genes encoding LEA protein in Tibetan hulless barley (*Hordeum vulgare* ssp. *vulgare*) and their relationship to drought tolerance. Mol Breeding. 2012;30(1):441–51.

47. Qian G, Han ZX, Zhao T, Deng GB, Pan ZF, Yu MQ. Genotypic variability in sequence and expression of *HVA1* gene in Tibetan hulless barley, *Hordeum vulgare* ssp. *vulgare*, associated with resistance to water deficit. Aust J Agr Res. 2007;58(5):425–31.

48. Zhang H, Song GQ, Ji WQ, Hu YG. Gene induction by drought stress in wheat variety Xiaoyan 22 and their expression analysis. J Agr Biotechnol. 2009;17(4):670–6.

49. Wang DL, Ye WW, Wang JJ, Song LY, Fan WL, Cui YP. Constructing SSH library of cotton under drought stress and analysis of drought associated genes. Acta Agron Sinica. 2010;36(12):2035–44.

50. Mortazavi A, Williams BA, McCue K, Schaeffer L, Wold B. Mapping and quantifying mammalian transcriptomes by RNA-Seq. Nat Methods. 2008;5(7):621–8.

51. Audic S, Claverie JM. The significance of digital gene expression profiles. Genome Res. 1997;7(10):986–95.

52. Benjamini Y, Hochberg Y. Controlling the false discovery rate: a practical and powerful approach to multiple testing. J Royal Stat Soc. 1995;57(1):289–300.

53. The Gene Onthology Consortium. The gene ontology project in 2008. Nucleic Acids Res. 2008;36 suppl 1:440–4.

54. Kanehisa M, Goto S. KEGG: Kyoto Encyclopedia of Genes andGenomes. Nucleic Acids Res. 2000;28(1):27–30.

55. Storey JD, Tibshirani R. Statistical significance for genomewide studies. Proc Natl Acad Sci U S A. 2003;100(16):9440–5.

56. Tarazona S, Garcia-Alcalde F, Dopazo J, Ferrer A, Conesa A. Differential expression in RNA-seq: a matter of depth. Genome Res. 2011;21:2213–23.

57. Mar JC, Wells CA, Quackenbush J. Defining an informativeness metric for clustering gene expression data. Bioinformatics. 2011;27(8):1094–100.

58. Ferdous J, Li Y, Reid N, Langridge P, Shi BJ, Tricker PJ. Identification of reference genes for quantitative expression analysis of MicroRNAs and mRNAs in barley under various stress conditions. PLoS One. 2015;10(3):e0118503.

CHAPTER 11

ROOT SPATIAL METABOLITE PROFILING OF TWO GENOTYPES OF BARLEY (HORDEUM VULGARE L.) REVEALS DIFFERENCES IN RESPONSE TO SHORT-TERM SALT STRESS

Megan C. Shelden, Daniel A. Dias, Nirupama S. Jayasinghe, Antony Bacic, and Ute Roessner

Keywords: Cell division, cereal, gas chromatography–mass spectrophotometry, metabolomics, osmotic stress, root elongation, root growth, salinity.

INTRODUCTION

Salt-affected soil is a major agricultural problem, with over 800 million ha of land estimated to be affected by salinity worldwide and 32 million ha of dryland agriculture estimated to be salt affected (FAO, 2015). Important cereal crops such as barley (*Hordeum vulgare*) and wheat (*Triticum* spp.) are growing in increasingly hostile environments, resulting in significant decreases in yield. The effect of salinity on plant growth has previously been described by a two-phase model (Munns, 2002). The early response to salt stress is referred to as an osmotic stress but is more accurately described as a shoot ion-accumulation independent stress as it occurs within minutes and continues over the duration of the salt stress (Roy *et al.*, 2014). Osmotic stress is a result of low water potential around the root and results in cell dehydration; loss of cell turgor; and inhibition of cell division, cell expansion, and photosynthesis; leading to a reduction in plant growth and yield (Munns and Tester, 2008). In contrast, ionic stress is a result of the accumulation of toxic concentrations of Na^+ and Cl^- in the cytoplasm of shoot tissues and usually occurs after long-term exposure to salt. High concentrations of Na^+ have been shown to accumulate in root cortical cells and cell walls, resulting in a decrease in cell turgor and root growth (Flowers and Hajibagheri, 2001), indicating that the roots may also elicit a salt-specific response.

Salinity has been shown to reduce root growth in many plant species, including *Arabidopsis* (Wu *et al.*, 1996), *Gossypium hirsutum* (cotton;Zhong and Lauchli, 1993), and the cereals *Oryza sativa* (rice; Lin and Kao, 1996), barley (Shelden *et al.*, 2013), wheat (Rahnama *et al.*, 2011), and *Zea mays* (maize; Bernstein and Kafkafi, 2002). Some plants maintain root elongation at salt concentrations (or in other low water potential media) that inhibit shoot growth (Spollen *et al.*, 1993; Yamaguchi and Sharp, 2010); this is an important adaptive mechanism ensuring seedling establishment and allowing an increase in soil root exploration for water and nutrient uptake (Yamaguchi and Sharp, 2010). In a study of eight genotypes of barley, increasing salinity resulted in a progressive reduction in seminal root elongation rate (Shelden *et al.*, 2013). Considerable variation in root elongation was observed among the genotypes, with wild barley identified as the most tolerant. A reduction in root growth in response to salt stress is often associated with an inhibition of cell division and cell expansion (Chatterjee and Majumder, 2010), as has been shown previously in barley (Tabur and Demir, 2009, 2010). Understanding the response of the root apical meristem, and of elongation and differentiation processes to environmental stresses will be important for understanding tolerance mechanisms in the root.

Citation: Megan C. Shelden, Daniel A. Dias, Nirupama S. Jayasinghe, Antony Bacic, and Ute Roessner, Root spatial metabolite profiling of two genotypes of barley (Hordeum vulgare L.) reveals differences in response to short-term salt stress, doi:10.1093/jxb/erw059 Advance Access publication 5 March 2016.

The response of roots to salinity is often transient; within hours, turgor can be restored by osmotic adjustment (Pritchard *et al.*, 1991). Osmotic adjustment occurs through the synthesis and accumulation of osmoprotectants (compatible solutes) and inorganic ions (Na^+, K^+, and Cl^-). In barley genotypes, root and shoot ion concentrations do not correlate with root elongation rates, suggesting Na^+ and K^+ concentrations are not directly influencing root growth (Shelden *et al.*, 2013). In the elongation zone of the maize primary root, osmotic adjustment occurs at low water potentials through an increase in proline deposition due to an increase in proline transport into the root tip (Voetberg and Sharp, 1991).

Recent advances in high-throughput functional genomics technologies make it possible to gain more insight into salt-tolerance mechanisms. Metabolism is likely to vary between tissues and cell types and these changes can often go undetected when bulk tissues are analysed. Thus, there is a need to use both spatial and temporal resolution to elucidate the molecular response to salinity (and other abiotic stresses). With the advances in analytical technologies it is now possible to measure metabolite changes in tissues or single cells (Wang and Bodovitz, 2010;Shelden and Roessner, 2013). In Arabidopsis, this technology has been used to produce a high-resolution metabolite map of specific cell types in the root using fluorescence-activated cell sorting (Moussaieff *et al.*, 2013). The metabolite responses to abiotic stress have been studied extensively in plants (Urano *et al.*, 2010; Obata and Fernie, 2012; Shelden and Roessner, 2013). The response to salinity has been studied in a number of agriculturally important plants, including barley (Widodo *et al.*, 2009; Wu*et al.*, 2013) and *Vitis vinifera* (grapevine; Cramer *et al.*, 2007); however, these studies were all conducted on bulk tissues (i.e. whole shoots or roots) and thus it is likely that some metabolites present in either specific tissues or cell types would have been undetected.

In terms of production, barley is the fourth most important cereal crop worldwide (after maize, rice, and wheat) producing 144 MT in 2013 (http://faostat3.fao.org/browse/rankings/commodities_by_regions/E) (FAO, 2015). It is more salt tolerant than other glycophytic cereals, such as wheat, making it a good model for studying salt-tolerance mechanisms. For this study, we used two barley genotypes, Clipper (malting barley) and Sahara (landrace), known to have contrasting salinity-tolerance mechanisms (Widodo *et al.*, 2009; Shelden *et al.*, 2013). Clipper (tolerant) has previously been shown to maintain root elongation in response to salt stress, whilst in Sahara (sensitive) root elongation was significantly reduced (Shelden *et al.*, 2013). In order to determine spatial changes in primary metabolites involved in osmotic adjustment in barley seminal roots in response to salt stress, we profiled primary metabolites involved in central metabolism in three different regions of the root: the root cap/cell division zone (R1), elongation zone (R2), and maturation zone (R3) (Fig. 1). A GC-MS metabolite profiling approach was used to monitor the metabolic changes between the different regions of the root and in response to short-term salt stress. To our knowledge, this is the first study to use spatially resolved metabolomics to gain insight into the short-term root response to salt stress.

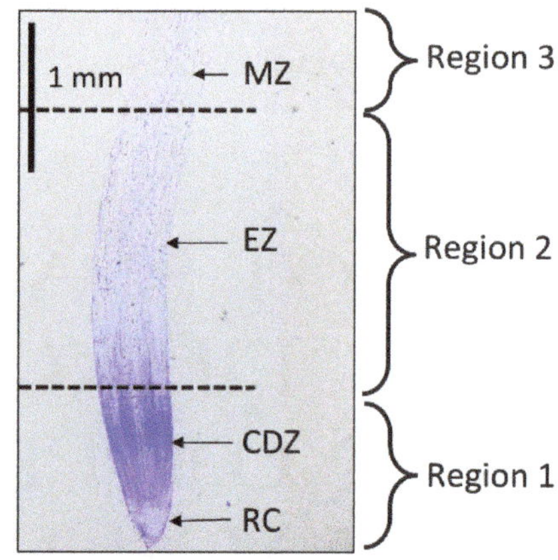

Figure 1: Longitudinal section through a 2-day-old barley cv. Clipper seminal root tip stained with toluidine blue. CDZ, cell division zone; EZ, elongation zone; MZ, maturation zone; RC, root cap.

MATERIALS AND METHODS

Plant material and growth conditions

Hordeum vulgare L. cv. Clipper (Australian malting variety) and the landrace Sahara (North Africa) were used for all experiments. Barley seeds were sourced from the Australian Centre for Plant Functional Genomics (Adelaide, South Australia). Growth conditions were as described previously (Shelden *et al.*, 2013). Briefly, seeds were germinated and transferred to nutrient agar medium for 48h. For salinity treatment, nutrient agar was supplemented with 100mM NaCl and additional $CaCl_2$ to a final activity of 1mM (Genc *et al.*, 2007; Shelden *et al.*, 2013). Seedlings were harvested for microscopy and metabolomics after 48h growth on agar media. Over this time period, seminal root growth has previously been shown to be linear and lateral roots are not yet formed (Shelden *et al.*, 2013).

Chemical fixation

Seminal roots of barley seedlings cv. Clipper and Sahara (n = 5) were harvested after 48h growth on agar media. A surgical blade was used to excise 4mm root tips from the root cap junction, which were then immediately immersed in 2.5% glutaraldehyde in 1× phosphate buffered saline (PBS; pH 7.4), and stored at 4°C. Excised root tips were fixed in 2.5% glutaraldehyde in PBS (pH 7.4) for 2h at room temperature. The samples were rinsed three times in fresh buffer for 10min each, before being dehydrated in increasing concentrations of ethanol consisting of 10, 30, 50, 70, 90, and 100% anhydrous ethanol for 30min at each step. Following dehydration the samples were infiltrated with increasing concentrations of LR white resin (Sigma) in ethanol consisting of 25, 50, 75, and 100% resin for 8h at each step. After a second change of 100% resin the tissues were embedded in fresh resin in gelatine capsules. The gelatine capsules were capped to exclude air and the resin polymerised in an oven at 60°C for 24h.

Light microscopy

Embedded root tips in resin blocks were sectioned longitudinally with a diamond knife on a Leica Ultracut-S microtome. Semi-thin sections (500nm) were collected onto glass slides and dried on a warm hotplate. Sections were stained with alkaline toluidine blue for 5min before being gently rinsed in a stream of distilled water until all the excess stain was removed. The stained sections on slides were dried and viewed with a DM2500 Leica compound microscope and digital images captured with a Leica 300F digital camera. Average cell lengths were calculated by measuring the length of 10–15 cortical cells at various positions from the root apex.

Sampling of 2-day-old roots for metabolomics

The apical region of the root was cut into three sections measured from the root cap junction (Fig. 1). R1 comprised the root cap and cell division zone, R2 the elongation zone, and R3 the maturation zone (defined as where the cortical cell length was uniform). Sections from the three longest seminal roots of 10 individual seedlings were pooled for each biological representative (n = 5). In total, 200 plants were used and the fresh weight (FW) for pooled samples ranged from 1.8 to 18.3mg. Owing to instrument sensitivity issues, we pooled tissue from each root region from multiple plants to ensure we had sufficient material to detect and identify as many metabolites as possible. The length of the root sections varied according to cultivar and treatment differences: Clipper (control) R1 0–1mm, R2 1–3.5mm, and R3 3.5–6.0mm; Clipper (100mM NaCl) R1 0–1.0mm, R2 1.0–3.0mm, and R3 3.0–5.0mm; Sahara (control) R1 0–1.0mM, R2 1.0–3.0mm, and R3 3.0–5.0mm; Sahara (100mM NaCl) R1 0–0.5mm, R2 0.5–2.5mm, and R3 2.5–5.0mm. Root sections were immediately transferred to aluminium boats that were floating in liquid N_2. The pooled root sections were immediately weighed to prevent thawing to determine the FW of each sample and subsequently snap frozen in 2mL Eppendorf tubes and stored at −80°C.

Extraction and derivatization

Approximately 10mg of the various sections of root tissue (R1, R2, and R3) was added to a Cryomill tube. Next, 150 μL methanol (100%) with 2 μL internal standards ($^{13}C_6$-sorbitol/$^{13}C_5$ ^{15}N-valine in water, 0.2mg mL^{-1}) was added. The sample was subsequently homogenized using a Cryomill (Bertin Technologies) using programme #2 (6100–3×45×45) at −10°C. The sample mixture was vortexed for 30s and then incubated for 15min at 70°C at 850rpm. Milli-Q water (150 μL) was added and then the sample was centrifuged at 13000rpm for 5min. A 90 μL aliquot was transferred into a glass insert and dried *in vacuo* for subsequent trimethylsilyl (TMS) and tri-*tert*-butyldimethylsilyl (TBS) polar metabolite derivatization.

Polar metabolite TMS and TBS derivatization

The dried samples were re-dissolved in 10 μL of 30mg mL^{-1}methoxyamine hydrochloride in pyridine and derivatized at 37°C for 120min with mixing at 500rpm using the on-line derivatization capability of the GERSTEL autosampler. One dried aliquot was then treated for 30min with 20 μL *N,O-bis*-(trimethylsilyl)trifluoroacetamide (BSTFA) and another for 45min with 20 μL *N-(tert*-butyldimethylsilyl*)-N*-methyltrifluoroacetamide (MTBSTFA). To both samples, 2.0 μL retention time standard mixture [0.029% (*v*/*v*) *n* dodecane, *n*-pentadecane, *n*-nonadecane, *n*-docosane, *n*-octacosane, *n*-dotriacontane, *n*-hexatriacontane dissolved in pyridine] was added, with mixing at 500rpm at 37°C (TMS) or 45°C (TBS). Each derivatized sample was allowed to rest for 60min prior to injection.

GC-MS analysis

The GC-MS analysis was conducted as described previously (Jacobs *et al.*, 2007; Hill *et al.*, 2013). For the polar TMS metabolite analysis, the temperature programme described in Hill *et al.* (2013) was used. For the TBS metabolite analysis, the following temperature programme was used: start injection at 100°C, hold for 1min, then a 12.5°C min^{-1} oven temperature increase to 325°C, with a final 6min of heating at 325°C. Both chromatograms and mass spectra were evaluated using the AnalyzerPro Deconvolution programme (Spectralworks, UK). Mass spectra of eluting compounds were identified using the public domain mass spectra library of Max Planck Institute for Plant Physiology, Golm, Germany (http://csbdb.mpimp-golm.mpg.de/csbdb/dbma/msri.html) and the in-house Metabolomics Australia mass spectral library. All matching mass spectra were additionally verified by determining the retention time by analysis of authentic standard substances. Resulting relative response ratios (area of analyte divided by area of internal standard, $^{13}C_6$-sorbitol for TMS-derivatized metabolites and/or $^{13}C_5$ ^{15}N-valine for TBS-derivatized metabolites) per sample FW (mg) for each analysed metabolite as described in Roessner *et al.* (2001). If a specific metabolite had multiple TMS or TBS derivatives, the metabolite with the greater detector response and improved peak shape within the dynamic range of the instrument was selected.

Statistical analysis

The metabolite data were analysed as described in Roessner *et al.* (2001)and are presented as fold changes relative to a reference that is set to 1. Statistical analysis of metabolite data was performed using Microsoft Excel. Differences between samples were validated by the Student's *t*-test incorporated in Excel. Differences were considered significant with a *t*-test value $P < 0.05$ (Bonferroni-corrected P value) and between the Bonferroni corrected value and 0.05. The heatmap was generated with the open-source software, MetaboAnalyst 3.0 (http://www.metaboanalyst.ca/MetaboAnalyst/faces/home.xhtml).

RESULTS

The root growth response and cellular anatomy is altered in response to salt

Previously we have shown a differential response in the root elongation of two cultivars of barley, Clipper and Sahara, to short-term salt stress (Shelden *et al.*, 2013). In order to further evaluate this difference, we examined the cortical cell length in the seminal roots to determine if the effects of 100mM NaCl on root elongation were due to an inhibition of cell expansion and/or cell division (Table 1). In the more salt-tolerant Clipper, the length of the cell division region was approximately 1mm and appeared to be unaffected by the addition of salt. In the elongation zone, cortical cells had variable lengths and thus were assumed to be undergoing cell expansion. In control plants, the elongation zone in Clipper was approximately 2mm in length. In a previous study we showed that in response to salt, growth rate was reduced by 20% compared to control plants (Shelden *et al.*, 2013). In this study we have shown the cortical cell length becomes uniform closer to the root apex, thus the elongation zone is shortened by salt treatment. The start of the maturation zone was defined as where the cortical cell length became uniform. Final cortical cell length was similar for both control and salt-treated Clipper roots (Table 1). In the more salt-sensitive Sahara, the cells started expanding closer to the root apex and the length of the elongation zone was unaltered in response to salt, most likely indicating an inhibition of cell division (Table 1). It was also observed that in response to salt treatment the final cortical cell length in Sahara roots was double that of control plants (Table 1).

Table 1: Spatial distribution of cortical cell length in Clipper and Sahara roots. The effect of 48h treatment with 100mM NaCl on the spatial distribution of cortical cell length in the apical 5mm of Clipper and Sahara seminal roots was examined compared with untreated roots

Genotype	Clipper		Sahara	
Treatment	Control	Salt	Control	Salt
Region 1 (mm)	0–1.0	0–1.0	0–1.0	0–0.5
Region 2 (mm)	1.0 – 3.5	1.0–3.0	1.0–3.0	0.5–2.5
Region 3 (mm)	3.5–5.5	3.0 – 5.0	3.0–5.0	2.5–4.5
Final cortical cell length (μm)	141.9±8.5[a]	143.4±11.6[a]	108.8±3.9[a]	220.5±15.0[b]

The length of each region was defined by the following: region 1, root cap and cortical cells dividing; region 2, cortical cells expanding; and region 3, cortical cells ceased expansion and the length became uniform. The final cortical cell length shown is the mean ± SEM (n = 8–10). Different superscript letters indicate significance between the salt treatment relative to control and between genotypes (two-way ANOVA, Bonferroni post-hoc test, $P < 0.001$).

Metabolite response differs spatially in the root

In order to precisely measure metabolic changes in the different regions of barley seminal roots, and in response to salt treatment, the roots were sectioned according to the estimates obtained from the cortical cell length measurements (Table 1). We employed GC-MS to measure a wide range of metabolites. In total, 98 metabolites were measured; of these, 76 were identified unambiguously and included 29 amino acids (AA) and amines, 20 organic acids (OA), and fatty acids (FA), and 19 sugars and sugar phosphates. Of those only the metabolites that changed significantly are further discussed below. We found that metabolite profiles changed spatially along the root and were different between the two genotypes and in response to salt stress.

Spatial metabolic profiles in the seminal roots of Clipper and Sahara

The heatmap analysis shows a clear separation between the three regions of the root (R1, R2, and R3) for both Clipper and Sahara (Fig. 2). The majority of metabolites detected changed significantly along the developmental gradient of the root in both genotypes. Significant fold-changes were described relative to R1 (Fig. 1). In Clipper roots, there were significant spatial changes to OAs, FAs, sugars, and sugar phosphates along the root. The general trend for AAs in Clipper was an increase in their relative response ratio, with levels highest in R3 and lowest in R1. The following AAs and amines increased significantly in R2 and/or R3 compared to R1: valine, isoleucine, leucine, alanine, tyrosine, phenylalanine, tryptophan, arginine, methionine, ornithine, putrescine, tyramine, histidine, glycine, proline, pyroglutamate, and glutamine. The largest increases in AAs were putrescine (R3; 12-fold), tyramine (R2; 6-fold), and tryptophan (R3; 9-fold). Conversely, the AAs aspartate, γ-aminobutyric acid (GABA), glutamate, homo-serine, and β-alanine, and the amines ethanolamine and guanine decreased with cell maturity, with levels highest in R1. Components of the tricarboxylic acid (TCA) cycle (citrate, succinate, fumarate, and α-ketoglutarate) were highest in the most apical region of the root tip (R1) and decreased shootward with levels lowest in R3. Components of lipid metabolism (adipic acid, palmitic acid, octadecanol, linoleic acid, stearic acid, and oleic acid) were also highest in R1. With the exception of shikimate, all other OA significantly decreased when compared to R1. The sugars maltose (51-fold, 29-fold), fructose (27-fold, 54-fold), glucose (10-fold, 23-fold), and erythrose (4.3-fold, 3-fold) increased in R2 and R3 respectively, when compared with R1. Raffinose increased only in R2 compared with R1 (2.5-fold). In R3, the largest significant increases were in fructose (54-fold) and glucose (23-fold). All other sugars measured decreased significantly relative to R1.

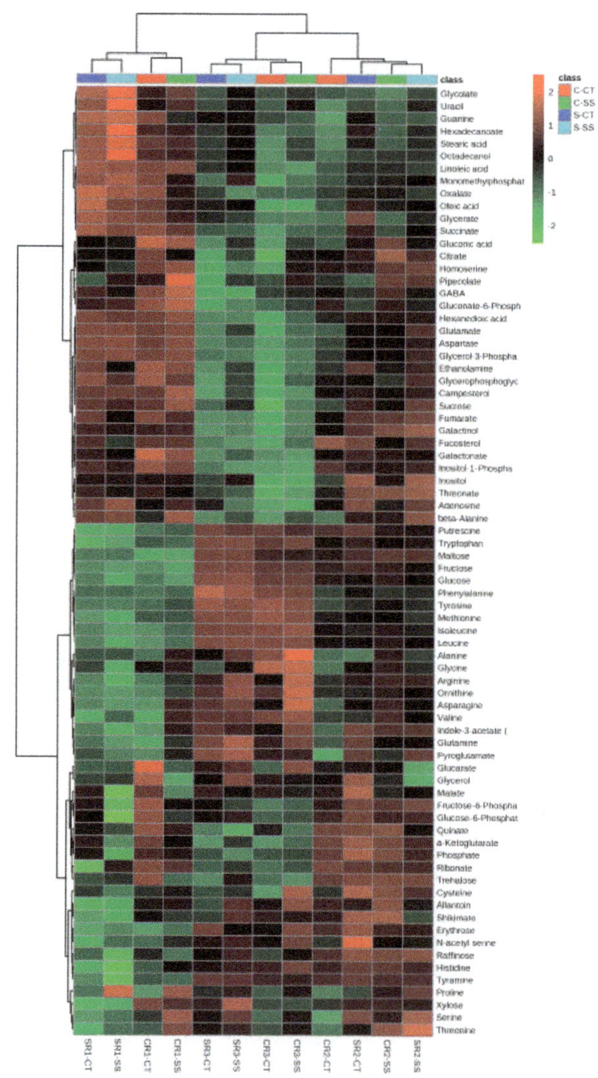

Figure 2: Clustered heatmap of the normalized metabolite log response between root zones/treatment and the measured metabolites in Clipper and Sahara. Clustering of the roots zones and salt treatments is depicted by the dendrogram at the top. Clustering of the metabo-

lites is depicted by the dendrogram at the left. Each coloured cell (red higher, green lower) on the map corresponds to a normalized log response value of the metabolite levels, with samples in columns and metabolites in rows. CRl CT, Clipper R1 Control; SR1 CT, Sahara R1 Control; CR1 SS, Clipper R1 salt-stressed; SR1 SS, Sahara R1 salt-stressed; CR2 CT, Clipper R2 Control; SR2 CT, Sahara R2 Control; CR2 SS, Clipper R2 salt-stressed; SR2 SS, Sahara R2 salt-stressed; CR3 CT, Clipper R3 control; SR3 CT, Sahara R3 control; CR3 SS, Clipper R3 salt-stressed; SR3 SS, Sahara R3 salt-stressed.

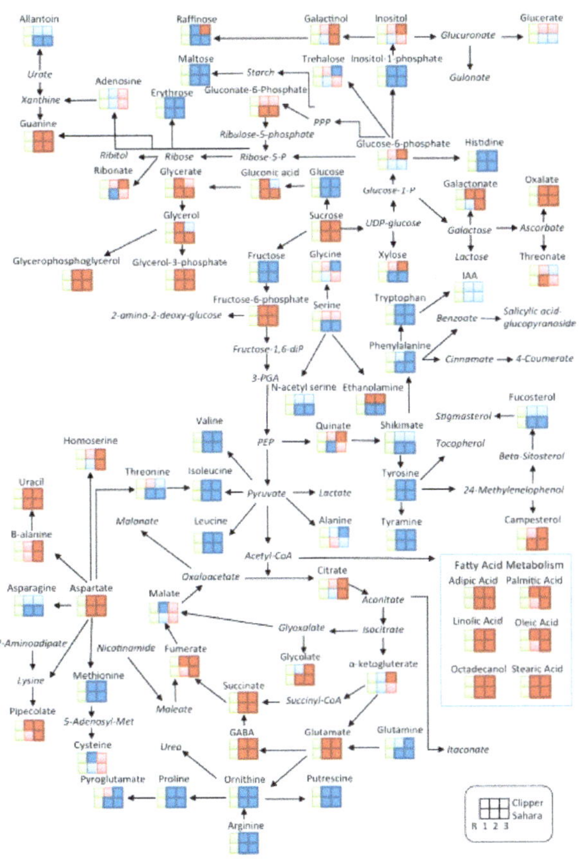

Figure 3: Analysis of metabolites in the root development zones (R1, R2, and R3) in barley genotypes Clipper and Sahara grown in control conditions. The identified metabolites are mapped onto a simplified metabolic network. Fold changes were determined for each root zone (R2, elongation zone; R3, maturation zone) relative to R1, cell division zone (pale green squares). Dark blue (dark red) squares indicate a significant fold-increase (decrease) in R2/R3 compared to R1. Light blue (light red) squares indicate an increase (decrease) in R2/R3 compared to R1 but these are not significant. Metabolite names in italics represent metabolites not detected in this study.

In Sahara, the relative response ratios of AAs generally increased spatially along the root towards the shoot (Fig. 4). The following AAs and amines increased significantly in R2 and/or R3 compared to R1: putrescine, tryptophan, tyramine, histidine, allantoin, threonine, *N*-acetyl serine, methionine, glutamine, arginine, leucine, asparagine, isoleucine, ornithine, phenylalanine, tyrosine, serine, pyroglutamate, valine, and proline. The largest fold changes in R2 and R3, respectively, were putrescine (25-fold, 31-fold), tryptophan (16-fold, 47-fold), tyramine (16-fold, 14-fold), and histidine (6.6-fold, 4.4-fold). The AAs β-alanine, glutamate, aspartate, guanine, GABA, and amine ethanolamine significantly decreased from R1 towards R3. The relative responses of measured components involved in FA metabolism (adipic acid, hexadeconoate, octadecanol, linoleic acid, and oleic acid) were highest in R1 and decreased shootwards towards R3, as seen in Clipper roots. Components of the TCA cycle (citrate, succinate, glycolate, and fumarate) were highest in the most apical region of the root tip (R1) and decreased shootwards with levels lowest in R3. The sugars fructose, glucose, and xylose increased significantly and were highest in R3, while raffinose, xylose, erythrose, ribonate, and glycerol were highest in R2.

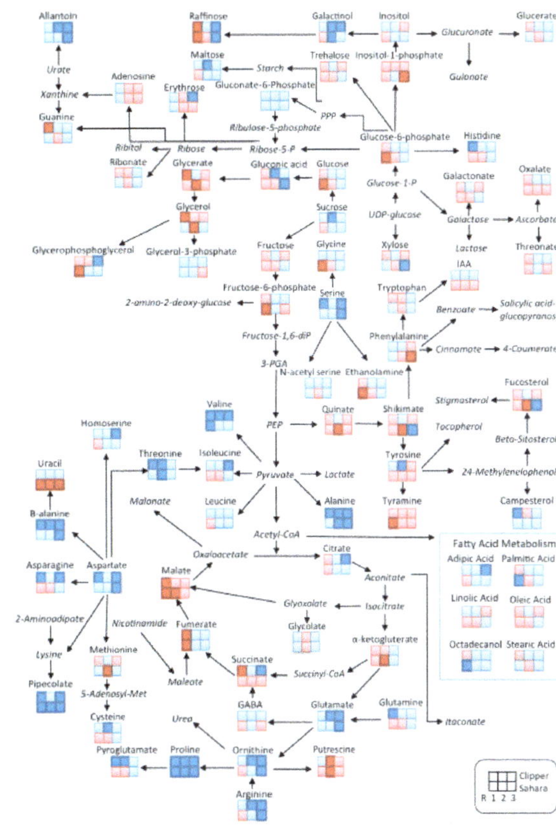

Figure 4: Analysis of metabolites in the root development zones (R1, R2, and R3) in barley genotypes Clipper and Sahara grown on 100mM salt. The identified metabolites are mapped onto a simplified metabolic network. Fold changes were determined for 100mM salt treatment compared to control for each root zone. Dark blue (dark red) squares indicate a significant fold-increase (decrease) in response to salt treatment compared to control conditions. Light blue (light red) squares indicate an increase (decrease) in response to salt treatment compared to control conditions but these are not significant. Metabolite names in italics were not detected in this study.

Although we identified many similarities in the metabolic response of Clipper and Sahara roots in control conditions, some important differences between genotypes were also observed. The sugars xylose, raffinose, ribonate, glucose-6-phosphate, galactonate, and inositol all had a differential response between genotypes. A number of AAs also differed, including asparagine, glycine, serine, threonine, and ethanolamine, and the OAs quinate and galactonate.

Metabolic changes in response to short-term salt stress

To determine the response of each genotype to salinity stress, we compared the level of each metabolite in each region of the root with the same metabolite measured in the same region of the control plants. The metabolic response was different for each genotype and in each region of the root. The heatmap analysis shows that for R1 and R3 there was clear separation between genotypes and salt treatment; however, in R2 the response was more variable and we did not observe a clear separation between genotypes and salt treatment (Fig. 2).

Metabolic changes in response to salt stress in Clipper roots

In response to salt treatment, there was a general trend seen in the roots of Clipper with increases in most AAs compared to the controls. The AAs alanine, proline, asparagine, β-alanine, and valine were significantly increased in all three regions of the root. In R1, there were significant increases in asparagine (3.8-fold), threonine (2.3-fold), alanine (1.8-fold), histidine (1.8-fold), glutamine (1.7-fold), serine (1.6-fold), isoleucine (1.3-fold), aspartate (1.3-fold), pyroglutamate (1.3-fold), and leucine (1.2-fold). Significant decreases in response to salt in R1 included the amine guanine (0.4-fold); OAs glycerate (0.8-fold), succinate (0.8-fold), fumarate (0.6-fold), and malate (0.5-fold); and the sugars raffinose (0.4-fold) and glycerol (0.3-fold).

In R2 (elongation zone), there was a general increase in AAs, sugars, sugar phosphates, and OAs in response to salt treatment in Clipper roots. Threonine (2.8-fold), proline (2.3-fold), β-alanine (2.1-fold), alanine (1.8-fold), allantoin (1.4-fold), glutamate (1.4-fold), arginine (1.8-fold), cysteine (1.6-fold), ornithine (1.6-fold), valine (1.6-fold), pyroglutamate (1.3-fold), tyrosine (1.5-fold), and the sugars galactinol (2-fold), maltose (1.9-fold), and sucrose (1.7-fold) all significantly increased compared to the control (Fig. 5A, B and C). The most significant increase in response to salt was the OA gluconate (5.9-fold) (Fig. 5B). There was only one metabolite, putrescine (0.6-fold), that significantly decreased in R2.

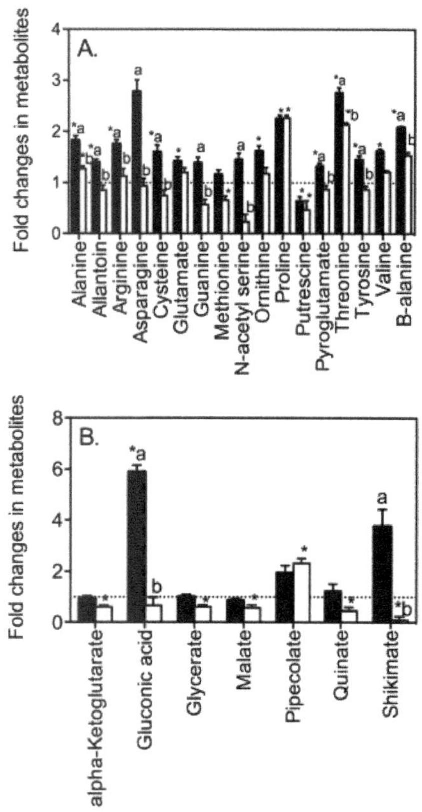

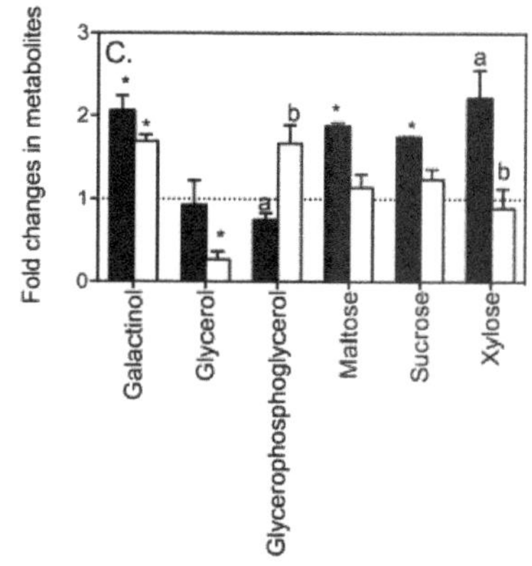

Figure 5: Comparison of metabolite changes in the elongation zone (R2) of Clipper and Sahara in response to 48h salt stress. Fold-changes for Clipper (black bars) and Sahara (white bars) were determined relative to the control for **A** amino acids, **B**organic acids, and **C** sugars. Only metabolites that changed significantly are shown. * denotes metabolites that are significantly different to a fold-change of 1 (student's *t*-test). Superscript letters indicate significant difference between genotypes Clipper and Sahara (two-way ANOVA, $P < 0.01$).

In R3 (maturation zone) of Clipper roots, there were small increases (up to 2-fold) in the following AAs and amines: allantoin, proline, alanine, β-alanine, glutamate, homo-serine, aspartate, isoleucine, serine, and valine. Asparagine (3.5-fold), arginine (2.9-fold), and orthinine (2.7-fold) also increased in response to salt stress. The sugars erythrose (2.2-fold), glycerophosphoglycerol (2.5-fold), raffinose (1.8-fold), and galactinol (1.4-fold) increased significantly. The OAs citrate (2.1-fold), pipecolate (1.7-fold), and succinate (1.3-fold) also increased, as did the FA hexanedioic acid (1.8-fold). R3 in Clipper showed no significant decreases in metabolites in response to salt.

Metabolic changes in response to salt stress in Sahara roots

In R1, the AAs proline (5.4-fold) and valine (1.5-fold) increased and glycine (0.8-fold), arginine (0.6-fold), and the amine tyramine (0.3-fold) decreased. The FAs octadecanol (1.7-fold) and hexadecanoate (1.5-fold) increased and the OAs fumarate (0.5-fold) and malate (0.5-fold) decreased, as did the sugars fructose-6-phosphate (0.3-fold), glucose 0.5-fold), glucose-6-phosphate (0.3-fold), and raffinose (0.3-fold). Campesterol and monomethylphosphate increased (1.7-fold) and ethanolamine (0.7-fold), fucosterol (0.5-fold), and uracil (0.4-fold) decreased.

In R2, the overall trend in AAs, OAs, and FAs in response to salt was a decrease (Fig. 5). The AAs proline (2.3-fold), threonine (2.1-fold), and alanine (1.3-fold) significantly increased and methionine (0.7-fold) and the amine putrescine (0.5-fold) decreased (Fig. 5A). Pipecolate increased by 2.3-fold and glycerate (0.6-fold), α-ketoglutarate (0.6-fold), malate (0.6-fold), phosphate (0.6-fold), quinate (0.5-fold), shikimate (0.1-fold), and uracil (0.6-fold) all significantly decreased in response to salt (Fig. 5B). The sugar alcohols galactinol (1.7-fold) increased and glycerol (0.3-fold) decreased (Fig. 5C).

In R3, there were significant increases in many metabolites in response to salt. The AAs and amines allantoin, ornithine, arginine, proline, β-alanine, glutamate, alanine, aspartate, and serine all increased significantly in response to salt, whilst phenylalanine decreased (0.8-fold). The OAs shikimate (3.9-fold), gluconic acid (2.5-fold), monomethylphosphate (2.2-fold), and pipecolate (1.8-fold) increased and phosphate decreased (0.8-fold). The sugars xylose (3.2-fold) and raffinose (1.6-fold) increased, while inositol-1-phosphate (0.8-fold) and uracil (0.8-fold) decreased.

Differential metabolite accumulation in the root zones of Clipper and Sahara

In response to short-term salt stress, there was a general decrease in OAs (specifically components of the TCA cycle) and sugars in R1 and an increase in AAs in R3 in both Clipper and Sahara roots. In both genotypes the known osmoprotectant proline increased in response to salt stress, with the highest accumulation in R1. The most pronounced differences in metabolite profiles between the two genotypes were in the elongation zone (R2), where there was a significant increase in some AAs, OAs, and sugars in Clipper and a decrease in these metabolites in Sahara (Fig. 5). Alanine, allantoin, arginine, asparagine, cysteine, guanine, N-acetyl serine, pyroglutamate, threonine, tyrosine, and β-alanine were all significantly higher in the elongation zone of Clipper than in Sahara (Fig. 5A). The fold-changes in gluconate, shikimate, and xylose relative to the control were significantly higher in the elongation zone of Clipper roots than Sahara roots (Fig. 5B, C). In Clipper, metabolites in the shikimate pathway were significantly increased in the elongation zone (quinate, shikimate, phenylalanine, and tyrosine) and cell division zone (tyramine) (Fig. 6) compared to Sahara. Salt stress resulted in a significant decrease in uracil in all root zones of Sahara compared with no significant change in Clipper.

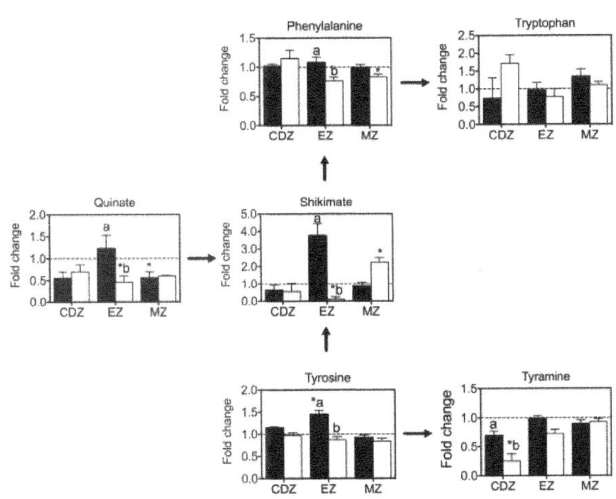

Figure 6: Changes in metabolites in the shikimate pathway of Clipper and Sahara in response to 48h salt stress. Fold-changes for Clipper (black bars) and Sahara (white bars) were determined relative to the control. * denotes metabolites that are significantly different to a fold-change of 1 (student's *t*-test). Superscript letters indicate significant difference between genotypes Clipper and Sahara (two-way ANOVA, $P < 0.01$).

DISCUSSION

Here we have shown the importance of performing metabolite analysis on specific regions of the root as a tool for understanding the changes in the growth zone. We have also shown that genotypes with contrasting root-growth response to short-term salt stress have different metabolite profiles. By carrying out GC-MS–based metabolomics we were able to assign specific metabolite changes to specific regions of the root. A plants' response to environmental stress occurs at all levels of organization and thus it is important to study the stress response at a cellular and tissue level. In a recent study in *Medicago truncatula,* a transcriptomics and metabolomics approach was used to gain insight into the metabolic differences between root tips and border cells (Watson *et al.*, 2015). Here we describe the first detailed metabolite profiling in different spatial regions of the barley root. The results indicate that the metabolome in the barley root is region and genotype specific as well as salt responsive.

In the root tip of the primary root, new cells are generated from cell division in the apical meristem for both the root cap and cells that will undergo further cell division, expansion, and differentiation. The process of cell division in the root is highly regulated both spatially and temporally (Benfey *et al.*, 2010). As the cells migrate away from the root tip and become fully expanded they undergo the complex process of cell differentiation into the epidermis, cortex, endodermis, and vasculature of the primary root.

Genetic variation in root phenotypes have previously been reported in barley (Bengough *et al.*, 2004) and wheat (Richards *et al.*, 2007; Wasson *et al.*, 2014) and in response to abiotic stresses (James *et al.*, 2008;Shavrukov *et al.*, 2010; Shelden *et al.*, 2013). We have previously shown variation in seminal root elongation in eight genetically diverse genotypes of barley (Shelden *et al.*, 2013). The root phenotype of Clipper is inherently different from the landrace Sahara, with Clipper having longer roots. The root phenotypes of Clipper and Sahara are altered in response to abiotic stresses including boron toxicity and salinity (Roessner *et al.*, 2006; Choi *et al.*, 2007; Widodo *et al.*, 2009; Shelden *et al.*, 2013). In order to gain a molecular understanding of the differences in root elongation in these two genotypes, we performed metabolite analyses by dividing the root tips into three segments corresponding with the apical meristem/cell division zone, elongation zone, and maturation zone (Fig. 1).

The metabolite analysis indicated that there were region-specific responses corresponding to root development (Figs 2 and 3). Both Clipper and Sahara showed very similar trends in their metabolic response along the developmental gradient under control conditions; however, there were some notable differences. Higher levels of TCA cycle intermediates in the cell division zone in both genotypes most likely indicate the increased energy requirement for cell division. FAs are a major component of membranes and were highest in the cell division zone where cell membrane synthesis is occurring. AAs increased along the developmental gradient towards the shoot, contributing to an increase in protein synthesis involved in cell elongation, maturation, and differentiation (Fig. 6).

The growth of the root system depends on the availability of carbon to the roots (Muller *et al.*, 1998). Cell division and cell expansion are thought to be two independent processes (Green, 1976), but both are affected by carbon availability. Cell division can be inhibited by carbon depletion (Muller *et al.*, 1998; Van't Hof, 1968). The process of cell elongation not only requires expansion of the cell wall but also an accumulation of intracellular solutes, including sugars, to generate an osmotic potential and facilitate water uptake into the cell (Cosgrove, 1999). Gradients in soluble sugars, including sucrose, glucose, and fructose, were seen along the root tip, with sucrose highest at the apex and glucose and fructose increasing further away from the apex and into the maturation zone. These trends are similar to what has been reported previously in maize (Sharp *et al.*, 1990; Muller *et al.*, 1998). This may be a result of either the termination of phloem in the maturation zone or the absence of sucrose-cleaving enzymes (invertases) in the root apex (Hellebust and Forward, 1962). Significant differences in both glucose and xylose were observed between Clipper and Sahara and potentially indicate differences in cell wall composition between the cultivated barley and the landrace; however, this needs to be evaluated directly by measuring cell wall composition and nucleotide sugars. In rice, genotype differences in glucose and xylose were shown to correlate with specific hemicellulosic polymers (Zhang *et al.*, 2012). The accumulation of the raffinose family oligosaccharides inositol, galactinol, and raffinose also showed differences between the two genotypes. These are important storage compounds and osmoprotectants and may reflect differences in response and adaptation to environmental stresses between cultivated barley and the landrace Sahara.

Interestingly, GABA showed a distinct profile in both genotypes, with higher levels in the cell division zone. GABA is believed to play an important role in many processes, including maintaining the carbon/nitrogen balance, pH regulation, signalling, and energy production (Fait *et al.*, 2008; Ramesh *et al.*, 2015), and has been shown to accumulate in response to both biotic and abiotic stresses (Kinnersley and Turano, 2000). In Arabidopsis, *pop2* mutants, which have a reduced ability to degrade GABA, accumulate high levels of GABA that result in an inhibition of cell elongation and represses cell wall synthesis genes (Renault *et al.*, 2011). The role of GABA in plant development is not well studied (Renault*et al.*, 2011) and to our knowledge there is no direct evidence of a role for GABA in root cell division.

The shikimate pathway (Fig. 6) is involved in biosynthesis of aromatic AAs, including tryptophan, tyrosine, and phenylalanine, and these serve as precursors for a wide range of secondary metabolites including lignin (Herrmann and Weaver, 1999; Maeda and Dudareva, 2012). Quinate synthesis branches from the main trunk of the shikimate pathway and is an ubiquitous plant building block for phytoalexins and UV protectants (Herrmann and Weaver, 1999). In plants, 20% of carbon is fixed through the shikimate pathway and this is linked to the biosynthesis of aromatic compounds (Herrmann, 1995). Increases in the aromatic AAs in both genotypes in the elongation zone and maturation zone are most likely owing to the differentiation of cells into specialized cell types requiring the synthesis of secondary metabolites and structural compounds, including lignin and suberin (Fan *et al.*, 2006; Barros *et al.*, 2015). The observed differences in the accumulation of quinate and shikimate in Clipper and Sahara roots may be a result of differences in the regulation of these pathways (Fig. 6).

Metabolite response to short-term salt stress in the roots

The ability of seedlings to maintain root elongation in saline soils is an important adaptation for seedling establishment and ensuring an adequate water and nutrient supply for shoot growth. The cellular response to salt stress can include changes in the cell cycle and division, and modifications to cellular membranes and cell walls (Shabala and Munns, 2012). Root elongation is often inhibited by osmotic stress and this can be due to an inhibition of cell division and/or cell expansion (Munns and Tester, 2008). The anatomical study demonstrated that the susceptible genotype Sahara has a significant inhibition in the cell division that is primarily responsible for a decrease in root elongation rate in response to salinity. Interestingly, cortical cell length in the root increased dramatically in response to salinity in Sahara roots. Cell enlargement in response to salinity has been reported previously in maize (Li *et al.*, 2014) and halophytes (Hajibagheri *et al.*, 1985) and may be a result of increased vacuolation—an adaptive mechanism to cope with increased intracellular ion concentration.

One mechanism to cope with osmotic stress is the sequestering of inorganic ions (Na^+ and Cl^-) into the vacuole. In response to abiotic stress, plants can also alter their metabolism and can synthesize compatible solutes (i.e., proline, glycine betaine, sucrose, and raffinose) to cope with the changes in water potential and maintain turgor pressure. Compatible solutes accumulate in the cytosol and organelles to help facilitate osmotic adjustment; however, the synthesis of compatible solutes is more energetically expensive than sequestering ions into the vacuole and thus this comes at a cost to the plant (Shabala and Munns, 2012). The concentrations of compatible solutes that accumulate are thought to be in the order of 10mM but can be higher when partitioned into the cytoplasm (Munns and Tester, 2008). This study provides novel insights into the metabolic response of barley seedlings exposed to salt stress by using spatial metabolite profiling of the root.

Salt-tolerant glycophytes (e.g. barley) can help mitigate the effects of salt stress by the synthesis of compatible solutes. Proline has been shown to act as an osmoprotectant in plants affected by both drought (Bowne *et al.*, 2012) and salinity (Ueda *et al.*, 2007). Increases in proline in response to salt stress have previously been reported in both barley roots and leaves (Widodo *et al.*, 2009; Wu *et al.*, 2013); however, proline levels do not always correlate with salt tolerance (Chen *et al.*, 2007). Interestingly, both Clipper and Sahara had similar increases in proline accumulation in response to salt stress, with proline levels highest in the most apical region of the root where cell division was occurring. Increased accumulation of proline in the apical region of the root tip (with values up to 160mM measured at the root apex) has previously been observed in maize in response to water stress (Voetberg and Sharp, 1991; Verslues and Sharp, 1999; Raymond and Smirnoff, 2002;Yamaguchi and Sharp, 2010) and is thought to contribute significantly to osmotic adjustment. In Arabidopsis, *P5CS* (Δ1-Pyrroline-5-carboxylase synthase), which regulates proline accumulation, has been shown to be induced by moderate salt stress, thus providing osmotic stress tolerance (Liu and Zhu, 1997). It been suggested that proline may also play a role in cell division in Arabidopsis (Mattioli *et al.*, 2009).

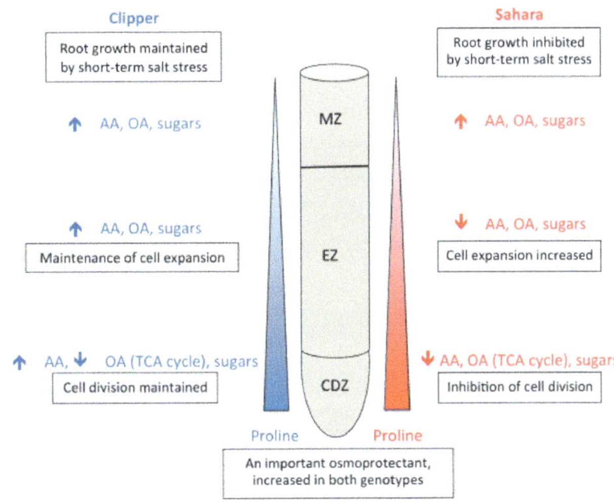

Figure 7: Schematic diagram showing the major trends in metabolic changes of AAs, OAs, and sugars in the maturation zone (MZ), elongation zone (EZ), and cell division zone (CDZ) of Clipper (blue) and Sahara (red) roots in response to short-term salt stress. Upward arrows indicate an increase in response to salt stress and downward arrows indicate a decrease in response to salt stress.

In Clipper, there were significant increases in AAs, OAs, and sugars in all three regions of the root, most likely contributing to the ability of Clipper roots to maintain elongation in response to salt stress (Fig. 7). Interestingly, the elongation zone of Clipper showed the most contrasting response compared to Sahara (Fig. 5) with significant increases in some AAs and amines (alanine, allantoin, asparagine, cysteine, guanine, N-acetyl serine, pyroglutamate, threonine, tyrosine, and β-alanine), sugars (xylose), and OAs (gluconate and shikimate). This corresponds with the ability of Clipper to maintain cell expansion (Table 1) and root elongation rate (Shelden *et al.*, 2013). Gluconate has previously been shown to increase in response to salt stress in barley roots (Wu *et al.*, 2013). Glucose dehydrogenase catalyses the oxidation of glucose to gluconate and this enzyme has been implicated in conferring tolerance to plant cells in response to stress (Witzel *et al.*, 2010).

Nitrogen-containing compounds (alanine, arginine, leucine, ornithine, serine, and valine), the amides (asparagine and glutamine), and polyamines (putrescine and spermine) have been shown to accumulate as part of the stress response in plants (Rabe, 1990). In two rice genotypes with different salt tolerance levels, most AAs and sugars significantly increased in the leaves and roots of both genotypes; however, the salt-tolerant genotype had a more prolonged response over time (Zhao *et al.*, 2014). Asparagine and glutamine have both been reported to accumulate in response to salt stress (Imamul Huq and Larher, 1983; Mansour, 2000). Asparagine has a high N:C and thus is an ideal storage compound that accumulates under stress conditions (Lea *et al.*, 2007). Asparagine levels reportedly increased in the roots of barley seedlings (Yamaya and Matsumoto, 1989) and in barley leaves (Garthwaite *et al.*, 2005) in response to NaCl. In this study, asparagine increased in the root of the salt-tolerant Clipper in all three regions but not in the salt-sensitive Sahara and thus is likely involved in the ability of Clipper to maintain root elongation. Glutamine is the primary AA involved in nitrogen assimilation in plants and the amine group donor for many other compounds. Glutamine increased in response to salt stress in the cell division zone of Clipper, potentially indicating an increase in nitrogen assimilation. Uracil, a pyrimidine derivative, decreased in all root zones of Sahara, possibly reflecting changes in the biosynthesis or degradation of RNA in response to salinity (Zhang *et al.*, 2011).

Polyamines, including spermidine, spermine, and their precursor putrescine, are involved in plant growth and development and have been suggested to play an important role in plant stress responses (Gill and Tuteja, 2010). Putrescine is synthesized from arginine via either arginine decarboxylase or orthinine decarboxylase. In Clipper and Sahara, putrescine significantly decreased in response to salt in the elongation zone. Differences in the accumulation of putrescine in response to salt stress in rice are contradictory, with some studies reporting increases in salt-sensitive rice cultivars (Katiyar and Dubey, 1990) and other studies reporting an accumulation in roots of a tolerant variety compared to a sensitive variety (Lefèvre *et al.*, 2001). Widodo *et al.* (2009) reported an increase in putrescine after 5 weeks of salt stress in barley leaves associated with leaf senescence and cell damage. In *Vigna radiata* (mung bean), free putrescine decreased in roots in response to salt stress but increased in the leaves (Friedman *et al.*, 1989). The contradictory results of putrescine accumulation in response to salt stress may be a reflection of the tissue sampling procedures and further emphasize the need for spatial metabolomics analyses.

Components of the shikimate pathway including quinate, shikimate, phenylalanine, and tyrosine increased significantly in the elongation zone of Clipper in response to salt stress compared to Sahara (Figs 5 and 6). A significant decrease in shikimate and quinate in the elongation zone of Sahara roots (Fig. 6) indicated a salt-specific response in metabolite accumulation, as has been reported previously in Sahara (Widodo *et al.*, 2009). Thus, we have shown the shikimate salt response is region and genotype dependent, highlighting the importance of conducting spatial metabolite profiling. Because shikimate is the precursor to aromatic AAs, these results may indicate differences between the two genotypes in the synthesis of structural compounds such as lignin, which is involved in cell wall biosynthesis (Barros *et al.*, 2015). In the elongation zone of maize, up-regulation of cell wall-related genes was associated with cell enlargement in response to salinity (Li *et al.*, 2014).

In response to salt stress, sugar accumulation was highest in the elongation zone and is most likely involved in the maintenance of turgor pressure and carbohydrates for cell wall synthesis. In both Clipper and Sahara, sucrose levels were lowest in the root apex and glucose was highest in the maturation zone. This is the same as has been previously reported in maize (Sharp *et al.*, 1990; Muller *et al.*, 1998).

The structure and fluidity of plant membranes is crucial in a plant's adaptation to abiotic stresses, and is affected by many factors including the FA composition of the lipids. Changes in the FA composition in response to salinity have been reported (Wu *et al.*, 1998; Chalbi *et al.*, 2013). In this study, FAs were found to be highest in R1 of the root, where cell division is occurring, most likely owing to the deposition of the cell membrane. The ratio of unsaturated to saturated FAs also altered in response to salt stress. In Sahara, an increase in the saturated FA palmitic acid in response to salt stress may be a mechanism to help restrict the uptake of Na^+ into the plant root. Octadecanol is a long chain unsaturated fatty alcohol previously found in Arabidopsis epidermal cell surfaces (Franke *et al.*, 2005) and in the epicuticular wax of *Allium ampeloprasum*(leeks; Gabriela-Anca Maier and Post-Beittenmiller, 1998).

CONCLUSION

We have used untargeted GC-MS metabolomics to identify differences along the development zones of the root in barley seedlings in response to salt stress. We have shown that the processes involved in root growth adaptation and the underlying coordination of metabolic pathways are controlled in a region-specific manner. In summary, the two genotypes had contrasting root growth responses to short-term salt stress and this was reflected in their metabolomes (Fig. 7). The synthesis of primary metabolites was spatially regulated in response to salt stress and differed between salt-tolerant and salt-sensitive genotypes. Clipper maintained cell division and root elongation through the synthesis of compatible solutes for osmotic adjustment and turgor maintenance. This study highlights the importance of utilizing 'omics' technologies for spatial profiling and provides us with potential metabolic pathways involved in root growth maintenance in response to salt stress. Further studies will involve matrix-assisted laser desorption/ionization imaging and transcriptional and genetic analyses to further elucidate the pathways involved in the root response.

ACKNOWLEDGEMENTS

This work was funded by grants to the Australian Centre for Plant Functional Genomics (ACPFG) from the Australian Research Council (ARC) and the Grains Research and Development Corporation (GRDC), the South Australian Government, The University of Adelaide, The University of Queensland, and The University of Melbourne. Metabolomics Australia is funded through Bioplatforms Australia Pty Ltd from Federal Government grants to the NCRIS/CRIS/EIF programmes, the Victorian State Government, and the University of Melbourne. The ARC and University of Adelaide have supported this research by funding MS (DE140100575). The authors wish to thank Dr Rosemary Manhire-Heath for technical assistance, Dr Simon Crawford for microscopy, and Mrs Terra Stark (Metabolomics Australia) for help with root sampling and GC-MS analysis. We also thank Professor Steve Tyerman for critical reading of the manuscript.

REFERENCES

1. Barros J Serk H Granlund I Pesquet E . 2015. The cell biology of lignification in higher plants. Annals of Botany 115, 1053–1074.
2. Benfey PN Bennett M Schiefelbein J . 2010. Getting to the root of plant biology: impact of the Arabidopsis genome sequence on root research. Plant Journal 61, 992–1000.
3. Bengough AG Gordon DC Al-Menaie H Ellis RP Allan D Keith R Thomas WTB Forster BP . 2004. Gel observation chamber for rapid screening of root traits in cereal seedlings. Plant and Soil 262, 63–70.
4. Bernstein N Kafkafi U . 2002. Root growth under salinity stress. In: Waisel Y Eshel A Kafkafi U , eds. Plant roots: the hidden half . New York: Marcel Dekker, Inc., 787–805.
5. Bowne JB Erwin TA Juttner J Schnurbusch T Langridge P Bacic A Roessner U . 2012. Drought responses of leaf tissues from wheat cultivars of differing drought tolerance at the metabolite level. Molecular Plant 5, 418–429.
6. Chalbi N Hessini K Gandour M Mohamed Salma N Smaoui A Abdelly C Youssef NB . 2013. Are changes in membrane lipids and fatty acid composition related to salt-stress resistance in wild and cultivated barley? Journal of Plant Nutrition and Soil Science 176, 138–147.
7. Chatterjee J Majumder AL . 2010. Salt-induced abnormalities on root tip mitotic cells of Allium cepa: prevention by inositol pretreatment. Protoplasma 245, 165–172.
8. Chen Z Cuin TA Zhou M Twomey A Naidu BP Shabala S . 2007. Compatible solute accumulation and stress-mitigating effects in barley genotypes contrasting in their salt tolerance. Journal of Experimental Botany 58, 4245–4255.
9. Choi EY Kolesik P McNeill A Collins H Zhang Q Huynh BL Graham R Stangoulis J . 2007. The mechanism of boron tolerance for maintenance of root growth in barley (Hordeum vulgare L.). Plant Cell and Environment 30, 984–993.
10. Cosgrove DJ . 1999. Enzymes and other agents that enhance cell wall extensibility. Annual Review of Plant Physiology and Plant Molecular Biology 50, 391–417.
11. Cramer GR Ergul A Grimplet J Tillett RL Tattersall EAR Bohlman MC Vincent D Sonderegger J Evans J Osborne C Quilici D Schlauch KA Schooley DA Cushman JC . 2007. Water and salinity stress in grapevines: early and late changes in transcript and metabolite profiles. Functional & Integrative Genomics 7, 111–134.
12. Fait A Fromm H Walter D Galili G Fernie AR . 2008. Highway or byway: the metabolic role of the GABA

shunt in plants. Trends in Plant Science 13, 14–19.

13. Fan L Linker R Gepstein S Tanimoto E Yamamoto R Neumann PM . 2006. Progressive inhibition by water deficit of cell wall extensibility and growth along the elongation zone of maize roots is related to increased lignin metabolism and progressive stelar accumulation of wall phenolics. Plant Physiology 140, 603–612.
14. FAO. 2015. FAO Land and Plant Nutrition Management Service . Rome: Food and Agriculture Organization of the United Nations.
15. Flowers TJ Hajibagheri MA . 2001. Salinity tolerance in Hordeum vulgare: ion concentrations in root cells of cultivars differing in salt tolerance. Plant and Soil 231, 1–9.
16. Franke R Briesen I Wojciechowski T Faust A Yephremov A Nawrath C Schreiber L . 2005. Apoplastic polyesters in Arabidopsis surface tissues – A typical suberin and a particular cutin. Phytochemistry 66, 2643–2658.
17. Friedman Ra Altman A Levin N . 1989. The effect of salt stress on polyamine biosynthesis and content in mung bean plants and in halophytes. Physiologia Plantarum 76, 295–302.
18.
19. Gabriela-Anca Maier C Post-Beittenmiller D . 1998. Epicuticular wax on leek in vitro developmental stages and seedlings under varied growth conditions. Plant Science 134, 53–67.
20. Garthwaite AJ von Bothmer R Colmer TD . 2005. Salt tolerance in wild Hordeum species is associated with restricted entry of Na+ and Cl− into the shoots. Journal of Experimental Botany 56, 2365–2378.
21. Genc Y McDonald GK Tester M . 2007. Reassessment of tissue Na+ concentration as a criterion for salinity tolerance in bread wheat. Plant Cell and Environment 30, 1486–1498.
22. Gill SS Tuteja N . 2010. Polyamines and abiotic stress tolerance in plants. Plant Signaling & Behavior 5, 26–33.
23. Green PB . 1976. Growth and cell pattern formation on an axis: Critique of concepts, terminology, and modes of study. Botanical Gazette 137, 187–202.
24. Hajibagheri MA Yeo AR Flowers TJ . 1985. Salt tolerance in Suaeda maritima (L.) dum. Fine structure and ion concentrations in the apical region of roots. New Phytologist 99, 331–343.
25. Hellebust JA Forward DF . 1962. The invertase of the corn radicle and its activity in successive stages of growth. Canadian Journal of Botany 40, 113–126.
26. Herrmann KM . 1995. The shikimate pathway - early steps in the biosynthesis of aromatic-compounds. Plant Cell 7, 907–919.
27. Herrmann KM Weaver LM . 1999. The shikimate pathway. Annual Review of Plant Physiology and Plant Molecular Biology 50, 473–503.
28. Hill CB Taylor JD Edwards J Mather D Bacic A Langridge P Roessner U . 2013. Whole-genome mapping of agronomic and metabolic traits to identify novel quantitative trait loci in bread wheat grown in a water-limited environment. Plant Physiology 162, 1266–1281.
29. Imamul Huq SM Larher F . 1983. Osmoregulation in higher plants: effects of NaCl salinity on non-nodulated Phaseolus aureus l. I. Growth and mineral content. New Phytologist 93, 203–208.
30. Jacobs A Lunde C Bacic A Tester M Roessner U . 2007. The impact of constitutive heterologous expression of a moss Na+ transporter on the metabolomes of rice and barley. Metabolomics 3, 307–317.
31. James RA von Caemmerer S Condon AGT Zwart AB Munns R . 2008. Genetic variation in tolerance to the osmotic stress component of salinity stress in durum wheat. Functional Plant Biology 35, 111–123.
32. Katiyar S Dubey RS . 1990. Changes in polyamine titer in rice seedlings following NaCl salinity stress. Journal of Agronomy and Crop Science 165, 19–27.
33. Kinnersley AM Turano FJ . 2000. Gamma aminobutyric acid (GABA) and plant responses to stress. Critical Reviews in Plant Sciences 19, 479–509.
34. Lea PJ Sodek L Parry MAJ Shewry R Halford NG . 2007. Asparagine in plants. Annals of Applied Biology 150, 1–26.
35. Lefèvre I Gratia E Lutts S . 2001. Discrimination between the ionic and osmotic components of salt stress in relation to free polyamine level in rice (Oryza sativa). Plant Science 161, 943–952.
36. Li H Yan S Zhao L Tan J Zhang Q Gao F Wang P Hou H Li L . 2014. Histone acetylation associated up-regulation of the cell wall related genes is involved in salt stress induced maize root swelling. BMC Plant Biology 14, 15.
37. Lin CC Kao CH . 1996. Proline accumulation is associated with inhibition of rice seedling root growth caused by NaCl. Plant Science 114, 121–128.
38. Liu J Zhu JK . 1997. Proline accumulation and salt-stress-induced gene expression in a salt-hypersensitive mutant of Arabidopsis. Plant Physiology 114, 591–596.
39. Maeda H Dudareva N . 2012. The shikimate pathway and aromatic amino acid biosynthesis in plants. Annual Review of Plant Biology . 63, 73–105.
40. Mansour MMF . 2000. Nitrogen containing compounds and adaptation of plants to salinity stress. Biologia Plantarum 43, 491–500.
41. Mattioli R Falasca G Sabatini S Altamura MM Costantino P Trovato M . 2009. The proline biosynthetic genes P5CS1 and P5CS2 play overlapping roles in Arabidopsis flower transition but not in embryo development. Physiologia Plantarum 137, 72–85.
42. Moussaieff A Rogachev I Brodsky L Malitsky S Toal TW Belcher H Yativ M Brady SM Benfey PN Aharoni A . 2013. High-resolution metabolic mapping of cell types in plant roots. Proceedings of the National Academy of Sciences 110, E1232–E1241.
43. Muller B Stosser M Tardieu F . 1998. Spatial distributions of tissue expansion and cell division rates are related to irradiance and to sugar content in the growing zone of maize roots. Plant, Cell and Environment 21, 149–158.
44. Munns R . 2002. Comparative physiology of salt and water stress. Plant Cell and Environment 25, 239–250.
45. Munns R Tester M . 2008. Mechanisms of salinity

tolerance. Annual Review of Plant Biology 59, 651–681.

46. Obata T Fernie AR . 2012. The use of metabolomics to dissect plant responses to abiotic stresses. Cellular and Molecular Life Sciences 69, 3225–3243.

47. Pritchard J Jones RGW Tomos AD . 1991. Turgor, growth and rheological gradients of wheat roots following osmotic-stress. Journal of Experimental Botany 42, 1043–1049.

48. Rabe E . 1990. Stress physiology - the functional-significance of the accumulation of nitrogen-containing compounds. Journal of Horticultural Science 65, 231–243.

49. Rahnama A Munns R Poustini K Watt M . 2011. A screening method to identify genetic variation in root growth response to a salinity gradient. Journal of Experimental Botany 62, 69–77.

50. Ramesh SA Tyerman SD Xu B Bose J Kaur S Conn V Domingos P Ullah S Wege S Shabala S Feijo JA Ryan PR Gillham M . 2015. GABA signalling modulates plant growth by directly regulating the activity of plant-specific anion transporters. Nature Communications 6, 7879.

51. Raymond MJ Smirnoff N . 2002. Proline metabolism and transport in maize seedlings at low water potential. Annals of Botany 89, 813–823.

52. Renault H El Amrani A Palanivelu R Updegraff EP Yu A Renou J-P Preuss D Bouchereau A Deleu C . 2011. GABA accumulation causes cell elongation defects and a decrease in expression of genes encoding secreted and cell wall-related proteins in Arabidopsis thaliana . Plant and Cell Physiology 52, 894–908.

53. Richards RA Watt M Rebetzke GJ . 2007. Physiological traits and cereal germplasm for sustainable agricultural systems. Euphytica 154, 409–425.

54. Roessner U Luedemann A Brust D Fiehn O Linke T Willmitzer L Fernie AR . 2001. Metabolic profiling allows comprehensive phenotyping of genetically or environmentally modified plant systems. The Plant Cell 13, 11–29.

55. Roessner U Patterson JH Forbes MG Fincher GB Langridge P Bacic A . 2006. An investigation of boron toxicity in barley using metabolomics. Plant Physiology 142, 1087–1101.

56. Roy SJ Negrão S Tester M . 2014. Salt resistant crop plants. Current Opinion in Biotechnology 26, 115–124.

57. Shabala S Munns R . 2012. Salinity stress: physiological constraints and adaptive mechanisms. In: Shabala S , ed. Plant Stress Physiology vol. 1 . Wallingford: CABI, 59–93.

58. Sharp RE Hsiao TC Silk WK . 1990. Growth of the maize primary root at low water potentials. 2. Role of growth and deposition of hexose and potassium in osmotic adjustment. Plant Physiology 93, 1337–1346.

59. Shavrukov Y Gupta NK Miyazaki J Baho MN Chalmers KJ Tester M Langridge P Collins NC . 2010. HvNax3-a locus controlling shoot sodium exclusion derived from wild barley (Hordeum vulgare ssp spontaneum). Functional and Integrative Genomics 10, 277–291.

60. Shelden MC Roessner U . 2013. Advances in functional genomics for investigating salinity stress tolerance mechanisms in cereals. Frontiers in Plant Science 4, 123.

61. Shelden MC Roessner U Sharp RE Tester M Bacic A . 2013. Genetic variation in the root growth response of barley genotypes to salinity stress. Functional Plant Biology 40, 516–530.

62. Spollen WG Sharp RE Wu Y Saab IN . 1993. Regulation of cell expansion in roots and shoots at low water potentials. In: Smith JAC Griffiths H , eds. Water Deficits: Plant Responses from Cell to Community . Oxford: Bios Scientific Publishers, 37–52.

63. Tabur S Demir K . 2009. Cytogenetic response of 24-epibrassinolide on the root meristem cells of barley seeds under salinity. Plant Growth Regulation 58, 119–123.

64. Tabur S Demir K . 2010. Protective roles of exogenous polyamines on chromosomal aberrations in Hordeum vulgare exposed to salinity. Biologia 65, 947–953.

65. Ueda A Yamamoto-Yamane Y Takabe T . 2007. Salt stress enhances proline utilization in the apical region of barley roots. Biochemical and Biophysical Research Communications 355, 61–66.

66. Urano K Kurihara Y Seki M Shinozaki K . 2010. 'Omics' analyses of regulatory networks in plant abiotic stress responses. Current Opinion in Plant Biology 13, 132–138.

67. Van't Hof J . 1968. Control of cell progression through the mitotic cycle by carbohydrate provision. I. Regulation of cell division in excised plant tissue. Journal of Cell Biology 37, 773–780.

68. Verslues PE Sharp RE . 1999. Proline accumulation in maize (Zea mays L.) primary roots at low water potentials. II. Metabolic source of increased proline deposition in the elongation zone. Plant Physiology 119, 1349–1360.

69. Voetberg GS Sharp RE . 1991. Growth of the maize primary root at low water potentials. 3. Role of increased proline deposition in osmotic adjustment. Plant Physiology 96, 1125–1130.

70. Wang D Bodovitz S . 2010. Single cell analysis: the new frontier in 'omics'. Trends in Biotechnology 28, 281–290.

71. Wasson AP Rebetzke GJ Kirkegaard JA Christopher J Richards RA Watt M . 2014. Soil coring at multiple field environments can directly quantify variation in deep root traits to select wheat genotypes for breeding. Journal of Experimental Botany 65, 6231–6249.

72. Watson BS Bedair M Urbanczyk-Wochniak E Huhman DV Yang DS Allen SN Li W Tang Y Sumner LW . 2015. Integrated metabolomics and transcriptomics reveal enhanced specialized metabolism in Medicago truncatula root border cells. Plant Physiology 167, 1699–1716.

73. Widodo Patterson JH Newbigin E Tester M Bacic A Roessner U . 2009. Metabolic responses to salt stress of barley (Hordeum vulgare L.) cultivars, Sahara and Clipper, which differ in salinity tolerance. Journal of Experimental Botany 60, 4089–4103.

74. Witzel K Weidner A Surabhi GK Varshney RK Kunze G Buck-Sorlin GH Borner A Mock HP . 2010. Comparative analysis of the grain proteome fraction in barley genotypes with contrasting salinity tolerance during germination. Plant Cell and Environment 33, 211–222.
75. Wu D Cai S Chen M Ye L Chen Z Zhang H Dai F Wu F Zhang G . 2013. Tissue metabolic responses to salt stress in wild and cultivated barley. PloS One 8, e55431.
76. Wu J Seliskar DM Gallagher JL . 1998. Stress tolerance in the marsh plant Spartina patens: impact of NaCl on growth and root plasma membrane lipid composition. Physiologia Plantarum 102, 307–317.
77. Wu SJ Ding L Zhu JK . 1996. SOS1, a genetic locus essential for salt tolerance and potassium acquisition. Plant Cell 8, 617–627.
78. Yamaguchi M Sharp RE . 2010. Complexity and coordination of root growth at low water potentials: recent advances from transcriptomic and proteomic analyses. Plant Cell and Environment 33, 590–603.
79. Yamaya T Matsumoto H . 1989. Accumulation of asparagine in sodium chloride stressed barley seedlings. Berichte des Ohara Instituts fuer Landwirtschaftliche Biologie Okayama Universitaet 19, 181–188.
80. Zhang J Zhang Y Du Y Chen S Tang H . 2011. Dynamic metabonomic responses of tobacco (Nicotiana tabacum) plants to salt stress. Journal of Proteome Research 10, 1904–1914.
81. Zhang SJ Song XQ Yu BS Zhang BC Sun CQ Knox JP Zhou YH . 2012. Identification of quantitative trait loci affecting hemicellulose characteristics based on cell wall composition in a wild and cultivated rice species. Molecular Plant 5, 162–175.
82. Zhao X Wang W Zhang F Deng J Li Z Fu B . 2014. Comparative metabolite profiling of two rice genotypes with contrasting salt stress tolerance at the seedling stage. PloS One 9, e108020.
83. Zhong HL Lauchli A . 1993. Spatial and temporal aspects of growth in the primary root of cotton seedlings - effects of NaCl and CaCl2 . Journal of Experimental Botany 44, 763–771.

CHAPTER 12

TRANSCRIPTIONAL ACTIVITY OF RRNA GENES IN BARLEY CELLS AFTER MUTAGENIC TREATMENT

Jolanta Kwasniewska, Joanna Jaskowiak

INTRODUCTION

Plant bioassays are commonly used to detect chromosome rearrangements induced by environmental, possibly mutagenic, factors. The application of modern cytogenetic techniques such as fluorescent *in situ* hybridization (FISH) permits more detailed analyses of chromosome aberrations and micronuclei. A cytogenetic study of physical location of 5S rRNA genes and genes encoding for 18S, 5.8S and 25S rRNAs (35S rRNA), in a few *Hordeum*species proved that rDNA provides useful chromosome landmarks [1]. A micronucleus (MN) test combined with FISH allowed a quantitative analysis of the involvement of specific chromosome fragments in micronuclei formation and thus permitted the possible origin of mutagen-induced micronuclei to be explained. 35S rDNA are genes that belong to the category of "housekeeping genes", which are present in every living cell and they have to have at least basal activity. In a previous study an attempt was made to identify the origin of micronuclei using FISH with rDNA as probes. A quantitative analysis of the involvement of specific chromosome fragments in micronuclei in barley cells was done using FISH. The characterization of micronuclei by identifying the rRNA genes involved in their formation in barley cells showed that rDNA-bearing chromosomes are often involved in micronuclei formation [2,3]. The commonly used histochemical silver staining, which is used to detect active ribosomal genes, can be useful in the analysis of the activity of the 35S rRNA genes in the micronuclei. Although the reuse of slides to localize rRNA genes and analyze the transcriptional activity of these genes has found wide application in plant cytogenetics [4,5], it has not yet been used in mutagenesis.

In this study, analyses of localization of 35S rRNA using FISH and the transcriptional activity of 35S rRNA genes with silver staining, were performed simultaneously. It permitted a determination whether the presence of rRNA genes in micronuclei is coupled with their activity. Maleic hydrazide (MH), which is a commonly known clastogenic agent that can lead to chromosome breaks as well as cause spindle fiber defects [2], was used as the model mutagen. It is worth investigating the activity of rRNA genes in micronuclei because they are eliminated during the next cell cycle. Additionally, an analysis of the influence of MH on the number of nucleoli and nuclear bodies in barley cells was applied.

MATERIAL AND METHODS

Material and Treatment

Barley (*Hordeum vulgare*, 2n = 14) seeds of the "Start" variety were presoaked in distilled water for 8 h and then treated with maleic hydrazide – 3 mM or 4 mM (MH; Sigma, CAS 123-3301) for 3 h. After treatment the seeds were washed three times in distilled water and then germinated in Petri dishes on wet filter paper at 21°C. The material was fixed in ethanol: glacial acetic acid (3:1) 72 h after treatment. The roots of M_1 seedlings were used as the source of the meristems for the analyses.

Chromosomal Preparations

For chromosome preparations the material was washed with a 0.01 mM sodium citrate buffer (pH 4.8) for 30 min and digested with an enzyme mixture containing 2% cellulase (w/v, Onozuka, Serva) and 20% pectinase (v/v, Sigma) for 2 h at 37°C. After digestion the material was washed again with a sodium citrate buffer for 30 min. Squash preparations

Citation: Kwasniewska J, Jaskowiak J (2016) Transcriptional Activity of rRNA Genes in Barley Cells after Mutagenic Treatment. PLoS ONE 11(6): e0156865. doi:10.1371/journal.pone.0156865.

were made in a drop of 45% acetic acid. After freezing and removing coverslips, the slides were dried.

Silver Staining

Silver staining was done using the modified method of Hizume et al. [6]. Prior to staining slides were treated with a borate buffer (pH 9.2; Merck) for 30 min. Then, a few drops of freshly prepared 50% $AgNO_3$ (Merck) in distilled water were applied to each preparation. Slides were covered with a nylon mesh and incubated in a humid chamber at 42°C for 70 min, washed in distilled water and air-dried. The photographs were taken with a Zeiss Axio Imager.Z.2 wide-field microscope equipped with an AxioCam Mrm monochromatic camera. The images were captured and processed using Adobe Photoshop 4.0.

After analysis of the silver-stained preparations, the immersion oil was removed from the preparations with 100% acetone (Poch). The slides were rinsed briefly in 4xSSC (0.3 M NaCl, 0.03 M sodium citrate) at 37°C to remove the coverslips and then for 5 min in 4xSSC at 37°C. To remove the silver staining, the preparations were placed in 30% hydrogen peroxide (Poch) for 20–30 s, rinsed 3 x 5 min in distilled water and air-dried. The slides were fixed in ethanol: glacial acetic acid (3:1) for 10 min and then 100% ethanol for 5 min. The slides were air-dried and then used for *in situ* hybridization.

Fluorescence in situ Hybridization (FISH)

Fluorescence *in situ* hybridization was applied according to the method described by Maluszynska and Heslop-Harisson [7] with some minor modifications. 25S rDNA, which had been isolated from *Arabidopsis thaliana* [8] and labeled with digoxygenin-11-dUTP using nick translation (Roche), was used as the probe.

Prior to FISH, the chromosome preparations were pretreated with RNase (1 mg/ml) for 60 min at 37°C and washed 3 times for 5 min in 2 x SSC at room temperature, dehydrated in methanol series and air-dried.

The hybridization mixture containing 2.5 µg/ ml of labeled DNA, 50% (v/v) formamide, 10% (w/v) dextran sulphate and 0.1 mg/µl salmon testes DNA in 2 x SSC was denaturated at 75°C for 10 min and immediately placed on ice for a few min. The hybridization mixture (38 µl) was added to the chromosome preparations and covered with a plastic coverslip. The chromosomes and DNA probes were denatured for 5 min at 70°C on a hot plate (Hybaid Thermal Cycler PCR). Hybridization was carried out in a moist chamber at 37°C for 20 h. After hybridization, the slides were washed for 4 min in 2 x SSC at 42°C, 2 x 4 min in 0.1 SSC at 42°C, 3 x 3 min in 2 x SSC at 42°C, 3 x 3 min in 2 x SSC at room temperature and in 0.2% Tween in 4 x SSC at room temperature for 5 min.

The digoxigenin-labeled probe was detected using FITC-conjugated anti-digoxigenin antibodies (Roche) and then the signal was amplified using a FITC-conjugated secondary antibody (FITC-conjugated Anti-Sheep; Dako). After three washes for 8 min in 0.2% Tween in 4 x SSC at 37°C and dehydration in an ethanol series, the slides were mounted in a Vectashield medium (Vector) containing 6 µg/ ml DAPI.

Preparations were examined with a Zeiss Axio Imager.Z.2 wide-field fluorescence microscope equipped with an AxioCam Mrm monochromatic camera. The images were captured and processed using Adobe Photoshop 4.0. The frequency of micronuclei with specific DNA signals and without signals was calculated. For each experimental group, 50 cells with micronuclei on three slides, each made from three meristems, were evaluated. The results of the analyses were pooled for all of the concentrations of MH. The total frequency of micronuclei was estimated on the same slides – 2000 cells in the control and the treated material were analyzed. The data on frequencies of nuclei were analyzed using a nonparametric Kruskal-Wallis one-way analysis of variance by rank test. If a significant K-value of $P<0.05$ was obtained, a Mann-Whitney U test was performed.

RESULTS AND DISCUSSION

In the present study, the activity of the 35S rRNA genes in MH-treated *H. vulgare* (2n = 14) cells was evaluated using the silver-staining method and fluorescence *in situ* hybridization. *H. vulgare* is a very good model for such investigations due to its convenient karyotype features—large chromosomes, two of which display the presence of 35S rRNA genes, together with activity of all of them are convenient karyotype features that made barley a suitable experimental model for this study (Fig 1). The transcriptional activities of the all of the 35S rRNA gene loci in the barley cells permitted an analysis of their activities in the micronuclei, which are eliminated from the cells in the next cell cycle. The use of the silver-staining method permitted an analysis of the activity of 35S rRNA genes in micronuclei that were induced by MH-treatment (Fig 2). The presence of the signals from histochemical staining in a micronucleus indicates that this micronucleus originated from the 35S rDNA region or the entire chromosome/chromosome arm and this confirmed the activity of these genes. The lack of signals in micronuclei could indicate the absence of 35S rRNA genes or their activity. The presence or lack of silver-staining signals in micronuclei was not correlated with the number of nucleoli or the size of micronuclei (Fig 2). The number of nucleoli in the *H. vulgare* Start var. control cell was one to four and correlated with two pairs of NOR chromosomes, which were transcriptionally active. However, the observed number of nucleoli is generally lower than the number of active rDNA loci due to their fusion during the cell cycle [9]. Treatment with MH influenced the number of nucleoli in the barley cells (Fig 3). The frequency of nuclei with one nucleolus increased significantly after treatment with MH–from 42.3% in the control to about 57% in the MH-treated cells. At the same time, the frequencies of the nuclei with two or three nucleoli were significantly lower after MH treatment in comparison with the control. The results of this study clearly indicated that MH could influence the fusion of the nucleoli in barley cells. This process occurred before the G1 of the cell cycle [10]. MH is responsible for disturbances in the biosynthesis of nucleic acids and proteins [11]. The nucleolus is very sensitive to stressful factors. The frequency of cells with one nucleus increases in response to heavy metals or changes of temperature [9]. In this study we have shown the influence of the MH on the number of nucleoli in plant cells for the first time. An increased size of the nucleoli was observed in *A. cepa* after treatment with another inhibitor of protein biosynthesis in eukaryotic cells by cycloheximide [12].

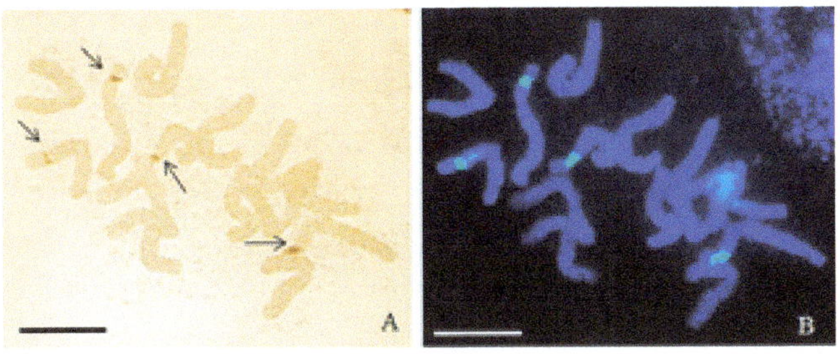

Figure 1: The metaphase chromosomes of *H. vulgare* **Start var. (2n = 14).**

(A) Silver-staining method showing the active loci of 35S rDNA (B) fluorescence *in situ*hybridization showing the distribution of all 35S rRNA genes (right). Bars represent 10 μm

http://dx.doi.org/10.1371/journal.pone.0156865.g001

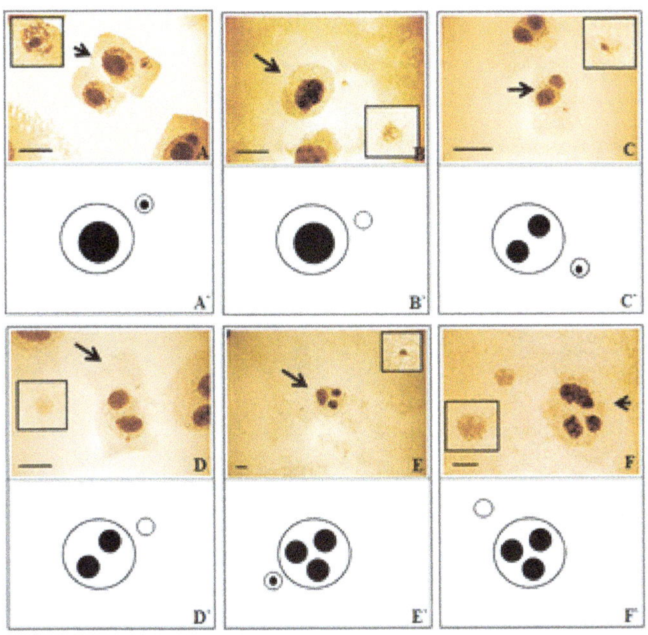

Figure 2: *H. vulgare* **root** meristematic cells with micronuclei after treatment with MH–silver-staining method.

(A-F) original images, arrows indicate cells chosen for drawings. Images of micronuclei are enlarged. (A`-F`) model drawings based on orginal images. Bars represent 10 μm.

http://dx.doi.org/10.1371/journal.pone.0156865.g002

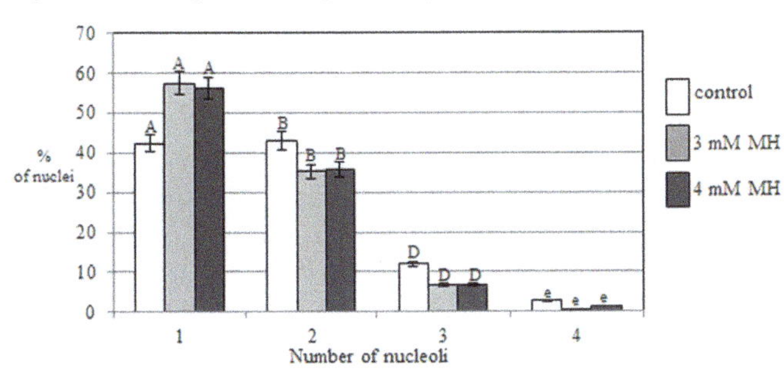

Figure 3: The frequencies of *H. vulgare* **nuclei with a different number of nucleoli (from 1 to 4) in root** meristematic cells in the control and after MH treatment.

The bars represent standard errors (SE). The capital letters A, B, D indicate that the frequencies of nuclei with 1, 2 or 3 nucleoli are significantly different (P<0.05) between the each of the experimental group: control, 3 mM MH and 4 mM MH. The lower case letter e indicates that the frequencies of nuclei with four nucleoli are not significantly different (P<0.05) between the control, 3 mM MH and 4 mM MH.

http://dx.doi.org/10.1371/journal.pone.0156865.g003

An analysis of the frequencies of the micronuclei with silver-staining signals was then conducted. The frequencies of micronuclei with signals that indicated the presence of active 35S rRNA genes were dependent on the MH concentration–38.2% after treatment with 3 mM MH and 29.5% after treatment with 4 mM MH (Fig 4). FISH with 25S rDNA as probe was used in the next experiments because the use of silver staining did not show the proportion of micronuclei with active 35S rRNA genes to the micronuclei that had these genes. Before the analysis of the micronuclei, an evaluation of the number of 35S rDNA signals using FISH together with an analysis of the number of nucleoli in the control cells without micronuclei were applied (Fig 5). Generally, four foci of 35S rDNA were observed in the nuclei (Fig 5A–5D). The number of 35S rDNA signals did not correlate with the number of nucleoli. Sporadically, the number of 35S rDNA signals in the nuclei was five (Fig 5E). The presence of five signals of 35S rDNA can indicate the probability of a duplication event including this region or of a break inside it. The sequential use of the silver-staining method and then fluorescence *in situ*hybridization with 25S rDNA as probe to the MH-treated root meristematic cells of barley with micronuclei permitted the simultaneous analysis of the presence of 35S rRNA genes and their activity (Fig 6). The number of 35S rDNA signals in the nuclei from the cells with micronuclei ranged from two to five. Nuclei with three signals of 35S rDNA and one signal in the micronucleus (Fig 6A) or nuclei with four signals but without any signals in the micronucleus (Fig 6B) were observed most often. Sporadically, nuclei with four signals of 35S rDNA and one signal in the micronucleus (Fig 6C) and nuclei with five signals and the absence of a signal in the micronucleus (Fig 6D) were also observed. However, cells with two FISH signals in the nucleus and two signals in the micronucleus were rarely observed (Fig 6E). Summarizing, all of the 35S rDNA foci that were present in the micronuclei were transcriptionally active. All of the 35S rDNA loci are active in *H. vulgare* Start var. cells, and therefore this result is predictable. However, it was interesting to determine whether this activity is maintained taking into account the fact that micronuclei are eliminated during the next cell cycle.

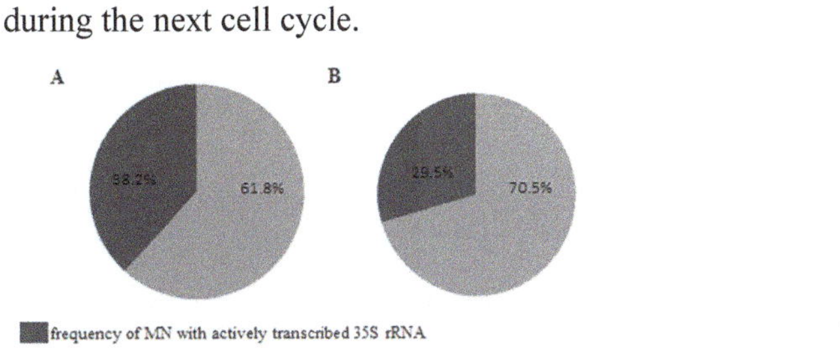

Figure 4: Frequencies of micronuclei with silver-staining signals (indicating the presence of active rRNA genes) or without silver-staining signals (indicating a complete lack of rRNA genes or a lack of active rRNA genes).

(A) 3 mM MH, (B) 4 mM MH

http://dx.doi.org/10.1371/journal.pone.0156865.g004

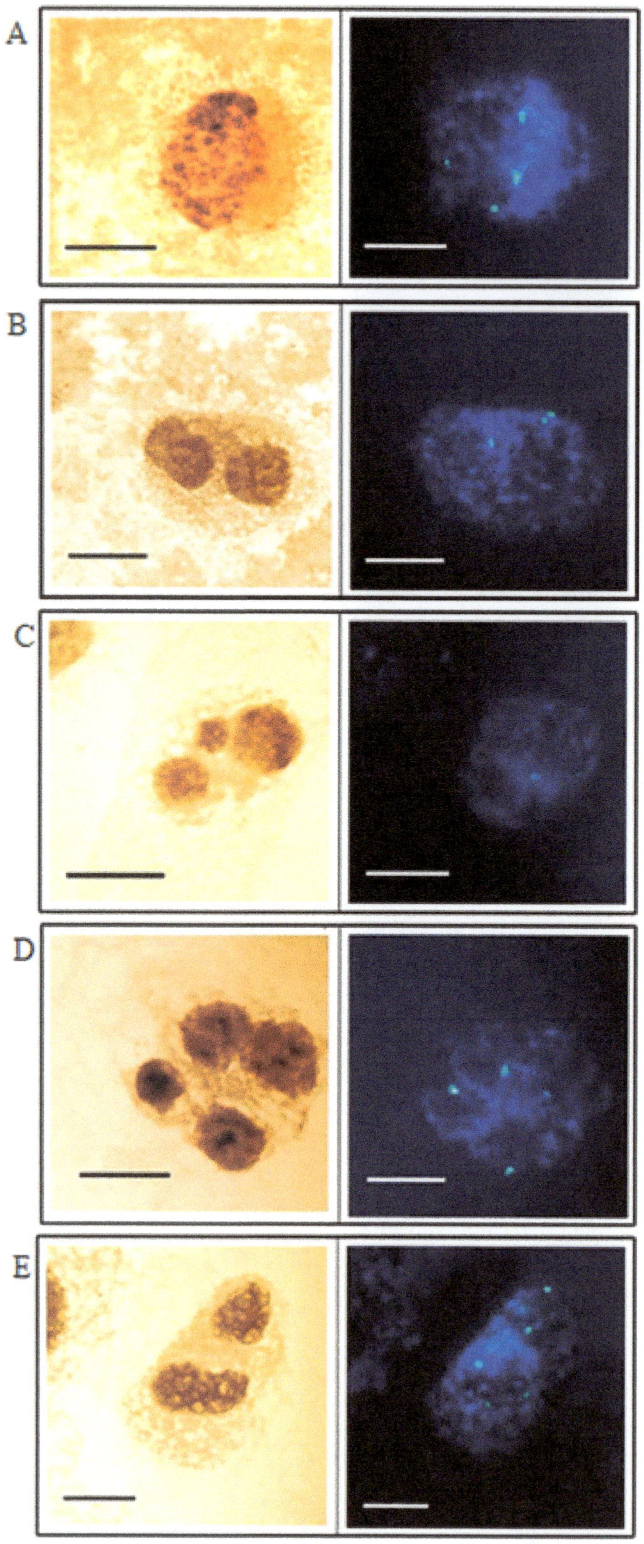

Figure 5: Untreated *H. vulgare* **interphase cells without micronuclei.**

Staining with the silver-staining method (left columns) and fluorescence *in situ*hybridization with 25S rDNA as the probe (right columns) applied to the same cells. Bars represent 10 μm

http://dx.doi.org/10.1371/journal.pone.0156865.g005

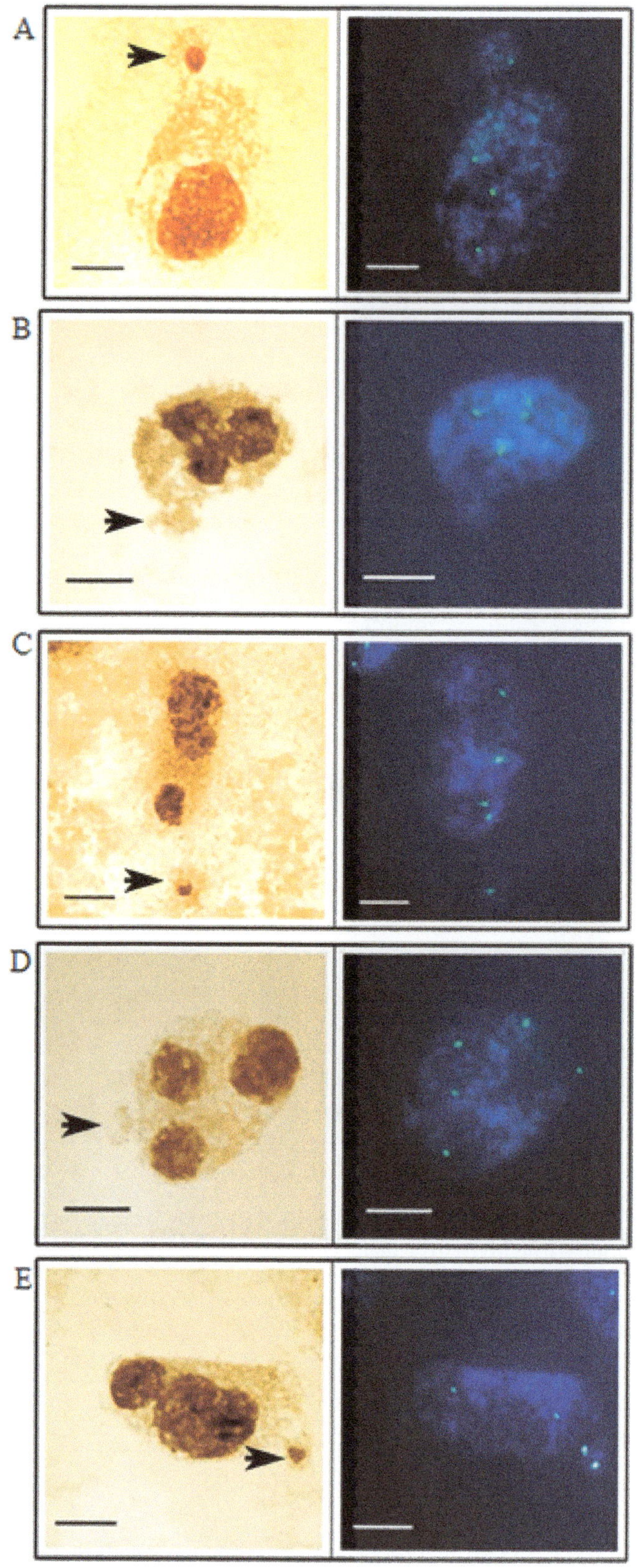

Fig 6 ***H. vulgare*** **interphase cells with micronuclei after treatment with MH.**

Staining with the silver staining-method (left columns) and fluorescence *in situ*hybridization with 25S rDNA as the probe (right columns) applied to the same cells. Bars represent 10 μm.

http://dx.doi.org/10.1371/journal.pone.0156865.g006

An analysis of the silver-stained cells treated with MH showed the presence of nuclear bodies. They were small, positively labeled structures (Fig 7). The size of the nuclear bodies was much smaller than the micronuclei. The number of nuclear bodies in the barley cells varied and ranged from one to three. Neither the presence nor the number of nuclear bodies correlated with the number of nucleoli (data not shown). Nuclear bodies were sporadically observed in the control cells (0.7%). Their frequency significantly increased after MH treatment and correlated with the concentration of the mutagen. Only 2.8% of cells that were treated with 3 mM MH and 10.1% of cells treated with 4 mM MH had nuclear bodies (Fig 8). Nuclear bodies have not been previously observed in any of the dividing or slowly dividing root cells of non-treated cells of *A.thaliana*, *B. napus*, *P. sativum* or *Z. mays* [13, 14]. Aluminum and lead are known as factors that induce nuclear bodies [15–17]. Although nuclear bodies have been observed in species that belong to grasses, e.g. *Oryza sativa* and *Tritcum aestivum* [13,18], their occurrence in *H.vulgare* was observed for the first time. Their random localization in barley cells will not allow to classified them as dense bodies (DB), which are predominantly localized near nucleoli [18,19]. FISH together with traditional histochemical staining provides insight into the mechanisms of the formation of chromosome aberrations during genotoxicity studies. Such an approach has commonly been used in studies of structure and function of genomes, e.g. in*Allium cepa* [20] and *Brassica* species [21]. FISH and the analysis of the transcriptional activity of genes using traditional staining has not previously been applied in mutagenesis and environmental genotoxicity studies. Although analyses of the presence of rDNA loci in the micronuclei induced by mutagens has been previously performed [2,3], in this study their activity was analysed for the first time.

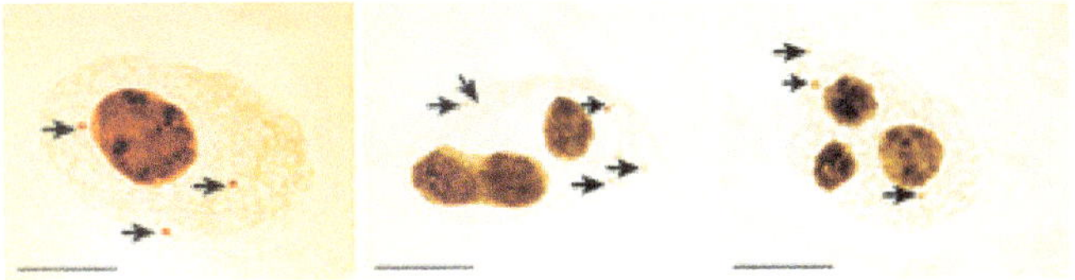

Fig 7: *H. vulgare* **nuclei with nuclear bodies after treatment with MH–silver-staining method.**

Bars represent 10 μm.

http://dx.doi.org/10.1371/journal.pone.0156865.g007

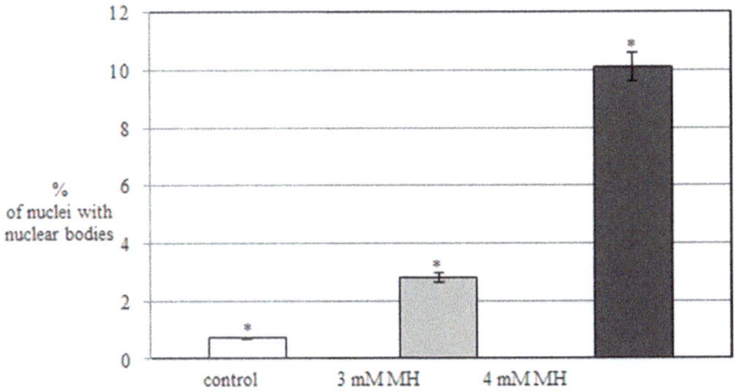

Figure 8: The frequencies of *H. vulgare* **nuclei with nuclear bodies in the control and MH-treated cells.**

The bars indicate standard errors (SE). An * indicates that the frequencies of nuclei with nuclear bodies are significantly different ($P<0.05$) between the experimental groups.

http://dx.doi.org/10.1371/journal.pone.0156865.g008

To conclude, this study is a first example of the application of the micronucleus test combined with analyses of the transcriptional activity of rRNA genes. Transcriptional activity is maintained in the micronuclei even taking into account the fact that they are eliminated during the next cell cycle. The analyses using this experimental model setup, including MH as a mutagen and barley as species, can be now extended to other abiotic stresses and plant species to study the response of plant cells to the environment.

AUTHOR CONTRIBUTIONS

Conceived and designed the experiments: JK JJ. Performed the experiments: JK JJ. Analyzed the data: JK JJ. Contributed reagents/materials/analysis tools: JK JJ. Wrote the paper: JK.

REFERENCES

1. Leitch IJ, Heslop-Harrison JS. Physical mapping of four sites of 5S rDNA sequences and one site of the alpha-amylase-2 gene in barley (*Hordeum vulgare*). Genome. 1993; 36: 517–523. pmid:18470006 doi: 10.1139/g93-071
2. Juchimiuk J, Hering B, Małuszyńska J. Multicolour FISH in an analysis of chromosome aberrations induced by N-nitroso-N-methylurea and maleic hydrazide in barley cells. J Appl Genet. 2007; 48(2): 99–106. pmid:17495342 doi: 10.1007/bf03194666
3. Juchimiuk-Kwasniewska J, Brodziak L, Małuszyńska J. FISH in analysis of gamma ray-induced micronuclei formation in barley. J Appl Genet. 2011; 52: 23–29. doi: 10.1007/s13353-010-0017-x. pmid:21136232
4. Hubbel HR. Silver staining as an indicator of active ribosomal genes. Biotech Histochem. 1985; 60(5): 285–294. doi: 10.3109/10520298509113926
5. Rogers SO, Bendich AJ. Ribosomal RNA genes in plants–variability in copy number and in the intergenic spacer. Plant Mol Biol. 1987; 9: 509–520. doi: 10.1007/BF00015882. pmid:24277137
6. Hizume M, Sato S and Tanaka A. A highly reproducible method of nucleolus organizer regions staining in plants. Stain Technology. 1980; 55: 87–90. pmid:6157230
7. Maluszynska J, Heslop-Harrison JS. Localization of tandemly repeated DNA sequences in *Arabidopsis thaliana*. Plant J. 1991; 1(2):159–166. doi: 10.1046/j.1365-313x.1991.00159.x
8. Unfriend I and Gruendler P. Nucleotide sequence of the 5.8S and 25S rRNA genes and of the internal transcribed spacers from *Arabidopsis thaliana*. Nucleic Acid Res. 1990; 18: 4011. pmid:2100998 doi: 10.1093/nar/18.13.4011
9. Gabara B, Krajewska M, Stecka E. Calcium effect on number, dimension and activity of nucleoli in cortex cells of pea (*Pisum sativum* L.) roots after treatment with heavy metals. Plant Sci. 1995; 111: 153–161. doi: 10.1016/0168-9452(95)04244-o
10. Dolezel J, Cihalikova J, Zakchlenjuk OV. Sequential estimation of nuclear DNA and silver staining of nucleoli in plant cells. Stain Technol. 1989; 64(1): 9–13. pmid:2472683 doi: 10.3109/10520298909108037

11. Jabee F, Ansari MYK, Shaban D. Studies on the effect of maleic hydrazide on root tip Cells and Pollen Fertility in *Trigonella foenum-graecum* L. Turk J Botany. 2008; 32: 337–344.
12. Guerrero F, De la Torre C, Garcia-Herdugo G. Control of nucleolar growth during interphase in higher plant meristem cells. Protoplasma. 1989; 152: 96–100. doi: 10.1007/bf01323067
13. Barlow PW. Nucleolus-associated bodies (karyosomes) in dividing and differentiating plant cells. Protoplasma. 1983a; 115: 1–10. doi: 10.1007/bf01293574
14. Małuszynska J, Hasterok R, Weiss H. rRNA genes–their distribution and activity in plants. W: Plant cytogenetics. Małuszynska J (Ed.) Wydawnictwo Uniwersytetu Śląskiego. Katowice. 1998. pp. 75–95.
15. Jiang W, Liu D. Effects of Pb^{2+} on Root Growth, Cell Division and Nucleolus of *Zea*mays L. Bull Environ Contam Toxicol. 2000; 65(6): 786–793. pmid:11080360 doi: 10.1007/s0012800191
16. Qin R, Jiao Y, Zhang S, Jiang W, Liu D. Effects of aluminum on nucleoli in root tip cells and selected physiological and biochemical characters in *Allium cepa* var. agrogarum L. Plant Biol. 2010; 10: 225.
17. Jiang Z, Zhang H, Qin R, Zou J, Wang J, Shi Q, Jiang W, Liu D. Effects of lead on the morphology and structure of the nucleolus in the root tip meristematic cells of *Allium cepa* L. Int J Mol Sci. 2014; 15: 13406–13423. doi: 10.3390/ijms150813406. pmid:25089875
18. Williams LM, Jordan EG, Barlow PW. The ultrastructure of nuclear bodies in interphase plant cell nuclei. Protoplasma. 1983; 118: 95–103. doi: 10.1007/bf01293065
19. Barlow PW. Changes in the frequency of two types of nuclear body during interphase of meristematic plant cells. Protoplasma. 1983b; 118: 104–113. doi: 10.1007/bf01293066
20. .Hasterok R, Małuszyńska J. Cytogenetic markers of *Brassica napus* L. chromosomes. J Appl Genet. 2000a; 41(1): 1–9.
21. Hasterok R, Małuszyńska J. Different rRNA genes expression in primary and adventitious roots of *Allium cepa* L. Folia Histochem Cytobiol. 2000b; 8(4): 181–184

CHAPTER 13

COMPARATIVE PROTEOMIC ANALYSIS OF TWO BARLEY CULTIVARS (HORDEUM VULGARE L.) WITH CONTRASTING GRAIN PROTEIN CONTENT

Baojian Guo, Haiye Luan, Shen Lin, Chao Lv, Xinzhong Zhang and Rugen Xu

Keywords: Barley, grain protein content, two-dimensional electrophoresis, differentially expressed protein, mass spectrometry

INTRODUCTION

Barley is the fourth largest cereal crop in the world, which is not only widely used for food and feed, but also used for malting and brewing. The former requires grain protein content (GPC) to be as high as possible, whereas malting barley requires GPC to be at the proper level (See et al., 2002;Clancy et al., 2003; Cai et al., 2013). In Europe, the acceptable protein content range for malting barley was 9.5–11.5% (Pettersson and Eckersten, 2007). Higher protein content levels resulted in excess extract yield, which could produce a beer with hazy appearance; while lower protein levels decreased enzyme activity (Weston et al., 1993; Eagles et al., 1995). In addition, a positive correlation has been observed between GPC and diastatic power (Wang et al., 2003; Cai et al., 2013).

Barley GPC revealed significant differences in different varieties. Cai et al. (2013) analyzed the GPC of 59 cultivated and 99 Tibetan wild barley accessions in 2008 and 2009. The results showed that the GPC ranged from 8.02 to 13.50% with a mean of 10.56% in 2008 and ranged from 8.28 to 14.45% with a mean of 10.87% in 2009. Tibetan wild barley was found to have a higher GPC than cultivated barley. Comparative GPC analysis performed in 10 two-rowed spring malting barley cultivars showed that Klages and Logan showed the highest (14.57%) and lowest (12.56%) GPC, respectively, (Qi et al., 2005). Remarkably, variation in the proportions of the individual B, C, and D hordeins was observed in barley cultivars (Howard et al., 1996; Qi et al., 2005). All the above results indicated that genetic background had significant effects on GPC.

Recently, Two-dimensional electrophoresis combined with tandem mass spectrometry (MS) have resulted in new insights into the molecular basis of grain filling and seed maturation in plants (Finnie et al., 2002; Ge et al., 2012; Guillon et al., 2012; Dong et al., 2014; Jiang et al., 2014). For example, two-dimensional gel electrophoresis was used for a time-resolved study of the changes in proteins that occur during seed development in barley with approximately 1,000 low-salt extractable protein spots detected on the two dimensional gels. Among which 19 protein spots were identified by using matrix-assisted laser-desorption ionization time of flight MS or nano-electrospray tandem MS/MS. Proteins were accumulated throughout grain filling and maturation stages, which is reported to be functional characteristics of barley cultivars (Finnie et al., 2002). Glutelin is a predominant compound in rice seed. Two-dimensional gel electrophoresis (2-DE) analysis revealed remarkable differences in protein profiles of the wild rice species and the two cultivated rice materials. A total of 35 different pattern of expression protein spots were found for glutelin acidic subunits, glutelin precursors and glutelin basic subunits in wild rice species. Among those, 18 protein spots were specific and 17 major spots were elevated (Jiang et al., 2014). Chinese bread wheat cultivars Jimai 20 and Zhoumai 16 with different quality properties were investigated by 2-DE and MALDI-TOF/TOF-MS. A total of 117 different pattern of expression protein spots representing 82 unique proteins, which included isocitrate dehydrogenase, triticin precursor, low-molecular-weight glutenin subunit, and replication factor C-like protein. Remarkably, Class II chitinase and peroxidase 1 with isoforms in developing grains

Citation: Guo B, Luan H, Lin S, Lv C, Zhang X and Xu R (2016) Comparative Proteomic Analysis of Two Barley Cultivars (Hordeum vulgare L.) with Contrasting Grain Protein Content. Front. Plant Sci. 7:542. doi: 10.3389/fpls.2016.00542.

were shown to be phosphorylated by Pro-Q Diamond staining and phosphorylated protein site prediction (Guo et al., 2012). Therefore, proteome analysis is a tool that can be used both to visualize and compare complex mixtures of proteins.

In the present study, we reported a comparative proteomics analysis between Yangsimai 3 (feed barley cultivar) and Naso Nijo (malting barley cultivar) by 2-DE and tandem MS. The main objectives were: (1) to obtain comparative information on proteins expression profiling between feed barley and malting barley; and (2) to identify potential candidate proteins that influence GPC and barley grain quality.

MATERIALS AND METHODS

Plant Material and Growth Conditions

Barley cultivars Yangsimai 3 and Naso Nijo were used in this study. Yangsimai 3 is a Chinese landrace of feed barley, two-rowed, with a high GPC. Naso Nijo is a Japanese two-rowed malting barley cultivar with a low GPC. Two barley cultivars were planted at the Yangzhou University Experimental Farm in the autumn of 2013. Twelve seeds of each cultivar were planted 10 cm apart with 20 cm between rows. The mature seeds were harvested from the middle region of the main spikelet, and then the seeds were frozen in liquid nitrogen and stored in -80°C for protein extraction. Three biological replicates were undertaken with 100 seeds per genotype being used for protein extraction and GPC measurement for each replication.

Grain Protein Content Measurement

The mature grains were dried to a constant weight at 80°C, and ground in a Cyclotec 1093 sample mill (Hoganas City, Sweden) and sieved through a 0.5 mm screen. The total nitrogen content in grain was quantified according to the Kjeldahl method by using FOSS Kjeltec ™ 2300 (Foss Analytical AB, Sweden; Kjeldahl, 1883). GPC was calculated by using the following formula: GPC = Nitrogen content × 5.83 × 100% (Sun et al., 2013). Statistical analysis of the differences in aerial part traits between cultivars was performed by using Student's *t*-test.

Protein Extraction

A 100 seeds of each genotype were pooled and milled to powder in liquid nitrogen. Approximately 0.1 g of flour was added into 1 mL of extraction buffer (2% SDS, 10% glycerol, 50 mM DTT, 40 mM Tris-HCl, pH 6.8) at 4°C (Hurkman and Tanaka, 2007). From this step onward, all manipulations were carried out at or below 4°C. The detailed method was as follows: the flour was extracted with stirring for 30 min and the insoluble material was removed by centrifugation at 20,000 *g* for 30 min (Fullerton City, CA, USA). The supernatant containing protein fractions were precipitated with four volumes of cold acetone containing 0.07% DTT. After 2 h incubation at -20°C, the extracts were centrifuged at 18,000 *g* for 30 min at 4°C and the supernatant was discarded. Protein pellets were resuspended with cold 80% acetone containing 0.07% DTT, incubated for 1 h at -20°C before centrifuging at 18 000 *g* for 15 min at 4°C (Kim et al., 2013). This step was repeated five times and the protein pellet was freeze-dried under vacuum. Protein pellets were solubilized and incubated in a protein buffer [7 M urea, 2 M thiourea, 2% CHAPS (powder to solution, w/v), 0.5% IPG buffer (v/v; pH 4-7 and 6-11; Fairfield City, OH, USA) and 36 mM DTT (5.6 mg/mL)] at room temperature for 1 h, vortexed every 10 min. The mixture was then centrifuged (20,000 *g*) for 15 min, and the supernatant was collected. Protein concentration was determined by Bradford assay (Bradford, 1976) with bovine serum albumin (BSA) used as a standard, and the R^2 of standard curve was 0.9974.

Two-dimensional Gel Electrophoresis and Image Analysis

Seed protein extract (200 μg) was loaded onto a GE Healthcare 18 cm IPG strip with a linear gradient of pH 4–7 and pH 6–11 during strip rehydration overnight. IEF was conducted using IPGPhorII (Fairfield City, OH, USA) at 20°C for a total of 65 kVh. Equilibration of the strips was performed immediately with 10 mL of two types of SDS equilibration buffer for 15 min each. Buffer 1 contained 1.5 M Tris-HCl (pH 8.8), 6 M urea, 30% glycerol, 2% SDS, and 1% DTT, and buffer 2 contained 1.5 M Tris-HCl (pH 8.8), 6 M urea, 30% glycerol, 2% SDS, and 2.5% iodoacetamide. The second dimension SDS-PAGE gels (12.5% linear gradient) were run on an Ettan DALTsix (Fairfield City, OH, USA), 0.5 h at 2.5 W per gel, then at 12 W per gel until the dye front reached the gel bottom. Upon electrophoresis, the protein spots were stained with silver nitrate according to the instructions of the PlusOne™ Silver Staining Kit for proteins (Fairfield City, OH, USA), which offered improved compatibility with subsequent mass spectrometric analysis. Briefly, gels were fixed in 40% ethanol and 10% acetic acid for 30 min, and then sensitized with 30% ethanol, 0.2% sodium thiosulfate (w/v), and 6.8% sodium acetate (w/v) for 30 min. Then gels were rinsed with distilled water three times; 5 min duration each time; then incubated in silver nitrate (2.5 g/L) for 20 min. Incubated gels were rinsed with distilled water and developed in a sodium carbonate solution (25 g/L) with formaldehyde (37%, w/v) added (300 μL/L) before use. Development was stopped with 1.46% EDTA-$Na_2 \cdot 2H_2O$ (w/v), and gels were stored in distilled water until they could be processed and reproducible spots were removed from them. Gel images were acquired using Labscan (Fairfield City, OH, USA). Image analysis was carried out with Imagemaster 2D Platinum Software Version 7.0 (Fairfield City, OH, USA). Three biological replicates of silver stained gels showed high reproducibility (>95%) by comparison using the Imagemaster 2D Platinum Software 7.0. Spot detection was performed automatically by the software used with the parameters smooth, minimum area, and saliency set to 2, 15, and 8, respectively, followed by manual spot editing, such as spot deletion, spot splitting, and merging. All the gels were matched to the reference gel in automated mode with Imagemaster 2D Platinum Software 7.0. The volume of each spot from three replicate gels was normalized and quantified against total spot volume in the Imagemaster 2D Platinum Software 7.0. Sequential k-nearest neighbor methods was used to impute missing values. Changes in the normalized spot volumes between experimental and control images were evaluated with a mixed linear mode. The spot number and normalized spot volume data were formatted in Excel. Student *t*-test analysis of protein expression was performed between Yangsimai 3 and Naso Nijo, and only those protein spots with the fold changes more than 1.5 and significant at $p < 0.05$ were considered as differentially expressed proteins. When comparing the different pattern of expression protein spots between Yangsimai 3 and Naso Nijo, both quantitative and qualitative differences were observed. The quantitative differences can be grouped into two categories: up-regulated or down-regulated protein spot in Yangsimai 3 compared with Naso Nijo. The qualitative differences can be grouped into two categories: (i) specific expressed in Yangsimai 3 (SEY), expression in Yangsimai 3 cultivar, but not in Naso Nijo cultivar; (ii) specific expressed in Naso Nijo (SEN), expression in Naso Nijo cultivar, but not in Yangsimai 3 cultivar. Student's *t*-test ($p < 0.05$) was used to calculate significant differences in relative abundances of protein spot features in the Yangsimai 3 compared with Naso Nijo. Spots with reproducible and significant variations, at least 1.5-fold up-regulated or down-regulated, were considered quantitative differentially expressed proteins.

In-gel Digestion of Proteins

Protein spots were excised manually and transferred to 1.5 mL microcentrifuge tubes, and proteins with lower abundance were removed from all the replicate gels to pool and digest in a single tube. Protein spots were destained twice with 30 mM potassium ferricyanide and 100 mM sodium thiosulfate, and then rinsed with 25 mM ammonium bicarbonate in 50% acetonitrile. Protein spots were dehydrated with 100% acetonitrile, dried under vacuum, and 10 μL trypsin (10 ng/μL) was added, imbibed 40 min on ice. Then protein spots were covered by using 25 μL 25 mM ammonium bicarbonate and incubated for 16 h at 37°C. The peptides were eluted by using 30 μL 0.1% TFA, shaken for 10 min, the digestion solution was transferred to a new 1.5 mL tube, and then the protein spots were eluted by using 70% v/v acetonitrile and 0.1% v/v trifluoroacetic acid twice, the digestion solution was then transferred to a new 1.5 mL tube once more, incorporating digestion solution, freeze-dried for 2 h, condensing the volume to 10 μL and stored in -80°C.

Identification of Proteins by Mass Spectrometry

The digestion solution was spotted on an MALDI target plate (1.0 μL) twice and recrystallized CHCA matrix dissolved in 0.1% TFA/70% ACN (0.5 μL). Mass Standards Kit for Calibration of SCIEX MALDI-TOF Instrument (Foster City, CA, USA) was used for Mass assignment. Each sample spot was desalted with 0.01% TFA, and completely dried. Protein identification was conducted using an SCIEX MALDI TOF-TOF™ 5800 Analyzer equipped with a neodymium. For the MS mode, peptide mass maps were acquired in positive reflection mode, and the 800–4,000 m/z mass range was used with 4,000 laser shots per spectrum. A maximum of 20 precursors per spot with a minimum S/N ratio of 20 were selected for MS/MS analysis in 2 kV Positive modes. The contaminant m/z peaks originating from trypsin auto-digestion, or matrix were excluded from MS/MS analysis.

A MS/MS results were analyzed by using ProteinPilot software (Foster City, CA, USA), and the results were searched using MASCOT software[1]. Matches to protein sequences from the Viridiplantae taxon (Green plants) in NCBInr database (updated 6 June 2014, 17893860 sequences) were considered acceptable if: (1) A protein score was obtained from MASCOT, which rates scores as significant if they are above the 95% significance threshold ($p < 0.05$); (2) At least two different predicted peptide masses matched the observed masses for an identification to be considered valid; (3) The coverage of protein sequences by the matching peptides should be higher than 5%; (4) A parent ion mass tolerance of ±0.2 Da and an MS/MS tolerance of ±0.1 Da; (5) Acetylation of the N-terminus, cysteine as carboxylamidomethyl cysteine, pyroglu formation of N-terminal Gln and methionine in an oxidized form were set as possible modifications. To understand the function of the proteins, the identified proteins were classified by using the MapMan ontology defined by Thimm et al. (2004).

Quantitative Real-Time PCR

Total RNA was isolated from the 10 days after flowering (DAF), 20 DAF, 30 DAF and mature seed using a TaKaRa MiniBEST Plant RNA Extration Kit (Tokyo City, Japan). cDNA was generated from the RNA with M-MLV reverse transcriptase (Tokyo City, Japan). Specific primers for quantitative real-time PCR (qRT-PCR) analysis were listed. Specificity of primers was checked by using NCBI database[2], and the PCR products were sequenced. Reaction was carried out in 20 μL reactions system containing 10 mM Tris-HCl (pH 8.5), 50 mM KCl, 2 mM $MgCl_2$, 0.4 μL DMSO, 200 mM dNTPs, specific PCR primers 10 pmol/μL, Taq DNA polymerase 1 U, SYBR GREEN I fluorescence dye 0.5 μL. qRT-PCR was performed in clear tubes using an Applied Biosystems ViiA™ 7 Real-Time PCR System (Carlsbad City, CA, USA) as follows: 94°C for 5 min, followed by 40 cycles at 94°C for 30 s, 58°C for 30 s, 72°C for 30 s, and a final extension of 72°C for 5 min. Actin was used as an internal control. All reactions were run in triplicate, Ct values were determined by the Applied Biosystems ViiA™ 7 software with default settings. Differences between the Ct values of target gene and Actin were calculated as $\Delta Ct = Ct_{target\ gene} - Ct_{Actin}$, and the relative expression levels of target genes were determined as $2^{-\Delta Ct}$. For each sample, PCR was performed with three biological replicates. The average values of $2^{-\Delta Ct}$ were used to determine difference in gene expression.

RESULTS

The Variation of Grain Protein Content between Barley Cultivar Yangsimai 3 and Naso Nijo

The GPC of Yangsimai 3 and Naso Nijo are 13.3 and 11.6%, respectively (Figure 1), which indicates the GPC of Yangsimai 3 is 14.7% higher than Naso Nijo. Analysis shows the difference of GPC between the two cultivars is statistically significant ($p < 0.01$; Figure 1).

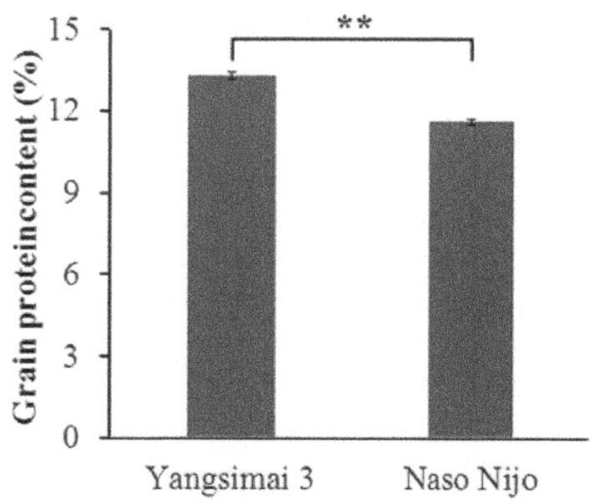

Figure 1: Differences in grain protein content between two barley cultivars.**$p < 0.01$.

Construction of a Differentially Expressed Protein Profiling of Grain Protein between Yangsimai 3 and Naso Nijo

To construct a 2-DE map of barley grain proteins, grain proteins were separated by 2-DE with three biological replicates from Yangsimai 3 and Naso Nijo, respectively. At linear gradient of pH 4–7, 456, 464, and 448 protein spots were observed on 2-DE gels of Yangsimai 3, while 448, 460, and 452 protein spots were detected on 2-DE gels of Naso Nijo. At linear gradient of pH 6–11, 59, 55, and 57 protein spots were detected on 2-DE gels of Yangsimai 3, correspondingly, 58, 54, and 60 protein spots were detected on 2-DE gels of Naso Nijo (Figure 2). Totally, 502 reproducibility protein spots were detected in both barley varieties, among which 41 (41/502, 8.17%) protein spots were found to be different pattern of expression between Yangsimai 3 and Naso Nijo by student's *t*-test at $p < 5\%$. When analyzing different pattern of expression between Yangsimai 3 and Naso Nijo, both quantitative and qualitative differences were observed. Student's *t*-test was used to calculate significant differences in relative abundance of protein spots in the Yangsimai 3 compared with Naso Nijo. The qualitative differences can be grouped into two categories, which were SEY (specific expressed in Yangsimai 3; 13 entries) and SEN (specific expressed in Naso Nijo; eight entries). The quantitative differences can be grouped into up-regulated or down-regulated in Yangsimai 3, with 13 and 7 protein spots in each category, respectively (Figure 3A).

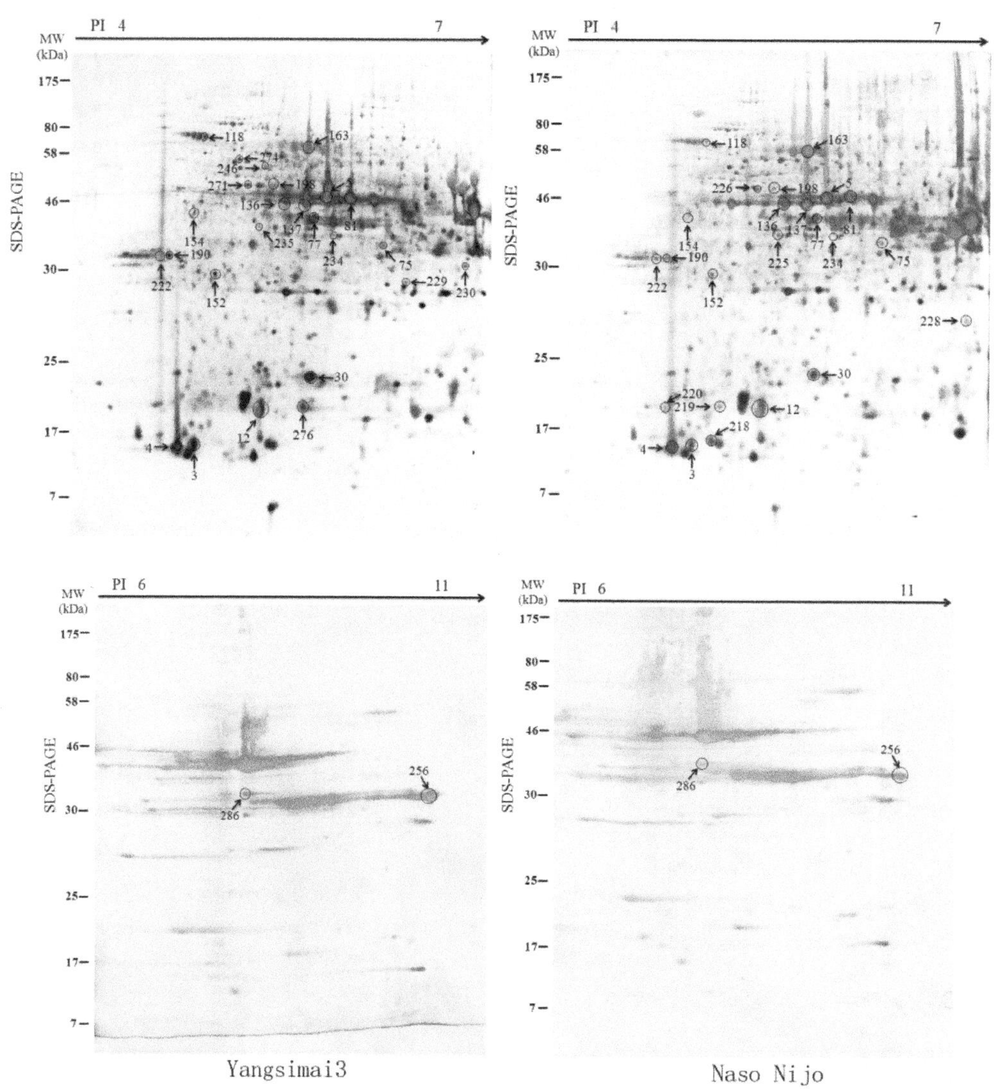

Figure 2: The differentially expressed protein profiling between Yangsimai 3 and Naso Nijo.

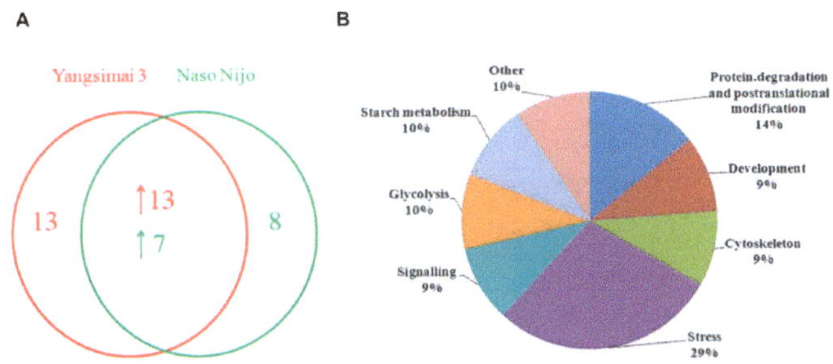

Figure 3: The number of differentially expressed protein spots and protein function category. (A) The number of differentially expressed protein spots; (B)Function category of differentially expressed proteins.

Identification of Differentially Expressed Protein Spots

All differentially expressed protein spots between Yangsimai 3 and Naso Nijo were excised from representative 2-DE gels for identification, among which 34 protein spots were successfully identified by tandem MS, corresponding to 23 unique proteins. These identified protein spots can be further grouped into eight categories according to their biological functions, and the category with the most proteins identified was stress (6/23, 26.09%). The other categories referred to protein degradation and post-translational modification (three entries), development (two entries), cell (two entries), signaling (two entries), glycolysis (two entries), starch metabolism (two entries), and other function (four entries; Figure 3B).

Further analysis revealed that these 34 identified protein spots derived from 23 different genes or gene families, among which 14 gel spots corresponding to 14 protein isoform identifications collapsed to three proteins (Table 1). For example, spot 12, 30, 219, and 220 identified as globulin (gi|167004); spot 137, 154, 225, and 235 identified as serpin-Z7 (gi|75282567); spot 5, 75, 77, 81, 136, and 234 identified as protein serpin-Z4 (gi|1310677). The isoforms matched to the same sequence, though they differed significantly with respect to their p*I*s and *Mr* (Figure 4). The number of isoforms for each protein ranged from 4 to 6. In the present study, four protein spots (spot 12, 30 219, and 220) were identified as globulin, which displayed an experimental Mr ranging from 20.380 to 22.147 kDa, less than the calculated Mr (637 aa, 72.551 kDa). MALDI-TOF MS/MS data which may be explained the Mr changes of the identified isoforms were summarized. All four protein spots were identified as C-terminal peptide sequences of globulin. In addition, some proteins have same protein name, but have different protein ID. For example, spot 190 (gi|22607) and 222 (gi|2492487) were identified as 14-3-3 protein, when aligned the two isoforms of 14-3-3 identified, it was shown that peptide 2 (LAEQAERYEEMVEFMEK) identified for gi|2492487 was common to both isoforms, whereas peptides 1 from each isoform (SAQDIALADLPTTHPIR and AAQEIALAELPPTHPIR) were unique and covered the same region of the proteins. Spot 198 (gi|326490934) and 274 (gi|326493636) were identified as enolase, both of them shared high homology (98.44%) and all of the peptides were common to both isoforms.

Table 1: Identified differentially expressed proteins between Yangsimai3 and Naso Nijo in mature barley grains

Spot no.	Protein name	Protein ID[a]	Experimental Mr(Da)/p*I*	Calculated Mr(Da)/p*I*	Species	Expression pattern[b]
Cell						
226	Actin	gi|326488133	42189/5.48	41849/5.23	*Hordeum vulgare*	SEN
228	Predicted protein	gi|297613620	27691/6.92	27303/4.93	*Oryza sativa*	SEN
Development						
229	Rab28	gi|326531218	31690/6.41	30439/5.23	*Hordeum vulgare*	SEY
12	Globulin	gi|167004	20412/5.45	72551/6.80	*Hordeum vulgare*	Up-regulated
30	Globulin	gi|167004	22147/5.72	72551/6.81	*Hordeum vulgare*	Up-regulated
219	Globulin	gi|167004	20386/5.23	72551/6.80	*Hordeum vulgare*	SEN
220	Globulin	gi|167004	20380/4.72	72551/6.80	*Hordeum vulgare*	SEN
Glycolysis						
198	Enolase	gi|326490934	42163/5.56	48601/5.39	*Hordeum vulgare*	Up-regulated
274	Enolase	gi|326493636	52435/5.31	48427/5.39	*Hordeum vulgare*	SEY
Protein degradation and post-translational modification						
5	Serpin-Z4	gi|1310677	40675/5.81	43307/5.61	*Hordeum vulgare*	Down-regulated
75	Serpin-Z4	gi|1310677	37625/6.24	43307/5.61	*Hordeum vulgare*	Down-regulated
77	Serpin-Z4	gi|1310677	38134/5.73	43307/5.61	*Hordeum vulgare*	Up-regulated
81	Serpin-Z4	gi|1310677	40580/6.01	43307/5.61	*Hordeum vulgare*	Up-regulated
136	Serpin-Z4	gi|1310677	40172/5.60	43307/5.61	*Hordeum vulgare*	Down-regulated
234	Serpin-Z4	gi|1310677	36214/5.92	43307/5.61	*Hordeum vulgare*	Up-regulated
137	Serpin-Z7	gi|75282567	40021/5.7	42851/5.45	*Hordeum vulgare*	Down-regulated
154	Serpin-Z7	gi|75282567	39041/4.92	42851/5.45	*Hordeum vulgare*	Up-regulated
225	Serpin-Z7	gi|75282567	36300/5.60	42851/5.45	*Hordeum vulgare*	SEN
235	Serpin-Z7	gi|75282567	37205/5.42	42851/5.46	*Hordeum vulgare*	SEY
230	Guanine nucleotide-binding protein subunit beta-like protein	gi|326491885	33456/6.90	36655/5.97	*Hordeum vulgare*	SEY
Signaling						
190	14-3-3 protein homolog	gi|22607	34126/4.77	29361/4.83	*Hordeum vulgare*	Up-regulated
222	14-3-3-like protein B	gi|2492487	34088/4.65	29787/4.67	*Hordeum vulgare*	Down-regulated
Starch metabolism						
163	Beta-amylase 1	gi|38349539	62321/5.70	57883/5.65	*Hordeum vulgare*	Down-regulated
271	Small subunit of ADP-glucose pyrophosphorylase	gi|27464770	42163/5.35	43861/4.91	*Hordeum vulgare*	SEY

Stress						
3	Alpha-amylase/trypsin inhibitor CMb	gi|585290	17054/4.91	17199/5.77	*Hordeum vulgare*	Up-regulated
4	CMd preprotein	gi|758343	16256/4.75	17894/5.24	*Hordeum vulgare*	Up-regulated
218	Alpha-amylase inhibitor BDAI-1	gi|123970	18374/5.20	17045/5.36	*Hordeum vulgare*	SEN
118	Heat shock cognate 70 kDa protein	gi|326497219	71675/5.13	72202/5.14	*Hordeum vulgare*	Up-regulated
9	Heat shock cognate 70 kDa protein	gi|326499406	70654/4.95	71629/5.09	*Hordeum vulgare*	Down-regulated
276	17 kDa class I small heat shock protein	gi|1536911	20653/5.68	16832/5.83	*Hordeum vulgare*	SEY
Others						
152	Lactoylglutathione lyase	gi|326493416	31.890/5.20	32811/5.34	*Triticum aestivum*	Up-regulated
246	ATP synthase beta subunit	gi|326492854	48524/5.51	59434/5.85	*Hordeum vulgare*	SEY
256	B hordein	gi|224386	31556/11.4	30850/8.26	*Hordeum vulgare*	Up-regulated
286	B hordein	gi|255348352	30782/7.55	30621/7.95	*Hordeum vulgare*	Up-regulated

[a]Protein ID against the NCBI database. [b]SEN, specific expressed in Naso Nijo; SEY, specific expressed in Yangsimai 3; Up-regulated: the abundance level of Yangsimai 3 is higher than Naso Nijo; Down-regulated: the abundance level of Yangsimai 3 is lower than Naso Nijo.

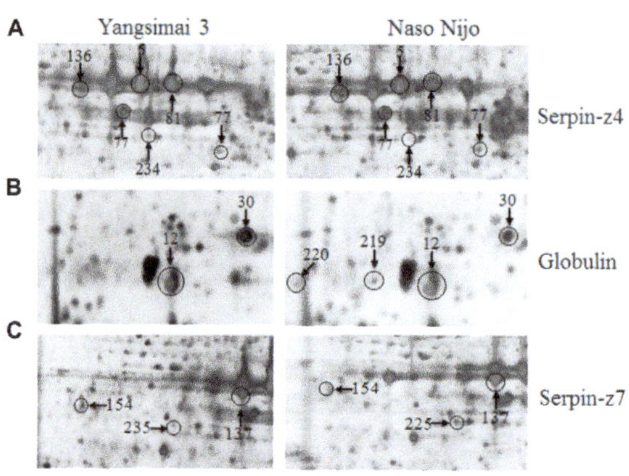

Figure 4: Protein isoforms were observed on 2-DE gels. Serpin-Z4 (A), Globulin(B), and Serpin-Z7 (C) isoforms which were observed on Yangsimai 3 and Naso Nijo 2-DE gels.

Spot 9 (gi|326499406) and 118 (gi|326497219) were identified as heat shock cognate 70 kDa protein, both of them shared 88.28% homology between two isoforms, all of the identified peptides were common to both isoforms except peptide 2 (STAGDTHLGGEDFDNR) from gi|326497219 was unique. Spot 256 (gi|224386) and 286 (gi|255348352) were identified as B hordein protein, both of them shared 67.29% homology between two isoforms, peptide 1 (TLPMMCSVNVPLYR and TLPTMCSVNVPLYR) and 2 (VFLQQQCSPVPVPQR and VFLQQQCSPVAMSQR) were unique and cover the same region of the protein isoforms, other peptides were common to both isoforms. Spot 3 (gi|585290) and 218 (gi|123970) were identified as alpha-amylase inhibitor, both of them contain a conserved AAI domain, all of the identified peptide for each isoforms were unique for each isoforms.

Expression Analysis of Three Differentially Expressed Proteins on RNA Level

To investigate whether differentially expressed proteins between Yangsimai 3 and Naso Nijo were also observed on the RNA level, genes encoding three proteins (spot 274, 218, and 271) were selected for qRT-PCR analysis (Figure 5). As shown in Figure 5, the expression levels of gene encoding ADP-glucose pyrophosphorylase (spot 271) increased with maturity, enolase (spot 274) increased and then decreased in mature seed, Alpha-amylase inhibitor BDAI-1 (spot 218) increased from 10 to 30 DAF but then dramatically decreased in mature seed. Comparison the expression pattern of three genes encoding differentially expression protein between protein level and RNA level in mature seed, genes encoding enolase, and small subunit of ADP-glucose pyrophosphorylase displayed higher expressed in Yangsimai3 than Naso Nijo on the RNA level, but specific expression in Yangsimai 3 in protein level (Figures 2 and 5). Remarkably, gene encoding Alpha-amylase inhibitor BDAI-1 showed the opposite expression patterns on the protein level compared with RNA level (Figures 2 and 5).

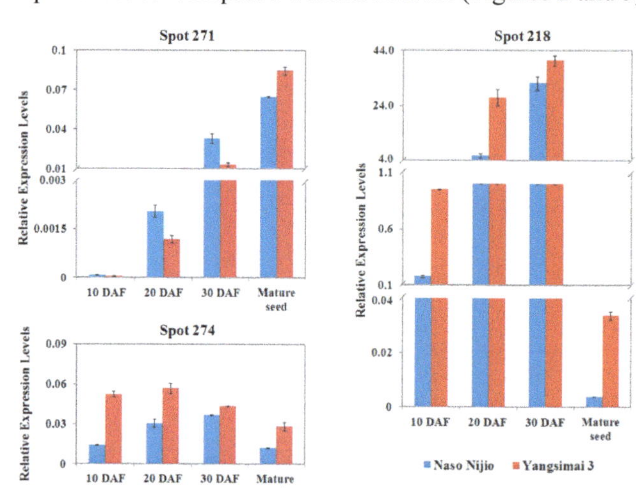

Figure 5: Quantitative real-time PCR of mRNA expression patterns for selected protein spots in 10 DAF (days after flowering), 20 DAF, 30 DAF and mature seed.

Integration with Genomic Information

To validate the potential implications of these functional annotations, the present results and those of a previous genetic analysis were integrated for GPC (Cai et al., 2013). Two proteins, including enolase and small subunit of ADP-glucose pyrophosphorylase (gi|326493636 and gi|27464770) were identified in the present study, both of which were mapped to on chromosomes 5^3 and positioned 44.17 and 47.01 cM in barley, respectively. Remarkably, both proteins were closed to a QTL of controling barley GPC whose linkage marker was bPb-3412 at 45.6 cM on Chr. 5 (Cai et al., 2013).

DISCUSSION

Barley cultivars used for malt should have a GPC not exceeding 11.5% (Pettersson and Eckersten, 2007). Actually, GPC is greatly affected by the growing environment. For instance, stresses caused by levels of nitrogen, drought, and/or heat may increase the GPC in barley (Savin and Nicolas, 1996; Qi et al., 2006). Therefore, it is important to breed cultivars with lower and less environmentally influenced GPC (Bertholdsson, 1999). In the current study, Yangsimai 3 was used as feed barley with GPC of 13.3%, which displayed significant difference in GPC compared to Naso Nijo which was used as malting barley with GPC of 11.6%, according to the data from 2013. Thus, these two cultivars provide good studying material to perform comparative proteome analysis and discover potential protein candidates involved in stable GPC expression and regulation.

Grain protein content is also influenced by genetic factors in barley. Genome-wide association study (GWAS) was performed by using 59 cultivated and 99 Tibetan wild barley genotypes for identifing molecular markers associated with GPC. Results showed a total of 10 DArT markers ($p < 0.01$) were associated with GPC in barley (Cai et al., 2013). Further analysis indicated that *HvNAM*genes could play a role in controlling barley GPC (Cai et al., 2013). Variation in protein expression profiles of barley cultivars reflected genetic variations, which was illustrated by the identification of different alleles of β-amylase in two protein spots (Finnie et al., 2002). For example, Barke and Morex are two malting cultivars different in seed maturation days. Comparative proteomics analysis revealed that differentially expressed proteins reflected the faster maturation of Morex seeds (Finnie et al., 2006). In the current study, a differentially expressed proteins profiling of barley mature seed was constructed. Among the 502 protein spots that were reproducibly detected, a total of 34 differentially expressed protein spots were identified by tandem MS, which corresponded to 23 different proteins.

ADP-glucose pyrophosphorylase (AGPase) catalyzes the conversion of Glc-1-P and ATP to PPi and ADP-glucose, and is a key regulatory enzyme of starch biosynthesis (Preiss et al., 1991; Tetlow et al., 2004; Geigenberger, 2011). In barley, AGPase gene generates two transcripts, one of which encodes the cytosolic small subunits of ADP-glucose pyrophosphorylase in the endosperm and another encodes the plastidial SSU in leaves (Beckles et al., 2001). The *Risø16* mutant of barley lacks cytosolic AGPase activity in the endosperm, which leads to decreased grain weight. Also, *Risø16*mutant contained 90% total N and proteins of wild type at the transcriptional level, down-regulated enolase and beta-amylase 1 (Faix et al., 2012). In the present study, protein spot 274 and 271 were identified as enolase (gi|326493636) and small subunit of ADP-glucose pyrophosphorylase (gi|27464770), respectively, both of which were higher expressed in Yangsimai 3 than in Naso Nijo on the protein and RNA levels. Genomic information verified that the two proteins are close to a QTL for GPC (bPb-3412) on Chr. 5 (Cai et al., 2013). Integrating the above genomic results, the comparative expressions from our current study indicated that the specific expression of enolase (spot 274) and small subunit of ADP-glucose pyrophosphorylase (spot 271) in barley Yangsimai 3 could contribute to the GPC in barley seed.

Prolamins is the major storage proteins in barley seed protein, which is specifically synthesized in the starchy endosperm and divided into B, C, D and γ hordeins according to their mobility in SDS-electrophoretic gels (Kreis and Shewry, 1992; Piston et al., 2005). Among the four types of hordeins, B fraction accounts for 70–80% of the total prolamins content in most barley cultivars. Previous studies reported that B hordein was significantly correlated with GPC, B hordein content increased as the sowing date was postponed and was significantly affected by nitrogen levels (Qi et al., 2005, 2006). In the present study, two protein spots (spot 256 and 286) were identified as B hordein, and also displayed up-regulated in Yangsimai 3, which would contribute to the difference of GPC in mature barley seeds.

Serpins are the most abundant proteins in beer 2DE gels, and characterized by their function in malting barley (Jin et al., 2013). It was widely believed that serpins are beer foam-positive proteins and improve malt filterability (Maceda et al., 1991; Jin et al., 2013). In barley, serpin was approximately 43-kDa polypeptide, which irreversibly inhibits the endogenous and exogenous target proteinases (Roberts and Hejgaard, 2008). Comparative proteomics based on fluorescent difference gel electrophoresis (DIGE) was employed to quantitatively analyze proteins in cultivars Dan'er and Metcalfe in China, and the results showed that serpin Z4 and Z7 were the most remarkable differentially expressed proteins, which played an important role in malt filterability (Jin et al., 2013). Interestingly, mutiple protein isoforms of serpin Z4 and Z7 were observed in barley seed proteome (Finnie et al., 2002; Jin et al., 2013). In the present study, 10 protein spots were identified as serpin (serpin Z4 and Z7) and displayed multiple expression patterns (Table 1). In addition, globulin, which undergoes post-translational processing, is an important seed storage protein in cereal crops and functions in glycosylation and partial endoproteolytic cleavage (Heck et al., 1993; Herman and Larkins, 1999). In wheat, globulin-3 was cleaved post-translationally in embryos. Five major polypeptide spots of globulin-3 were identified by MS and N-terminal sequencing, and each spot displayed different Mr and pI, these post-translational processing events that lead to the maturation of the globulin family of proteins observed seed protein fraction that could be associated with type 1 diabetes or celiac disease following endoproteolytic processing (Koziol et al., 2012). In barley, globulin contains 637 amino acids with one signal peptide detected by the SignalP v4.0 program, the peptide starts at position 1 and ends at position 27. Four protein spots (spot 12, 30, 219, and 220) were identified as barely globulin in the current study, the MALDI-TOF MS/MS result revealed that the four protein spots only contains several fragments of the globulin, it is suggesting that globulin undergo cleavage of signal peptide from precursor, or non-specific degradation pathway.

The proteinaceous barley alpha-amylase/subtilisin inhibitor (BASI) is synthesized during grain filling and is an abundant protein of the endosperm and the aleurone layers of the mature seed (Mundy et al., 1983; Lecommandeur et al., 1987; Leah and Mundy, 1989). Recently, alpha-amylase inhibitor (BDAI-1) and CMb component of tetrameric alpha amylase inhibitor (CMb) were identified by using MS analysis. All of them were differentially expressed among four haze samples; further analysis indicated that BDAI-1, CMb were not predominant haze-active proteins, but growth factors of beer colloidal haze (Iimure et al., 2009). Two alpha-amylase inhibitors were identified in the current study, which were Alpha-amylase inhibitor BDAI-1 (spot 218) and CMb (spot 3) displayed specific expressed and down-regulated pattern in Naso Nijo, respectively. Dramatically, Alpha-amylase inhibitor BDAI-1 displayed higher expression in Yangsimai 3 than in Naso Nijo on the RNA level. These data suggested that different members in the protease inhibitor family may have different functions in barley cultivars, and also the mRNA of BDAI-1 maybe undergo post-transcriptional processing in the seed development. However, detailed studies on alpha-amylase inhibitor may facilitate a better understanding of the mechanisms involved in malting barley.

CONCLUSION

In the present study, the GPC was significant difference in Yangsimai 3 and Naso Nijo. A differentially expressed profiling of grain proteins was established by using a combination of 2-DE and tandem MS. In total, 41 protein spots were detected differentially expressed between Yangsimai 3 and Naso Nijo, among which 34 protein spots corresponding to 23 different proteins were identified. In particular, differentially expressed proteins from the seed were mainly related to stress, protein degradation and post-translational modification, development, cell, signaling, glycolysis and starch metabolism. Seeds proteome explicitly displays that differentially expressed proteins are involved in GPC and malting quality. However, a malting barley cultivar is quite different from a feed barley cultivar and there is an interest to develop a greater knowledge of the determinants of malting and feed quality at the protein level, in order to improve the evaluation of new barley varieties. The quality of the barley is determined in part by the proteins content produced during grain filling, meanwhile the accumulation of storage proteins plays an important role in seed development by regulating the appearance of proteins in different development stages. Therefore, an understanding of the mechanism of how storage proteins accumulation initiates during seed development is required to be further investigated.

AUTHOR CONTRIBUTIONS

BG and HL carried out the seed proteins preparation, 2-DE analysis, mass spectrometry analysis and interpretation of the 2-DE data, and wrote portions of the manuscript. SL, CL, and XZ carried out image analysis and 2-DE data. BG and RX conceived of the project and wrote a draft of the manuscript and provided conceptual framework. All authors read and approved of the final manuscript.

FUNDING

This research was funded by the National Natural Science Foundation of China 31401370, National Barley and Highland Barley Industrial Technology Specially Constructive Foundation of China (CARS-05), A Project Funded by the Priority Academic Program Development of Jiangsu Higher Education Institutions, Jiangsu Agriculture Science and Technology Innovation Fund [CX(14)5093], Jiangsu Provincial Key Laboratory Open Project 205031310.

ACKNOWLEDGMENT

The authors wish to thank Dr. Yuyang Wang for his assistance in mass spectrometry analysis.

REFERENCES

1. Beckles, D. M., Smith, A. M., and ap Rees, T. (2001). A Cytosolic ADP-glucose pyrophosphorylase is a feature of graminaceous endosperms, but not of other starch-storing organs. *Plant Physiol.* 125, 818–827. doi: 10.1104/pp.125.2.818
2. Bertholdsson, N. O. (1999). Characterization of malting barley cultivars with more or less stable grain protein content under varying environmental conditions. *Eur. J. Agron.* 10, 1–8. doi: 10.1016/S1161-0301(98)00043-4
3. Bradford, M. M. (1976). A rapid and sensitive method for the quantitation of microgram quantities of protein utilizing the principle of protein-dye binding. *Anal. Biochem.* 72, 248–254. doi: 10.1006/abio.1976.9999
4. Cai, S. G., Yu, G., Chen, X. H., Huang, Y. C., Jiang, X. G., Zhang, G. P., et al. (2013). Grain protein content variation and its association analysis in barley. *BMC Plant Biol.* 13:35. doi: 10.1186/1471-2229-13-35
5. Clancy, J. A., Han, F., and Ullrich, S. E. (2003). The North American barley genome project: comparative mapping of β-amylase activity QTLs among three barley crosses. *Crop Sci.* 43, 1043–1052. doi: 10.2135/cropsci2003.1043
6. Dong, M. H., Gu, J. R., Zhang, L., Chen, P. F., Liu, T. F., Deng, J. H., et al. (2014). Comparative proteomics analysis of superior and inferior spikelets in hybrid rice during grain filling and response of inferior spikelets to drought stress using isobaric tags for relative and absolute quantification. *J. Proteom.* 109, 382–399. doi: 10.1016/j.jprot.2014.07.001
7. Eagles, H. A., Bedggood, A. G. J., Panozzo, F., and Martin, P. J. (1995). Cultivar and environmental effects on malting quality in barley. *Aust. J. Agric. Res.* 46, 831–844. doi: 10.1071/AR9950831
8. Faix, B., Radchuk, V., Nerlich, A., Hümmer, C., Radchuk, R., Emery, R. J. N., et al. (2012). Barley grains, deficient in cytosolic small subunit of ADP-glucose pyrophosphorylase, reveal coordinate adjustment of C: N metabolism mediated by an overlapping metabolic-hormonal control. *Plant J.* 69, 1077–1093. doi: 10.1111/j.1365-313X.2011.04857.x
9. Finnie, C., Bak-Jensen, K. S., Laugesen, S., Roepstorff, P., and Svensson, B. (2006). Differential appearance of isoforms and cultivar variation in protein temporal profiles revealed in the maturing barley grain proteome. *Plant Sci.* 170, 808–821. doi: 10.1016/j.plantsci.2005.11.012
10. Finnie, C., Melchior, S., Roepstorff, P., and Svensson, B. (2002). Proteome analysis of grain filling and seed

maturation in barley. *Plant Physiol.* 129, 1308–1319. doi: 10.1104/pp.003681

11. Ge, P., Ma, C., Wang, S., Gao, L., Li, X., Guo, G., et al. (2012). Comparative proteomic analysis of grain development in two spring wheat varieties under drought stress. *Anal. Bioanal. Chem.* 402, 1297–1313. doi: 10.1007/s00216-011-5532-z
12. Geigenberger, P. (2011). Regulation of starch biosynthesis in response to a fluctuating environment. *Plant Physiol.* 155, 1566–1577. doi: 10.1104/pp.110.170399
13. Guillon, F., Larré, C., Petipas, F., Berger, A., Moussawi, J., Rogniaux, H., et al. (2012). A comprehensive overview of grain development in *Brachypodium distachyon* variety Bd21. *J. Exp. Bot.* 63, 739–755. doi: 10.1093/jxb/err298
14. Guo, G. F., Lv, D. W., Yan, X., Subburaj, M., Ge, P., Li, X. H., et al. (2012). Proteome characterization of developing grains in bread wheat cultivars (*Triticum aestivum* L.). *BMC Plant Biol.* 12:147. doi: 10.1186/1471-2229-12-147
15. Heck, G. R., Chamberlain, A. K., and Ho, T. H. D. (1993). Barley embryo globulin-1 gene, beg1: characterization of cDNA, chromosome mapping and regulation of expression. *Mol. Gen. Genet.* 239, 209–218. doi: 10.1007/BF00281620
16. Herman, E., and Larkins, B. (1999). Protein storage bodies and vacuoles. *Plant Cell* 11, 601–613. doi: 10.1105/tpc.11.4.601
17. Howard, K. A., Gayler, K. R., Eagles, H. A., and Halloran, G. M. (1996). The relationship between D hordein and malting quality in barley. *J. Cereal Sci.* 24, 47–53. doi: 10.1006/jcrs.1996.0036
18. Hurkman, W. J., and Tanaka, C. K. (2007). Extraction of wheat endosperm proteins for proteome analysis. *J. Chromatogr. B*849, 344–350. doi: 10.1016/j.jchromb.2006.11.047
19. Iimure, T., Nankaku, N., Watanabe-Sugimoto, M., Hirota, N., Tiansu, Z., and Kihara, M. (2009). Identification of novel haze-active beer proteins by proteome analysis. *J. Cereal Sci.* 49, 141–147. doi: 10.1016/j.jcs.2008.08.004
20. Jiang, C. M., Chengm, Z. Q., Cheng, Z., Yu, T. Q., Zhong, Q. F., Shen, Q. X., et al. (2014). Proteomic analysis of seed storage proteins in wild rice species of the *Oryza* genus. *Proteome Sci.* 12:51. doi: 10.1186/s12953-014-0051-4
21. Jin, Z., Li, X. M., Gao, F., Sun, J. Y., Mu, Y. W., and Jian, L. (2013). Proteomic analysis of differences in barley (*Hordeum vulgare*) malts with distinct filterability by DIGE. *J. Proteom.* 93, 93–106. doi: 10.1016/j.jprot.2013.05.038
22. Kim, Y. J., Lee, H. M., Wang, Y. M., Wu, J. N., Kim, S., Kang, K. Y., et al. (2013). Depletion of abundant plant rubisco protein using the protamine sulfate precipitation method. *Proteomics* 13, 2176–2179. doi: 10.1002/pmic.201200555
23. Kjeldahl, J. Z. (1883). A new method for the determination of nitrogen in organic matter. *Z. Anal. Chem.* 22, 366–382. doi: 10.1007/BF01338151
24. Koziol, A. G., Loit, E., McNulty, M., MacFarlane, A. J., Scott, F. W., and Altosaar, I. (2012). Seed storage proteins of the globulin family are cleaved post-translationally in wheat embryos. *BMC Res.* 5:385. doi: 10.1186/1756-0500-5-385
25. Kreis, M., and Shewry, P. R. (1992). "The control of protein synthesis in developing barley seeds," in *Barley: Genetics, Molecular Biology and Bio/Technology*, ed. P. R. Shewry (Wallingford: CAB International).
26. Leah, R., and Mundy, J. (1989). The bifunctionala-amylase/subtilisin inhibitor of barley: nucleotide sequence and patterns of seed-specific expression. *Plant Mol. Biol.* 12, 673–682. doi: 10.1007/BF00044158
27. Lecommandeur, D., Lauriére, C., and Daussant, J. (1987). α-Amylase inhibitor in barley seeds: localization and quantification.*Plant Physiol. Biochem.* 25, 711–715.
28. Maceda, K., Yokoi, S., Kamada, K., and Kaminura, M. (1991). Foam stability and physicochemical properties of beer. *J. Am. Soc. Brew. Chem.* 49, 14–18.
29. Mundy, J., Svendsen, I., and Hejgaard, J. (1983). Barley α-amylase/subtilisin inhibitor: I. *Isolation and characterization. Carlsberg Res. Commun.* 48, 81–91. doi: 10.1007/BF02906171
30. Pettersson, C. G., and Eckersten, H. (2007). Prediction of grain protein in spring malting barley grown in northern Europe. *Eur. J. Agron.* 27, 205–214. doi: 10.1016/j.eja.2007.04.002
31. Piston, F., Martin, A., Dorado, G., and Barro, F. (2005). Cloning and molecular characterization of B-hordeins from *Hordeum chilense* (Roem. *et Schult). Theor. Appl. Genet.* 111, 551–560. doi: 10.1007/s00122-005-2046-0
32. Preiss, J., Ball, K., Smith-White, B., Iglesias, A., Kakefuda, G., and Li, L. (1991). Starch biosynthesis and its regulation. *Biochem. Soc. Trans.* 19, 539–547. doi: 10.1042/bst0190539
33. Qi, J. C., Chen, J. X., Wang, J. M., Wu, F. B., Cao, L. P., and Zhang, G. P. (2005). Protein and hordein fraction content in barley seeds as affected by sowing date and their relations to malting quality. *J. Zhejiang Univ. Sci.* 6B, 1069–1075. doi: 10.1631/jzus.2005.B1069
34. Qi, J. C., Zhang, G. P., and Zhou, M. X. (2006). Protein and hordein content in barley seeds as affected by nitrogen level and their relationship to beta-amylase activity. *J. Cereal Sci.* 43, 102–107. doi: 10.1016/j.jcs.2005.08.005
35. Roberts, T. H., and Hejgaard, J. (2008). Serpins in plants and green algae. *Funct. Integr. Genom.* 8, 1–27. doi: 10.1007/s10142-007-0059-2
36. Savin, R., and Nicolas, M. E. (1996). Effects of short periods of drought and high temperature on grain growth and starch accumulation of two malting barley cultivars. *Funct. Plant Biol.* 23, 201–210. doi: 10.1071/PP9960201
37. See, D. K., Kephart, V., and Blake, K. (2002). Mapping genes controlling variation in barley grain protein concentration. *Crop Sci.* 2002, 680–685. doi: 10.2135/cropsci2002.0680
38. Sun, H. Y., Cao, F. B., Wang, N. B., Zhang, M., Ahmed,

I. M., Zhang, G. P., et al. (2013). Differences in grain ultrastructure, phytochemical and proteomic profiles between the two contrasting grain Cd-accumulation barley genotypes. *PLoS ONE*8:e79158. doi: 10.1371/journal.pone.0079158

39. Tetlow, I. J., Morell, M. K., and Emes, M. J. (2004). Recent developments in understanding the regulation of starch metabolism in higher plants. *J. Exp. Bot.* 55, 2131–2145. doi: 10.1093/jxb/erh248

40. Thimm, O., Bläsing, O., Gibon, Y., Nagel, A., Meyer, S., Krüger, P., et al. (2004). MAPMAN: a user-driven tool to display genomics data sets onto diagrams of metabolic pathways and other biological processes. *Plant J.* 37, 914–939. doi: 10.1111/j.1365-313X.2004.02016.x

41. Wang, J. M., Zhang, G. P., Chen, J. X., Ding, S. R., and Zhou, T. Y. (2003). Variation of grain and malt qualities in barley as affected by cultivars and environments. *Agric. Sci. China* 2, 699–705.

42. Weston, D. T., Horsley, R., Schwarz, P. B., and Goos, R. J. (1993). Nitrogen and planting date effects on low-protein spring barley. *Agron. J.* 85, 1170–1174. doi: 10.2134/agronj1993.00021962008500060015x

CHAPTER 14

THE MALTASE INVOLVED IN STARCH METABOLISM IN BARLEY ENDOSPERM IS ENCODED BY A SINGLE GENE

Vasilios M. E. Andriotis, Gerhard Saalbach, Robbie Waugh, Robert A. Field, Alison M. Smith

INTRODUCTION

Starch degradation in the endosperm of barley seedlings provides the glucose substrate for the early growth of the seedling. Conversion of starch polymers to glucose involves four classes of enzyme, α-amylase, β-amylase, limit-dextrinase and maltase. While the first three of these classes are well understood at the gene, transcriptional and protein level in the endosperm [1–9] and in the context of brewing and distilling [10–13], relatively little is known about maltase. Its nature, its role in endosperm starch degradation and its importance for brewing and distilling remain subjects of debate.

Maltases are specialised α-glucosidases (exo-acting enzymes that release α-d-glucose from the non-reducing end of their substrates) that act preferentially on maltose, producing glucose. Maltases are also able to release α-d-glucose from the non-reducing end of generic α-glucosidase substrates such as 4-nitrophenyl α-D-glucopyranoside (pNPG), but at rates lower than that for maltose hydrolysis [14]. It has long been known that germinating barley grains contain one or more enzymes capable of the hydrolysis of maltose [15]. At least some of this maltase activity is attributable to Agl97, an α-glucosidase of glycoside hydrolase family GH31 with a high affinity for maltose [16–19]. There is evidence for the presence of other maltases in the endosperm. Fractionation by ion exchange chromatography of soluble proteins from germinating grains consistently revealed the presence of at least two separable maltase activities [20–23]. Agl97 encodes a protein with a molecular mass of 97 kDa [16,19] predicted to be processed by removal of an N-terminal signal peptide to a mature mass of 92 kDa. Although a protein of this mass was purified from barley endosperm by Frandsen and colleagues [18], maltases with masses of 66 and 14 kDa [20], and 33 kDa [23] have also been partially purified. Henson and colleagues reported the occurrence in barley of four distinct genes encoding α-glucosidases (named Agl1 to Agl4 [24] see http://www.cazy.org/GH31), although it is not clear whether any of the gene products preferentially hydrolyse maltose. Thus the number and nature of maltases in the endosperm of barley seedlings remains to be determined.

The role of maltases in the endosperm of barley seedlings is also poorly understood. Sun and Henson reported from in vitro experiments that two maltase activities purified from barley endosperm could hydrolyse soluble and granular starch as well as maltose in vitro, and that these enzymes strongly and synergistically promoted the degradation of granules in the presence of α-amylases [21,25]. A subsequent study [23] failed to reproduce these strong synergistic effects, and the authors proposed that previous maltase preparations [21] were contaminated with other hydrolytic enzymes.

The quality of barley malt–the substrate for the brewing and distilling industries–is largely determined by the activities in the endosperm of enzymes of starch degradation. During the malting process, barley grains are hydrated and then allowed to germinate, typically over three to four days. Maltsters aim to maximise the activities of enzymes while minimising the loss of starch and the growth of the embryo. The barley amylases are crucial for conversion of starch to sugars in the subsequent mashing stage of alcohol production, but maltase is heat labile

Citation: Andriotis VME, Saalbach G, Waugh R, Field RA, Smith AM (2016) The Maltase Involved in Starch Metabolism in Barley Endosperm Is Encoded by a Single Gene. PLoS ONE 11(3): e0151642. doi:10.1371/journal.pone.0151642.

and much of it may be denatured at the high temperatures imposed during mashing [26,27]. Some authorities argue that introduction of a thermostable maltase would reduce potential inhibition of amylases by maltose and–if maltases can indeed synergistically promote starch granule degradation by amylases—could potentially increase the rate of conversion of starch to fermentable sugars [21,27–29]. However, others propose that the presence of maltase is detrimental during the mashing stage of brewing—in part because the resulting elevated glucose levels can potentially inhibit maltose uptake and lead to over-production of flavour compounds by brewer's yeast [26,30,31]. Full evaluation of the potential for improving malt quality by manipulation of maltases thus requires more information on the nature and roles of the endogenous maltases [12].

We previously used transgenic and chemical approaches to shed further light on the maltases of barley endosperm [32]. We showed that seedlings of lines carrying an RNA interference silencing cassette for *HvAgl97* had reductions of up to 50% in α-glucosidase activity (measured with the generic α-glucosidase substrate pNPG) in the endosperm. The ratio of glucose to maltose in the endosperm of seedlings was strongly reduced in these lines, but the patterns of starch degradation and seedling growth were unaffected. Different effects were seen when wild-type seedlings were grown in the presence of deoxynojirimicin (DNJ), an established inhibitor of α-glucosidases [33] that has been shown to inhibit Agl97 [32]. The glucose to maltose ratio was strongly reduced as in the transgenic lines, but both starch degradation and seedling growth were also inhibited.

These results suggested first that Agl97 is important for conversion of maltose to glucose but may play no role in degradation of starch granules in vivo, and second that the endosperm of barley seedlings contains DNJ-sensitive enzymes other than Agl97 that are required for normal rates of starch degradation. Since DNJ does not inhibit amylases [32], we speculated that its effects on starch degradation are either via inhibition of unidentified enzymes that act on starch granules, or via inhibition of other unknown enzymes indirectly required for normal starch degradation [32].

In this paper we investigate the possibility that the barley endosperm contains unidentified maltases that may be necessary for normal rates of starch degradation. We use protein fractionation and identification techniques on wild-type seedlings and seedlings of transgenic lines with reduced expression of Agl97 to investigate the extent to which Agl97 accounts for the maltase activity in the endosperm.

RESULTS

To determine the optimal time at which to study maltase activity, we assayed maltase and general α-glucosidase activity (using maltose and the generic α-glucosidase substrate pNPG, respectively) during the first ten days post imbibition (dpi) of barley grains (cv NFC Tipple). Both activities increased sharply in the endosperm from about 3 dpi (Fig 1). Whereas maltase activity was undetectable prior to this point and was almost completely inhibited by inclusion of 0.5 mM DNJ in the assay, α-glucosidase activity was present from the point of imbibition. This early activity was not inhibited by DNJ, but it was increasingly susceptible to DNJ inhibition from about six days onwards (Fig 1). Based on these results, we performed subsequent experiments on barley endosperm at 10 dpi. At this point the α-glucosidase and the maltase activities are maximal, more than 75% of the starch originally present in the grain has been degraded, and levels of both maltose and glucose in the endosperm have peaked ([32], and our observations).

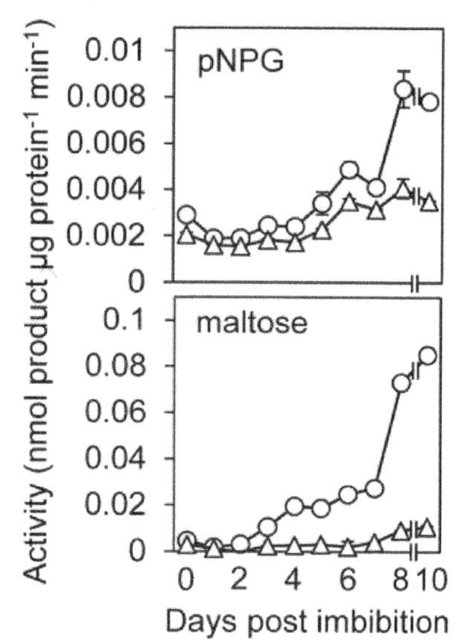

Figure 1: Time course of development of maltase and α-glucosidase activity in the endosperm.

Extracts were prepared from endosperm pooled from five seedlings at each time point. Values are means of three technical replicates ± SE. Circles, without inhibitor; triangles, with 0.5 mM DNJ in the assay. Upper graph, assay with pNPG as the substrate. Lower graph, assay with maltose as the substrate.

http://dx.doi.org/10.1371/journal.pone.0151642.g001

To identify maltases present in the endosperm, proteins that precipitated from endosperm extracts between 20% and 80% (w/v) ammonium sulfate (a fraction that contained essentially all of the maltase activity) were fractionated on a MonoQ anion exchange column. Under our chromatographic conditions, α-glucosidase and maltase activities had very similar profiles. Both appeared in the unbound fraction (peak I) and during elution of bound proteins with an increasing NaCl gradient (peaks II and III: Fig 2A and 2B).

To discover which peak(s) are attributable to the previously-identified maltase Agl97, we repeated these experiments with extracts from endosperm of transgenic lines of barley (cv Golden Promise) carrying an RNA interference (RNAi) silencing cassette for *Agl97* (lines 21 and 23 [32]). These transgenic lines carry a single insert of the transgene. Relative to an out-segregant control line they have greatly reduced α-glucosidase activity but normal rates of starch degradation in the endosperm and normal seedling growth [32]. The elution profile on a MonoQ column of maltase activity from endsperm of the out-segregant control line closely resembled that of extracts from barley cv NFC Tipple. However, in extracts from the transgenic lines 21 and 23 the activity in all peaks (I-III) was strongly reduced (Fig 2C). This result indicates that maltase activity in the endosperm is largely attributable to Agl97, and/or to enzyme(s) encoded by genes highly similar to *Agl97* and thus susceptible to silencing by the RNAi cassette.

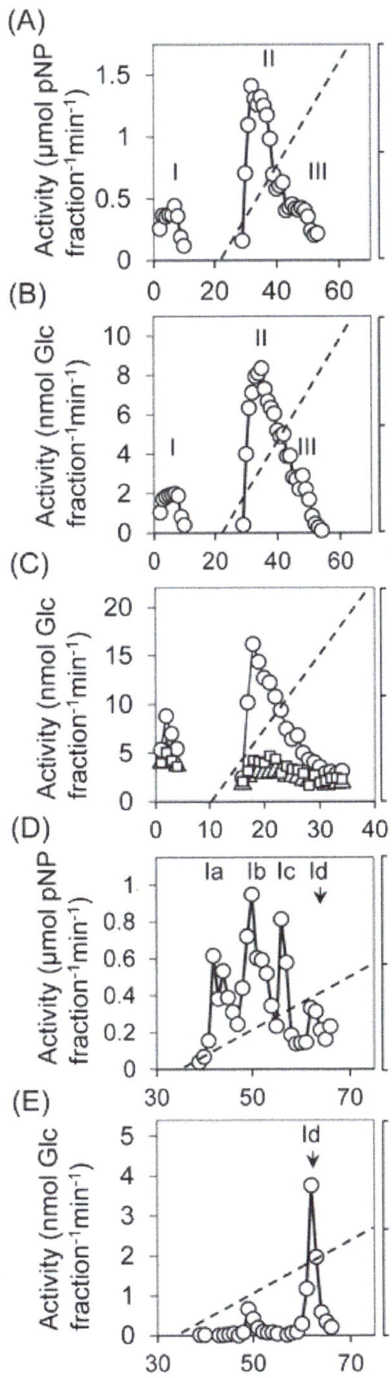

Figure 2: Fractionation on a MonoQ anion exchange column of maltase activity from endosperm.

All fractionations were reproduced at least twice, on independent batches of seedlings. Representative examples are shown. (A), (B). Extracts of endosperm of seedlings of cv NFC Tipple at 10 dpi were applied to a MonoQ column by FPLC. Proteins binding the column were eluted with a gradient of increasing NaCl concentration (dashed line). One-mL fractions were collected and assayed for α-glucosidase activity with pNPG (A), or maltase activity with maltose (B). I-III indicate the three peaks of activity. (C). Extracts were prepared, fractionated and assayed for maltase as in (B), except that seedlings were either transgenic lines in a cv Golden Promise background carrying an RNA interference (RNAi) silencing cassette for *Agl97* [lines 21 and 23 [31] (squares)], or an out-segregant line (i.e. not carrying the silencing cassette) from the same transformation experiment (circles). Chromatographic conditions were the same as in (B) except that the column wash to remove unbound proteins was reduced from 20 to 10 mL. Note that peak II maltase activity elutes at the same NaCl concentration in the experiments shown in (B) and (C). (D), (E). Fractionation on a MonoS column of proteins not binding to a MonoQ column [peak I in (A) and (B)]. Proteins binding to the MonoS column were eluted with a gradient of increasing NaCl concentration (dashed line). One-mL fractions were collected and assayed for α-glucosidase activity with pNPG (D), or for maltase activity with maltose (E). Peaks of activity eluting from the column are designated Ia-Id.

http://dx.doi.org/10.1371/journal.pone.0151642.g002

To fractionate the unbound activity in peak I, this material was applied to a MonoS cation exchange column at pH 5. Under these conditions, the α-glucosidase activity bound to the column and eluted in four distinct peaks at 150, 250, 350 and 400 mM NaCl [peaks Ia, Ib, Ic and Id respectively (Fig 2D)]. By contrast, maltase activity eluted as a single peak that corresponded to peak Id of α-glucosidase activity (Fig 2E). Maltase activity in peak Id was purified 170-fold relative to that in crude extracts of endosperm, with a recovery of ~10% (Table 1).

Table 1: Purification of maltase from endosperm extracts.

http://dx.doi.org/10.1371/journal.pone.0151642.t001

	Protein (mg)	Specific activity (nmol μg^{-1} min^{-1})	Total activity (μmol min^{-1})	Recovery (%)	Fold purification
Crude	45	0.1	3.6	100	1
20–80% ammonium sulfate	30	0.1	4	111	1.7
MonoQ (unbound)	6	0.4	2	56	4.5
MonoS, peak Id	0.025	13.5	0.34	9.5	168

Recovery is the percentage of the initial (crude extract) activity remaining at each step of the purification. Fold purification is the fold increase in specific activity relative to the crude extract at each stage of the purification.

doi:10.1371/journal.pone.0151642.t001

MALDI-ToF mass spectrometry was used to identify proteins in peak Id. Tryptic digestion of proteins in peak Id prior to fractionation on SDS-PAGE yielded 13 unique peptides that matched with the following known or predicted α-glucosidases: Agl97 (Q9LLY2 [16,18], coverage of 24%), an α-glucosidase identical to Agl97 (D1MDV2, Agl1), and a predicted protein essentially identical to Agl97 except for an additional 53 amino acids N-terminal to the Agl97 sequence (F2DV72; predicted from mRNA AK367790.1). This N-terminal peptide is not predicted to be a secretion or transit peptide, or to have a cleavage site (PSORT, SignalP;http://expasy.org/resources/). It is not found in predicted Arabidopsis or rice proteins that are most similar to Agl97 (At5g11720 and Os06g46284 respectively; seehttps://ics.hutton.ac.uk/morexGenes/view_gene2.cgi?seq_name=MLOC_66806&dataset=assembly3_WGSMorex_rbca.fasta#homologies). Searches of translated nucleotide databases with the 53 amino acid peptide did not identify other plant glycosyl hydrolases containing this sequence.

Tryptic digestion of proteins from a band of 95–110 kDa from an SDS-polyacrylamide gel of peak Id contained 37 peptides that matched the same α-glucosidase proteins, with a coverage of Agl97 of 47% (Fig 3A). Other proteins identified in the tryptic digests are listed.

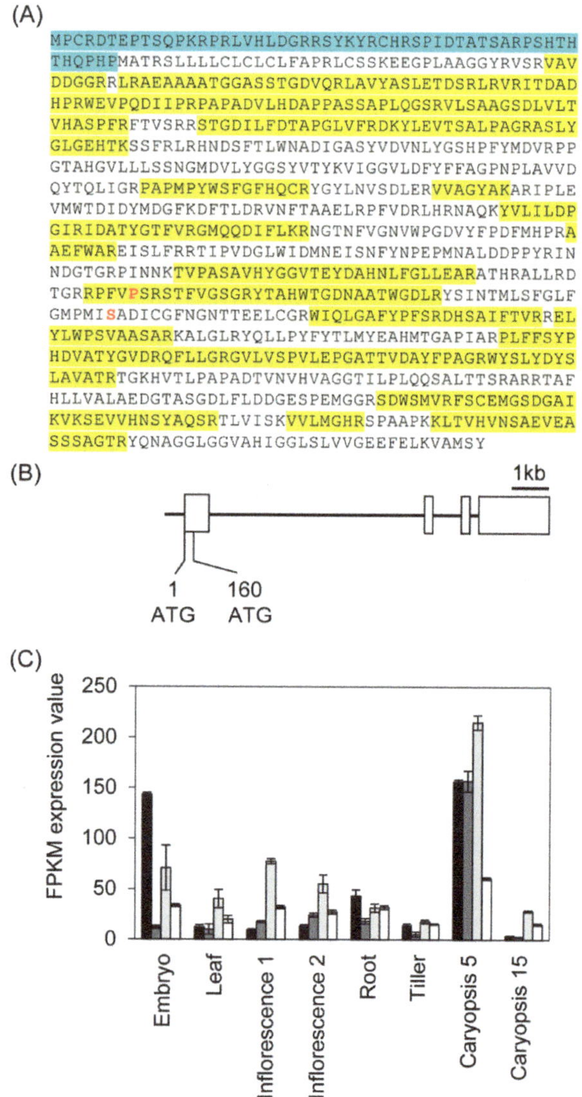

Figure 3: Identification of Maltase-Like Proteins Following Purification, and Structure of the Gene Encoding Them.

(A) Predicted amino acid sequence of maltase-like proteins Agl97 (Q9LLY2), Agl1 (D1MDV2) and predicted protein (F2DV72), and coverage of the sequence by tryptic peptides identified by mass spectrometry in peak Id from Fig 2E. These three separately-annotated barley proteins have near-identical predicted amino acid sequences. Agl97 and Agl1 are identical, whereas the predicted protein F2DV72 has an additional 53 amino acids at the N-terminus (blue), and also differs with respect to the P and S residues shown in red: in Agl1 and Agl97 these are respectively L and G. Sequences in yellow were identified among tryptic peptides from peak Id of maltase activity recovered after purification. A full list of the maltase peptides identified. (B) Structure of the gene encoding the three maltase-like proteins. Boxes represent exons. The first exon encodes the N-terminal amino acid sequence unique to F2DV72 (from bp 1 to bp 159), and a further 73 amino acids (from an ATG at bp 160) that are common to all three predicted maltase-like proteins. The remaining three exons encode amino acid sequences common to the three proteins encoded by the remaining three exons. (C) Expression pattern of the *HvAgl* genes. Results are average values from three biological replicates per tissue (± SE) and were obtained from the International Barley Genome Sequencing Consortium [34]. Data are in FPKM expression values (fragments per kilobase of exon per million fragments mapped). Black bars: *Agl1/Agl97*; grey bars: *Agl2*; light grey bars: *Agl3*, open bars: *Agl4*. Tissues analysed were: embryo (from germinating grains at 4 dpi, including root, coleoptile and scutellum); leaf (10 cm shoot stage); inflorescence 1 (young inflorescence at 5 mm); INF2: inflorescence 2 (inflorescences at 1–1.5 cm); root (from seedlings at 10 cm shoot stage); tiller (developing tillers at six leaf stage, 3rd internode); caryopsis 5 developing grain at 5 days post anthesis); caryopsis 15 (developing grain at 15 days post anthesis).

http://dx.doi.org/10.1371/journal.pone.0151642.g003

Of the peptides that matched Agl1/Agl97, three (of seven, seven and six amino acids: Fig 3A) also matched the predicted product of a barley cDNA annotated as a GH31 α-glucosidase (BAK05419; http://www.cazy.org/GH31_eukaryota; [35]). This predicted protein is only 69% identical in amino acid sequence to Agl1/Agl97, and no other peptides from the tryptic digests matched it. All three matching peptides are also found in numerous other proteins with various functions from many species, and two [(R)VVAGYAK(A) and (R)QFLLGR(G)] are common to numerous α-glucosidases from microorganisms, animals and plants (information in CAZy GH31 family; www.cazy.org/GH31.html). There is thus no strong reason to believe that the product of BAK05419 was present in peak Id. Neither tryptic digest contained peptides that matched the predicted products of barley *Agl2* (GenBank: BAG72143; MLOC_57756, chromosome 3H), *Agl3* (GenBank: BAG72144; MLOC_77920.1, chromosome 1H) or *Agl4*(GenBank: BAG72145.1; MLOC_36871.1, chromosome 4H), identified by Tanaka et al. [24], apart from a seven amino-acid peptide common to Agl97, Agl4 and many other α-glucosidases [(R)RPFVLSR(S)). The level of amino acid identity between Agls 2, 3 and 4 and Agl1/Agl97 is only 50% or less (Table 2). No other proteins predicted to be glucosidases were identified in the tryptic digests.

Table 2: Percentage of amino acid sequence identity between *Agl97*, *Agl1*, *Agl2*,*Agl3* **and** *Agl4*.

Genbank accession numbers are shown.

http://dx.doi.org/10.1371/journal.pone.0151642.t002

	Agl1	*Agl97*	*Agl2*	*Agl3*	*Agl4*	Locus	Chromosome
Agl1_ACZ37245	100.0	99.0	50.1	32.5	31.1		7H
Agl97_AAF76254.1		100.0	49.9	32.4	30.9	MLOC_66806	7H
Agl2_BAG72143			100.0	30.6	30.4	MLOC_57756	3H
Agl3_BAG72144				100.0	30.8	MLOC_77920.1	1H
Agl4_BAG72145					100.0	MLOC_36871.1	4H

doi:10.1371/journal.pone.0151642.t002

We found that both Agl97 and the longer predicted protein F2DV72 are encoded by a single gene (MLOC_66806) on the long arm of barley chromosome 7H. The same locus also encodes the α-glucosidase D1MDV2 (Agl1). The N-terminal peptide unique to the predicted protein F2DV72 is encoded by 159 nucleotides starting with an ATG translation initiation codon upstream of and in frame with the previously-identified *Agl1/Agl97* ATG codon (Fig 3B). Small sequence differences in clones obtained by different research groups (for example AAB02985.1 [16] and AAF76254.1 [18]) are almost certainly the result of use of different barley cultivars, rather than the existence of multiple, near-identical loci encoding Agl1/Agl97-like proteins. Analysis of publicly-available data showed that *Agl1/Agl97* is expressed in several organs and developmental stages of the barley plant. Of the tissues tested, expression is strongest in the early stages of grain development, and in the embryo following germination (Fig 3C). *Agl2*,*Agl3* and *Agl4* are most highly expressed in the early stages of grain development. *Agl3* is also relatively highly expressed in the embryo following germination, although at a lower level than*Agl1*/*Agl97*.

DISCUSSION

Our results indicate that all of the maltase activity in the endosperm of barley seedlings is attributable to the product of a single gene. In line with previous studies, we found that maltase activity from seedling endosperm could be separated into distinct fractions with different charge properties. However, two lines of evidence indicate that the activity in all of the fractions is a function of the same or highly similar proteins. First, for a given amount of endosperm protein, maltase activity in all of the fractions was substantially higher in extracts of endosperm from wild-type plants than in extracts from two transgenic lines carrying an RNAi silencing cassette for *Agl97* [32]. This result implies that all of the maltase activity was a function of *Agl97* expression, or of expression of *Agl97* and other gene(s) very closely related in sequence to *Agl97* over the 411 bp region used in the RNAi construct.

Second, sequencing by mass spectrometry of proteins in the maltase-containing fraction from the MonoS column revealed the presence of the previously-characterised maltase Agl97. Tryptic peptides covered 47% of the predicted Agl97 protein. The same peptides are also encoded by *Agl1* and by a third sequence known from a cDNA AK367790.1. The products of*Agl97* and *Agl1* are identical in amino-acid sequence, and this protein has been shown experimentally to have maltase activity [19]. The predicted protein encoded by AK367790.1, F2DV72, is almost identical in sequence to Agl1/Agl97, apart from an N-terminal extension of 53 amino acids. A single, unique locus in the barley genome accounts for these sequences (MLOC_66806, chromosome 7).

The pattern of expression of Agl1/Agl97 is consistent with a role for the encoded maltase specifically during seed germination. In addition to publicly available transcript data showing a high level of expression in germinating grain at 4 dpi, transcript levels were shown by Northern blotting to increase strongly in germinating grains between 2 and 3 dpi–the point at which maltase activity appeared in our experiments–and to be elevated by treatment with gibberellin [16].

It seems likely that the multiple, separable maltase activities reported from barley endosperm in this and previous studies [20,22] are all derived by post-translational modification of Agl1/Agl97. Consistent with this assumption, Finnie et al. [36] reported that the Agl97 gene product appeared as five separate spots on 2-D gels of proteins secreted from isolated barley aleurone layers. We propose two possible explanations for the multiple activities/proteins derived from Agl1/Agl97. First, expression of the *Agl1*/*Agl97* gene may produce both a protein of 97 kDa and a 103 kDa protein with an additional N-terminal domain of 53 amino acids, using the two ATG codons in the first exon: both products would be expected to be processed to smaller sizes by removal of an N-terminal signal peptide upon export from the aleurone to the endosperm. The mature maltase protein of 92 kDa purified from endosperm by Frandsen et al. [18] may originate from the 97 kDa protein, which is predicted with confidence to have a cleavage site for an N-terminal signal peptide between amino acids 23 and 24. A transcript encoding the 103 kDa protein has been identified, but it is not known whether it is translated. Tibbot et al. [17] detected proteins estimated to be of 95 and 101 kDa in endosperm of germinating seeds using an antiserum raised to recombinant Agl97. However, the predicted 103 kDa protein has no apparent signal peptide cleavage site in the first 53 amino acids (SignalP 4.1, http://www.cbs.dtu.dk/services/SignalP-4.1) so it is not clear whether it could give rise to mature proteins of the masses detected by Tibbot et al.

Second, it seems likely that the product(s) of *Agl1*/*Agl97* is subject to extensive proteolysis following secretion from the aleurone into the endosperm. Several studies have reported the purification from barley endosperm of maltases with apparent masses considerably smaller than 97 kDa (e.g. [20,23]). Tibbot et al. [17] observed that the principal proteins in endosperm recognised by an Agl97 antiserum were of 101 and 95 kDa at 4 dpi, but of only 81 kDa two days later. Other possible modifications include variable levels of glycosylation and non-enzymic glycation, both of which have been proposed to account for the occurrence of multiple forms of other endosperm proteins [37,38].

There are at least four GH31 α-glucosidase-like proteins in addition to Agl1/Agl97 encoded in the barley genome ([24] and see http://www.cazy.org/GH31). These are less than 70% identical to Agl1/Agl97, and their substrate specificities are not known. Some of them may account for the peaks Ia, Ib and Ic of α-glucosidase activity that eluted from the MonoS column, since these fractions showed activity with the generic α-glucosidase substrate pNPG but not with maltose.

Our demonstration that the maltase activity of the barley endosperm is the product of a single gene opens the way for engineering this activity. For brewing purposes, it will be interesting to test whether maltase-free grain (obtainable by gene silencing, genome editing or mutant selection for the *Agl1*/*Agl97* gene) will give rise to lower glucose to maltose ratios in wort, and hence improved fermentation by yeast. Given that the maltase is apparently not required for normal germination and growth of barley [32], a maltase-free variety should be indistinguishable from normal barley in the field.

MATERIALS AND METHODS

Plant Material and Grain Germination

Grain of the UK elite barley variety NFC Tipple (2011 harvest) was a kind gift from Dr. Phil Howell (National Institute of Agricultural Botany, Cambridge, UK). Transgenic lines carrying an RNAi interference silencing cassette for *HvAgl97* (RNAi$_{Agl97}$) were generated in cv Golden Promise [32]. Grains were surface sterilised according to Stanley et al. [32] and germinated on two filter papers (Whatman No. 1) moistened with 4 mL water in 9 cm petri dishes (10 grains per dish), at 17°C in the dark.

Purification of α-Glucosidases with Maltase Activity

Procedures were at 4°C unless otherwise stated. The roots

and coleoptile of seedlings at 10 dpi were removed and the remaining tissue (referred to as the endosperm), was rapidly frozen and stored at -80°C until use. Tissue was ground under liquid nitrogen, homogenized in medium A [100 mM MOPS/KOH (pH 7.2), 1 mM EDTA, 1 mM DTT, 10% (v/v) ethanediol, with protease inhibitors (plant cocktail, Sigma-Aldrich, UK)], and filtered through two layers of muslin. After clarification by centrifugation, (20,000*g* for 30 min) ammonium sulfate was added to the filtrate and protein precipitating between 20 and 80% saturation was collected by centrifugation as above. The pellet was resuspended in medium B [20 mM HEPES/KOH (pH 7.5), 5 mM $CaCl_2$, 5% (v/v) ethanediol], desalted with Sephadex G25 (PD-10 columns, GE Healthcare) equilibrated with the same medium, then applied to a MonoQ 5/50 GL anion exchange column (GE Healthcare) equilibrated with medium B, on an AKTA FPLC system (GE Healthcare). The column was washed with medium B at a flow rate of 1 mL min^{-1}, then eluted with a 40-mL linear gradient of 0–1 M NaCl in medium B. Proteins in the initial wash (unbound MonoQ fraction) were desalted with Sephadex G25 equilibrated with medium C [25 mM Na acetate (pH 4.5), 5 mM $CaCl_2$, 5% (v/v) ethanediol], then applied to a MonoS 5/50 GL cation exchange column equilibrated with medium C. The column was washed with medium C at a flow rate of 1 mL min^{-1}, and eluted with a 40-mL linear gradient of 0–0.5 M NaCl in medium C.

Mass Spectrometric Analysis and Mascot Search Parameters

Fractions collected during column chromatography were subjected to solid-phase extraction using OMIX C4 pipette tips (Agilent Technologies, UK), re-dissolved in 0.1 M triethylammonium bromide buffer (TEAB; Sigma-Aldrich), 0.1% (w/v) Rapigest (Waters Ltd, UK), reduced and alkylated with dithiothreitol and iodoacetamide respectively and digested with trypsin (Promega, UK) at 37°C for 16 h. Protein bands after SDS-PAGE were excised from colloidal Coomassie-stained gels and treated as described by Shevchenko et al. [39]. Peptides were extracted with 5% (v/v) formic acid, 50% (v/v) acetonitrile, dried down, and re-dissolved in 0.1% (v/v) TFA.

For LC-MS/MS analysis, samples were applied via a nanoAcquity™ (Waters Ltd) UPLC™-system at 250 nL min^{-1} to an LTQ-Orbitrap™ mass spectrometer (Thermo Fisher, MA), trapped with a Symmetry® C18 pre-column (5 μm, 180 μm x 20 mm, Waters Ltd) in-line to an analytical BEH C18 column (1.7 μm, 75 μm x 250 mm, Waters Ltd) and eluted with a gradient of 3–37% (v/v. aqueous) acetonitrile, 0.1% (v/v) formic acid. The column was connected to a 10 μm SilicaTip™ nanospray emitter (New Objective, USA) attached to a nanospray interface (Proxeon, Denmark). MS was in positive ion mode at 200°C capillary temperature. The source and focusing voltages were tuned for the transmission of MRFA peptide (m/z 524) (Sigma-Aldrich, USA). Analysis was performed in Orbitrap-IT parallel mode (CID fragmentation of the five most abundant ions in each cycle, resolution of 30,000 over the MS range from m/z 350 to m/z 1800, MS target of 1xe6 and 1 s maximum scan time, collision energy 35, isolation width 2). Only mono-isotopic 2+ and 3+ charged precursors were selected for MS2 (minimal signal of 1000, AGC target of 3xe4 ions and 150 ms scan time, dynamic exclusion 1 count and 30 s exclusion, exclusion mass window ± 20 ppm).

Raw files were processed with MaxQuant 1.3.0.5 (http://maxquant.org) [40] to generate re-calibrated peaklist-files used for database searches. The searches were performed using Mascot® 2.3 Server (Matrix Science Limited, UK) on the SPtrEMBL fasta database (Sprot_sptrembl20120711) with taxonomy set to *Hordeum vulgare* (23,891 sequences) and a common contaminants database using trypsin/P with two missed cleavages and oxidation (M) and acetylation (N-term) as variable modifications. Mass tolerances were 6 ppm for precursor ions and 0.6 Da for fragment ions. Results were evaluated in Scaffold 3.6.4 (proteomsoftware.com, Portland, USA) with thresholds of 99% and 95% for proteins and peptides respectively, resulting in a false discovery rate of 0% for both peptides and proteins.

Enzyme Assays

Endosperm extracts were prepared as for enzyme purification. Samples were desalted with Sephadex G25 (coarse) columns equilibrated with 50 mM sodium acetate (pH 5). α-Glucosidase and maltase were assayed with pNPG and maltose respectively as described by Stanley et al. [32].

ACKNOWLEDGMENTS

We are grateful to Dr. Phil Howell (National Institute of Agricultural Botany, Cambridge, UK) for the kind gift of grain of barley NFC Tipple.

AUTHOR CONTRIBUTIONS

Conceived and designed the experiments: VMEA AMS RAF. Performed the experiments: VMEA GS. Analyzed the data: VMEA GS. Contributed reagents/materials/analysis tools: GS RW RAF. Wrote the paper: VMEA AMS.

REFERENCES

1. Kristensen M, Lok F, Planchot V, Svendsen I, Leah R, Svensson B. Isolation and characterization of the gene encoding the starch debranching enzyme limit dextrinase from germinating barley. Biochim Biophys Acta 1999; 1431: 538–546. pmid:10350630 doi: 10.1016/s0167-4838(99)00077-1
2. Clark SE, Hayes PM, Henson CA. Effects of single nucleotide polymorphisms in β-amylase1 alleles from barley on functional properties of the enzymes. Plant Physiol Biochem. 2003; 41: 798–804. doi: 10.1016/s0981-9428(03)00118-9
3. Ritchie S, Swanson SJ, Gilroy S. From common signalling components to cell specific responses: insights from the cereal aleurone. Physiol Plant. 2002; 115: 342–351. pmid:12081526 doi: 10.1034/j.1399-3054.2002.1150303.x
4. Muralikrishna G, Nimala M. Cereal α-amylases—an overview. Carbohydr Polym. 2005; 60: 163–173. doi: 10.1016/j.carbpol.2004.12.002
5. Robert X, Haser R, Gottschalk TE, Ratajczak F, Driguez H, Svensson B et al. The structure of barley α-amylase isozyme 1 reveals a novel role of domain C in substrate recognition and binding: A pair of sugar tongs. Structure 2003; 11: 973–984. pmid:12906828 doi: 10.1016/s0969-

2126(03)00151-5

6. Xie Z, Zhang ZL, Zou X, Yang G, Komatsu S, Shen QJ. Interactions of two abscisic-acid induced *WRKY* genes in repressing gibberellin signaling in aleurone cells. Plant J. 2006; 46: 231–242. pmid:16623886 doi: 10.1111/j.1365-313x.2006.02694.x
7. Moreno-Risueno MA, Díaz I, Carrillo L, Fuentes R, Carbonero P. The HvDOF19 transcription factor mediates the abscisic acid-dependent repression of hydrolase genes in germinating barley aleurone. Plant J. 2007; 51: 352–365. pmid:17565581 doi: 10.1111/j.1365-313x.2007.03146.x
8. Finnie C, Svensson B. Barley seed proteomics from spots to structures. J Proteom. 2009; 72: 315–324. doi: 10.1016/j.jprot.2008.12.001
9. Vester-Christensen MB, Hachem MA, Svensson B, Henriksen A. Crystal structure of an essential enzyme in seed starch degradation: barley limit dextrinase in complex with cyclodextrins. J Mol Biol. 2010; 403: 739–750. doi: 10.1016/j.jmb.2010.09.031. pmid:20863834
10. Bamforth CW. Current perspectives on the role of enzymes in brewing. J Cereal Sci. 2009; 50: 353–357. doi: 10.1016/j.jcs.2009.03.001
11. Evans DE, Li C and Eglinton JK. Improved prediction of malt fermentability by measurement of the diastatic power enzymes β-amylase, α-amylase and limit dextrinase: I. Survey of the levels of diastatic power enzymes in commercial malts. J Am Soc Brew Chem. 2008;66: 223–232. doi: 10.1094/asbcj-2008-0909-02
12. Evans DE, Li C, Eglintorr JK. The properties and genetics of barley malt starch degrading enzymes. In: Zhang G, Li C, editors. Genetics and Improvement of Barley Malt Quality. Hangzhou: Springer, Zhejiang University Press; 2009. pp. 143–189.
13. Bathgate GN, Bringhurst TA. Update on knowledge regarding starch structure and degradation by malt enzymes (DP/DU and limit dextrinase). J Inst Brew. 2011; 117: 33–38. doi: 10.1002/j.2050-0416.2011.tb00440.x
14. Frandsen TP, Svensson B. Plant alpha-glucosidases of the glycoside hydrolase family 31. Molecular properties, substrate specificity, reaction mechanism, and comparison with family members of different origin. Plant Mol Biol. 1998; 37: 1–13. pmid:9620260
15. Ling AR, Nanji DR. On the presence of maltase in germinated and ungerminated barley. Biochem J. 1923; 17: 593–596. pmid:16743257 doi: 10.1042/bj0170593
16. Tibbot BK, Skadsen RW. Molecular cloning and characterization of a gibberellin-inducible, putative alpha-glucosidase gene from barley. Plant Mol Biol. 1996; 30: 229–241. pmid:8616248 doi: 10.1007/bf00020110
17. Tibbot BK, Henson CA, Skadsen RW. Expression of enzymatically active, recombinant barley α-glucosidase in yeast and immunological detection of α-glucosidase from seed tissue. Plant Mol Biol. 1998; 38: 379–391. pmid:9747846
18. Frandsen TP, Lok F, Mirgorodskaya E, Roepstorff P, Svensson B. Purification, enzymatic characterization, and nucleotide sequence of a high-isoelectric-point alpha-glucosidase from barley malt. Plant Physiol. 2000; 123: 275–286. pmid:10806244 doi: 10.1104/pp.123.1.275
19. Naested H, Kramhøft B, Lok F, Bojsen K, Yu S, Svensson B. Production of enzymatically active recombinant full-length barley high pI α -glucosidase of glycoside family 31 by high cell-density fermentation of *Pichia pastoris* and affinity purification. Protein Expr Purif. 2006; 46: 56–63. pmid:16343940 doi: 10.1016/j.pep.2005.10.008
20. Stark JR, Yin XS. Evidence for the presence of maltase and α-glucosidase isoenzymes in barley. J Inst Brew. 1987; 93: 108–112. doi: 10.1002/j.2050-0416.1987.tb04485.x
21. Sun Z, Henson CA. Degradation of native starch granules by barley α-glucosidases. Plant Physiol. 1990; 94: 320–327. pmid:16667704 doi: 10.1104/pp.94.1.320
22. Sissons MJ, MacGregor AW. Hydrolysis of barley starch granules by α-glucosidases from malt. J Cereal Sci. 1994; 19: 161–169. doi: 10.1006/jcrs.1994.1022
23. Im H, Henson CA. Characterization of high pI α-glucosidase from germinated barley seeds: substrate specificity, subsite affinities and active-site residues. Carbohydr Res. 1995; 277: 145–159. doi: 10.1016/0008-6215(95)00212-c
24. Tanaka Y, Duke S, Henson CA. Alpha-glucosidases from the glycoside hydrolase family 31 in germinating seeds and seedling leaves of barley. In: Proceedings of the XXIIIrd International Carbohydrate Symposium. Whistler, Canada: American Chemical Society/National Research Canada; 2006 (USDA publication #196842;http://www.ars.usda.gov/research/publications/publications.htm?seq_no_115=196842&pf=1.
25. Sun Z, Henson CA. A quantitative assessment of the importance of barley seed α-amylase, β-amylase, debranching enzyme, and α-glucosidase in starch degradation. Arch Biochem Biophys. 1991; 284: 298–305. pmid:1824915 doi: 10.1016/0003-9861(91)90299-x
26. Osman AM. The advantage of using natural substrate-based methods in assessing the roles and synergistic and competitive interactions of barley malt starch-degrading enzymes. J Inst Brew. 2002; 108: 204–214. doi: 10.1002/j.2050-0416.2002.tb00542.x
27. Muslin EH, Karpelenia CB, Henson CA. The impact of thermostable alpha-glucosidase on the production of fermentable sugars during mashing. J Am Soc Brew Chem. 2003; 61: 142–145.
28. Muslin EH, Clark SE, Henson CA. The effect of proline insertions on the thermostability of a barley α-glucosidase. Prot Eng Design Select. 2002; 15: 29–33. doi: 10.1093/protein/15.1.29
29. Clark SE, Muslin EH, Henson CA. Effect of adding and removing *N*-glycosylation recognition sites on the thermostability of barley α-glucosidase. Prot Eng Design Select. 2004; 17: 245–249. doi: 10.1093/protein/gzh028
30. Phaweni P, O'Connor-Cox ESC, Pickerell ATW, Axcell

BC. Influence of adjunct carbohydrate spectrum on the fermentative activity of a brewing strain of *Saccharomyces cerevisiae*. J Am Soc Brew Chem. 1993; 51: 0010.

31. Hazell BW, Attfield PV. Enhancement of maltose utilisation by *Saccharomyces cerevisiae* in medium containing fermentable hexoses. J Ind Microbiol Biotechnol. 1999; 22: 627–632. pmid:10455490 doi: 10.1038/sj.jim.2900672
32. Stanley D, Rejzek M, Naested H, Smedley M, Otero S, Fahy B, et al. The role of α-glucosidase in germinating barley grains. Plant Physiol. 2011; 155: 932–943. doi: 10.1104/pp.110.168328. pmid:21098673
33. Asano N. Glycosidase inhibitors: update and perspectives on practical use. Glycobiol. 2003; 13: 93R–104R. doi: 10.1093/glycob/cwg090
34. International Barley Genome Sequencing Consortium. A physical, genetic and functional sequence assembly of the barley genome. Nature 2012; 491: 711–716. doi: 10.1038/nature11543. pmid:23075845
35. Matsumoto T, Tanaka T, Sakai H, Amano N, Kanamori H, Kurita K, et al. Comprehensive sequence analysis of 24,783 barley full-length cDNAs derived from 12 clone libraries. Plant Physiol. 2011; 156: 20–28. doi: 10.1104/pp.110.171579. pmid:21415278
36. Finnie C, Andersen B, Shahpiri A, Svensson B. Proteomes of the barley aleurone layer: A model system for plant signalling and protein secretion. Proteomics 2011; 11: 1595–1605. doi: 10.1002/pmic.201000656. pmid:21433287
37. Barba-Espín G, Dedvisitsakul P, Hägglund P, Svensson B, Finnie C. Gibberellic acid-induced aleurone layers responding to heat shock or tunicamycin provide insight into the N-glycoproteome, protein secretion, and endoplasmic reticulum stress. Plant Physiol. 2014; 164: 951–965. doi: 10.1104/pp.113.233163. pmid:24344171
38. Bobálová J, Petry-Podgórska I, Laštovicková M, Chmelík J. Monitoring of malting process by characterization of glycation of barley protein Z. Eur Food Res Technol. 2010; 230: 665–673. doi: 10.1007/s00217-009-1205-y
39. Shevchenko A, Tomas H, Havli J, Olsen JV, Mann M. In-gel digestion for mass spectrometric characterization of proteins and proteomes. Nature Protoc. 2007; 1: 2856–2860. doi: 10.1038/nprot.2006.468
40. Cox J, Mann M. MaxQuant enables high peptide identification rates, individualized p.p.b.-range mass accuracies and proteome-wide protein quantification. Nature Biotechnol. 2008; 26: 1367–1372. doi: 10.1038/nbt.1511

CHAPTER 15

ASSESSMENT OF GENETIC DIVERSITY IN BRAZILIAN BARLEY USING SSR MARKERS

Jéssica Rosset Ferreira, Jorge Fernando Pereira, Caroline Turchetto, Euclydes Minella, Luciano Consoli and Carla Andréa Delatorre

Keywords: Domesticated barley, genetic diversity, *Hordeum vulgare* ssp. vulgare, *Hordeum vulgare* ssp. spontaneum, microsatellite markers

INTRODUCTION

Barley (*Hordeum vulgare*) was one of the first cereal species domesticated by humans and, since then, it is cultivated in several environments (Ladizinsky 1998; Feuillet *et al.*, 2007). It is considered one of the best adapted cereals due to its tolerance to salinity, low temperatures and lower water demand (Baik and Ullrich, 2008;Goyal and Ahmed, 2012), although it is highly sensitive to aluminum (Bian *et al.*, 2013). A broad range of end use, such as human consumption, malt in brewing and distilling industry and animal feeding, makes barley one of the most important cereal crops in the world ranking as the fourth most produced cereal after wheat, maize, and rice.

Cultivated barley (*Hordeum vulgare* ssp. *vulgare*) and its wild progenitor (*Hordeum vulgare*ssp.*spontaneum*) belong to the barley primary gene pool (Minella, 2005). Greater allelic diversity has been detected in wild than in cultivated barley (Saghai Maroof *et al.*, 1994; Ledovskoy *et al.*, 2010; Nandha and Singh, 2014).

The wild progenitor is of great interest for crop breeding because it carries a rich source of genes, enabling the adaptation to different biotic and abiotic stresses (Nevo *et al.*, 2012). Moreover, these species preserved their intercrossing ability. The genetic diversity found in cultivated barley varies among different studies, some of them reporting high genetic diversity (Baik and Ullrich, 2008;Yahiaoui *et al.*, 2008; Khodayari *et al.*, 2012). Previous studies also showed different levels of genetic diversity in accessions of different geographic origins. For instance, genotypes from Europe showed lower genetic diversity compared to those of other continents (Malysheva-Otto *et al.*, 2006).

In the last years, Brazil imported around 350 thousand tons yearly to supply the domestic needs, and 75% of the barley has been destined to produce malt for beer brewing (de Mori and Minella, 2012). Area expansion, and consequently production increase, would require the development of cultivars adapted to the diverse Brazilian environments. Hence, knowledge about the allelic composition of materials may facilitate the parental selection for breeding to specific environments. However, up to now, few genetic variability studies have been conducted for Brazilian barley. SDS-PAGE analysis was employed to analyze the hordein polypeptide patterns, which revealed a surprisingly extensive intravarietal polymorphism for Brazilian barley varieties (Echart-Almeida and CavalliMolina, 2000). The intravarietal variability was also detected by RAPD and isoenzymes, although in lower frequency (Selbach and Cavalli-Molina, 2000; Maris AF, 1992, Bachelor's thesis, Universidade Federal do Rio Grande do Sul, Porto Alegre, RS). However, a great genetic similarity was found among cultivars (Echart-Almeida and Cavalli-Molina, 2000; Kroth *et al.*, 2005; Maris AF, 1992, Bachelor's thesis, Universidade Federal do Rio Grande do Sul, Porto Alegre, RS). Another good option for cultivar characterization is the use of microsatellite markers (SSR) that possess high levels of polymorphism, are multiallelic and spread throughout the genome (Gupta *et al.*, 1996; Oliveira *et al.*, 2006). Although SSR markers have been used to characterize barley from different countries (Saghai Maroof *et al.*, 1994; Karakousis *et al.*, 2003; Baik and Ullrich, 2008; Yahiaoui*et al.*, 2008; Ledovskoy *et al.*, 2010;Khodayari *et al.*, 2012; Nandha and Singh, 2014), no

Citation: Jéssica Rosset Ferreira, Jorge Fernando Pereira, Caroline Turchetto, Euclydes Minella, Luciano Consoli and Carla Andréa Delatorre, Assessment of genetic diversity in Brazilian barley using SSR markers, doi: 10.1590/1678-4685-GMB-2015-0148.

studies evaluating the molecular diversity of Brazilian barley using this kind of markers have been conducted.

The objectives of the present study were to describe the molecular diversity of Brazilian barley, as well as to compare that diversity with the one found for foreign and wild genotypes. The aim was to detect new genetic variability and thus to assist the breeding programs in the development of new cultivars.

MATERIAL AND METHODS

Plant Material and Genomic DNA Isolation

Seeds of 64 barley accessions, provided by the Active Germplasm Bank of CNPT-Embrapa, Brazil, were cultivated under controlled conditions. These accessions represent two- and six-row materials including six wild (*Hordeum vulgare* ssp. *spontaneum*) genotypes, 35 cultivars and 16 breeding lines developed by different breeding programs in Brazil, and seven cultivars from other countries (Table 1). The Brazilian accessions used here have spring growth habit and are representative of the germplasm cultivated in Brazil in the last four decades. Because we were interested in evaluating the genetic variability among the different Brazilian breeding programs (development institution), the wild and domesticated accessions, the breeding lines and cultivars, and among the Brazilian and foreign genotypes, we grouped the 64 barley accessions in seven sets according to development institution, country, and type as detailed in Table 2.

Table 1: Barley genotypes used in this study

Genotype	Row	Institution	Country	Year	Pedigree
Alpha	2	Foreign	USA	-	`Manchuria'/'Champion of Vermont'
Antarctica 01	2	Cervejaria Antarctica Paulista	Brazil	1960's	Selection of Breun Volla
Antarctica 04	2	Cervejaria Antarctica Paulista	Brazil	1999	Alpha1959/Union
Antarctica 05	2	Cervejaria Antarctica Paulista	Brazil	1999	Unkown
BRS 180	6	CNPT/Embrapa	Brazil	1999	73Ab2199/Karla
BRS 195	2	CNPT/Embrapa	Brazil	2001	Defra/BR 2
BRS 224	2	CNPT/Embrapa	Brazil	2002	Embrapa 43/PFC 9114
BRS 225	2	CNPT/Embrapa	Brazil	2002	PFC 9103/Defra
BRS Brau	2	CNPT/Embrapa	Brazil	2009	MN 698/3/BRS 195//Schooner/Embrapa 129
BRS Cauê	2	CNPT/Embrapa	Brazil	2008	BRS Borema/BRS 195
BRS Elis	2	CNPT/Embrapa	Brazil	2008	BRS 195/Scarlett
BRS Greta	2	CNPT/Embrapa	Brazil	2006	Krona/PFC 9219//PFC 9204
BRS Itanema	2	CNPT/Embrapa	Brazil	2013	BRS 195//PFC8590/PFC9205
BRS Lagoa	2	CNPT/Embrapa	Brazil	2005	PFC 9215/PFC 9288
BRS Marciana	2	CNPT/Embrapa	Brazil	2005	PFC 9240/PFC 9211
BRS Mariana	2	CNPT/Embrapa	Brazil	2005	PFC 88137/PFC 8905//PFC 9205
BRS Sampa	2	CNPT/Embrapa	Brazil	2008	BRS 195//PFC 8590/PFC 9205
BRS Suábia	2	CNPT/Embrapa	Brazil	2006	BRS 195/MN 698
Cevada BR1	2	CNPT/Embrapa	Brazil	1987	Selection of W 5586 =Binder/Opal//Balder/kenia/3/Alpha
Cevada BR2	2	CNPT/Embrapa	Brazil	1989	FM 424/TR 206
Dayton	6	Foreign	USA	1955	Composite cross X (CI 6625) selection
Embrapa 127	2	CNPT/Embrapa	Brazil	2000	BR2/Alexis
Embrapa 128	2	CNPT/Embrapa	Brazil	1999	LM 844/PFC 84148//BR2
Embrapa 129	2	CNPT/Embrapa	Brazil	1999	LM 844/MN610//Cevada BR 2
Embrapa 43	2	CNPT/Embrapa	Brazil	1999	IPB 194//C 2146/TR 208
FM 404	2	Maltaria Navegantes (Filial Maltaria)	Brazil	1970's	Selection from a cross probably involving Alpha
FM 424	2	Maltaria Navegantes (Filial Maltaria)	Brazil	1970's	Quinn/Malteria Heda//W 5746
FM 434	2	Maltaria Navegantes (Filial Maltaria)	Brazil	1970's	Quinn/Malteria Heda//FM 424
FM 437	2	Maltaria Navegantes (Filial Maltaria)	Brazil	1985	Alpha 1959/2*Mansholt Twerijge Zomergerst
FM 519	2	Maltaria Navegantes (Filial Maltaria)	Brazil	1985	Kr 1/Union//Volla/3/Kr 2/Volla//Wisa/4/Alpha
Golden Promise	2	Foreign	England	1966	Maythorpe Gamma-Ray Mutant
Hspo 584	2	-	-	-	-
Hspo PI 282590	2	-	Israel	-	-
Hspo PI 466338	2	-	Israel	-	-
Hspo PI 466394	2	-	Israel	-	-
Hspo PI 466396	2	-	-	-	-
Hspo PI 466381	2	-	Israel	-	-
IAC 74310	2	Instituto Agronômico de Campinas	Brazil	1974	Unkown
IPB 1219	2	International Plant Breeders	Brazil	1980	Vada/Zephyr
IPB 194	2	International Plant Breeders	Brazil	1980	Mazurka/Nackta
MN 599	2	Maltaria Navegantes	Brazil	1990	Ariana/Volla//FM 462
MN 6021	2	Maltaria Navegantes	Argentina	2012	Dominique/Quilmes Ayelen
MN 656	2	Maltaria Navegantes	Brazil	1993	SG 4279/FM 404//Bacco/Union/3/ FM 434
MN 684	2	Maltaria Navegantes	Brazil	2001	Antarctica 05/MN 577
MN 698	2	Maltaria Navegantes	Brazil	2001	MN 599/MN 635
MN 743	2	Maltaria Navegantes	Brazil	2004	MN 681/Gimpel
Murasakimochi	6	Foreign	Japan	-	-

Paraí-I	6	CNPT/Embrapa	Brazil	-	Unkown
PFC 7802	2	CNPT/Embrapa	Brazil	7802	Binder/Opal//Balder/Kenya
PFC 8115	2	CNPT/Embrapa	Brazil	1981	Volla*3/Wpgm 626-46-25
PFC 8153	2	CNPT/Embrapa	Brazil	1981	FM 424*2//Volla/Wpgm 626-46-25
PFC 8280	2	CNPT/Embrapa	Brazil	1982	Volla/C 2146
PFC 84148	2	CNPT/Embrapa	Brazil	1984	Volla*3/Wpgm 626-46-25//MAGNIF 131
PFC 8601	2	CNPT/Embrapa	Brazil	1986	Seleção em TR 207 em solo corrigido (1/8 SMP)
PFC 8610	2	CNPT/Embrapa	Brazil	1986	PFC 8144/Park
PFC 86125	2	CNPT/Embrapa	Brazil	1986	PFC 8153/IPB 194
PFC 8644	2	CNPT/Embrapa	Brazil	1986	Antarctica 05*3/Park
PFC 88209	6	CNPT/Embrapa	Brazil	1988	Selection of FM 70
PFC 88210	6	CNPT/Embrapa	Brazil	1988	Selection of FM 71
PFC 88211	6	CNPT/Embrapa	Brazil	1988	Selection of FM 80
PFC 88212	6	CNPT/Embrapa	Brazil	1988	Selection of FM 80
Quench	2	Foreign	England	-	Sebastian/Drum
Quest	6	Foreign	USA	-	MN Brite/Zhedar 1
Vacaria	6	CNPT/Embrapa	Brazil	-	Unkown

Table 2: Groups of genotypes separated according to institution, country and barley type

	CNPT/Embrapa cultivars	Companhia Antarctica Paulista	Maltaria Navegantes	Breeding lines from other institutions[1]	Breeding lines from CNPT/Embrapa[2]	Foreign genotypes	Wild barley
Number of genotypes	22	3	10	3	13	7	6
Total of alleles	146	61	105	63	130	123	145
Alleles per locus	4.29	1.79	3.09	1.85	3.82	3.62	4.26
PIC	0.46	0.27	0.43	0.32	0.52	0.57	0.63

[1] Lines from Instituto Agronômico de Campinas and International Plant Breeders.
[2] Genotypes identified in Table 1 as "PFC".

Young leaves of the seedlings were collected, transferred to 2 mL Eppendorf tubes containing three stainless steel beads (2.3 mm diameter) and immediately frozen in liquid nitrogen. Tissue was powdered using a Mini-Beadbeater™ (Biospec Products) platform and total genomic DNA was extracted using a CTAB protocol (Doyle and Doyle, 1987). The DNA quality and quantity was estimated on 0.8% agarose gels. DNA was stored at −20 ° C for further analyses.

Microsatellite (SSR) Amplification

Genomic DNA was amplified using 34 previously described SSR markers (Liu *et al.*, 1996; Röder *et al.*, 1998; Ramsay *et al.*, 2000; Varshney *et al.*, 2007;Soto-Cerda *et al.*, 2013) that are scattered throughout the barley genome. The details of the 34 SSR primers are available in Table S1.

PCR was conducted based on a three primer system, as described by Schuelke (2000), where one of the locus-specific primers (forward or reverse) was extended with a non-labeled M13-tail (TGTAAAACGAC GGCCAGT), and a M13 primer labeled with a fluorescent dye (FAM, NED, PET or VIC). The reaction mixture was prepared in a final volume of 10 µL containing 1x buffer (RBC Bioscience), 0.2 µM primer, 0.02 µM M13 tailed primer and 0.2 µM fluorescence-labeled M13 primer. Other reagents were optimized for each primer and added as mixA [1.5 mM $MgCl_2$ (RBC Bioscience), 0.2 mM of each dNTP (Thermo Scientific), 0.5 U of *Taq* polymerase (RBC Bioscience)], mix C [2.5 mM $MgCl_2$, 0.2 mM of each dNTP, 0.75 U of *Taq* polymerase], or mixD [2.5 mM $MgCl_2$, 0.35 mM of each dNTP, 0.75 U of *Taq* polymerase]. Amplification was performed in a GenAmp® PCR System 9700 (Applied Biosystems) with two programs set according to the melting temperatures of the primers (Table S1): TD60-50 (94 °C for 30 s, 60 °C for 30 s, and 72 °C for 30 s, followed by 10 cycles at decreasing annealing temperatures of 1 °C per cycle and then 30 cycles at 94 °C for 30 s, 50 °C for 30 s, 72 °C for 30 s), or TD60-55 (94 °C for 30 s, 60 °C for 30 s, and 72 °C for 30 s, followed by five cycles at decreasing annealing temperatures of 1°C per cycle and then 35 cycles at 94 °C for 30 s, 55 °C for 30 s, 72 °C for 30 s). After amplification, the reactions from up to four primer combinations containing different fluorescent dyes were multiplexed, diluted, mixed with Hi-Di formamide and GeneScan 500 LIZ size standard (Applied Biosystems), denatured and run on an ABI 3130xL Genetic Analyzer containing a 36 cm capillary array with POP7 polymer. The program GeneMapper v3.5 was used to determine allele sizes.

Data Analysis

FSTAT 2.9.3.2 software (Goudet, 2002) was used to evaluate summary statistics, such as the number of alleles per locus (*A*) and number of private alleles (*E*) for each locus and groups of accessions. Polymorphism information content (PIC) was estimated using the PowerMarker software (Liu and Muse, 2005) according to Andersonet *al.* (1993): PIC = $1 - \sum \mathbf{P}_{ij}^{\ 2}$, where P_{ij} is the frequency of the*i*th allele at the *j*th marker, to evaluate the diversity level of each SSR marker.

In addition, we implemented analyses of molecular variance (AMOVA; Excoffier *et al.*, 1992) using ARLEQUIN 3.5.1.2 software (Excoffier and Lischer, 2010) among seven groups of accessions. Grouping of the accessions was established according to the institution that developed the cultivars or breeding lines (Table 1). The genetic diversity among Brazilian genotypes developed in different decades was estimated based on the number of alleles, frequency of alleles and PIC values.

To investigate the genetic similarity between accessions, we carried out a Principal Coordinates Analysis in GENALEX 6.4 software (Peakall and Smouse, 2006, 2012). A matrix of distance based on shared alleles among accessions was used to depict relationships among all accessions; the original data were bootstrapped 10,000 times using PowerMarker software (Liu and Muse, 2005). In addition, an Unweighted Pair Group Method with Arithmetic Mean (UPGMA) tree was constructed based on the matrix of shared microsatellite alleles among the 64 accessions calculated from 34 SSRs using the PowerMarker software (Liu and Muse, 2005). A Bayesian clustering analysis as implemented in STRUCTURE 2.3 software (Pritchard *et al.*, 2000) was performed to compare to UPGMA and PCoA results. This approach uses a Bayesian clustering analysis to assign individuals to clusters (K) without prior knowledge of their population affinities. The parameters were correlated to allele frequencies (Falush*et al.*, 2003) and no prior accession information was used. The number of groups (K) was evaluated from one to 20, with 10 independent runs per K value, to determine the maximum value of the posterior likelihood [lnP(D)] and the best value of K. Each run was performed using 2.5 10^5 burn-in periods and 10^6 Markov chain Monte Carlo (MCMC) repetitions after burn-in, and the convergence was checked. The optimal value of K was calculated using the maximum value of ΔK (Evanno *et al.*, 2005) implemented in STRUCTURE HARVESTER 0.6.93 (Earl and von Holdt, 2012). CLUMPP 1.1.2 software was used to summarize the results of the optimal K value on the basis of the average pairwise similarity of individual assignments across runs using Greedy's method and the G' statistic (Jakobsson and Rosenberg, 2007). DISTRUCT 1.1 software (Rosenberg, 2004) was used to visualize the STRUCTURE results after processing with the CLUMPP software.

RESULTS

Genetic Variation

We have evaluated 34 SSR loci in 64 barley accessions including two- and six-row genotypes as well as wild and domesticated barley. Although 11 genotypes from six countries are among the ones studied here, the majority of accessions represent genotypes developed in Brazil. The 34 markers were spread through the barley genome with at least four markers on each chromosome. We detected 280 alleles with an average of eight alleles per locus. The number of alleles per locus varied from one for markers XGWM6, HvLOX and Bmac0251 (monomorphic loci) to 18 for the marker Bmag0032 (Table 3). Sixteen markers detected at least 10 alleles and only eight amplified less than five alleles. For only one marker (Bmac0167), a null allele was observed. An association between the number of motif repeats and the number of alleles could be established. On average, loci containing 15 or more motif repeats generated 10 alleles per marker while the ones with less than 15 motif repeats showed five alleles per marker. Among the monomorphic loci, two contained less than 15 motif repeats while one (XGWM6) possessed the highest number (40 repeats) among the evaluated markers. The highest number of alleles was found in the 22 cultivars developed by CNPTEmbrapa (146 alleles averaging four alleles per locus) and in the six wild accessions (145 alleles averaging four alleles per locus). However, considering the number of accessions within each set, the foreign genotypes also generated a high number of alleles (Table 2).

Table 3: Repeat motif, number of alleles, allele size range, and polymorphism information content (PIC) of the 34 SSR markers used

Marker	Repeat motif	Number of alleles	Allele size range	PIC
Bmac0090	(AC)20	10	207-247	0.75
Bmag0032	(AC)7T(CA)15(AT)9	18	211-319	0.78
Bmag211	(CT)16	6	193-207	0.54
HVM20	(GA)19	6	150-172	0.66
Bmag0125	(AG)19	10	136-162	0.77
Bmag0378	(AG)14	8	150-170	0.26
Bmag0749	(AG)11	4	180-186	0.62
HVM36	(GA)13	6	124-166	0.65
HVM54	(GA)14	7	166-182	0.58
Bmac0067	(AC)18	9	180-250	0.72
Bmag0013	(CT)21	12	147-183	0.81
Bmag225	(AG)26	10	155-185	0.81
HVM60	(AG)11(GA)14	10	101-133	0.77
Bmac0030	(AC)22	11	150-180	0.83
Bmac0310	(CT)11(AC)20	11	159-199	0.76
Bmag0353	(AG)21	11	105-145	0.69
EBmac0669	(AC)8	2	236-238	0.07
EBmag0781	(CT)21	4	162-182	0.26
HVM68	(GA)22	12	196-240	0.77
HvML03	(CTT)6	3	251-266	0.33
wms165	(GA)20	8	215-263	0.23
XGWM6	(GA)40	1	196	0.00
Bmac0113	(AT)7(AC)18	10	208-236	0.81
Bmac096	(AT)6(AC)16	11	180-212	0.77
Bmag0387	(AG)16	8	119-161	0.53
HvLOX	(AG)9	1	170	0.00
Bmac0316	(AC)19	14	142-200	0.66
Bmac251	(AC)12A(AC)13	1	161	0.00
Bmag0173	(CT)29	8	156-180	0.79
HVM65	(GA)10	3	139-143	0.30
Bmac0031	AC(28)	10	172-211	0.53
Bmac0167	AC(20)	10[1]	180-212	0.76
Bmag0120	(AG)15	17	230-290	0.86
Bmag0135	(AG)10GG(AG)12	8	133-179	0.79

[1] Null allele was observed.

Private alleles were detected in all sets except from Companhia Antarctica Paulista (Table S2). In the wild accessions, 69 private alleles were detected in 28 different loci (Table S2). Among the foreign genotypes and the cultivars developed by CNPT-Embrapa we found 23 and 21 private alleles, respectively. Only one private allele was visualized in the breeding lines developed by other institutions. The Bmag0032 marker was the locus with the highest number of alleles and presented the highest number of private alleles (Table S2).

Furthermore, indicators of locus diversity (polymorphism information content - PIC) were calculated (Table 3). Large variation was observed among the polymorphic loci; PIC values ranged from 0.07 (EBmac0669) to 0.86 (Bmag120) with an average of 0.57. Not considering the monomorphic loci, only five markers originated PIC values below 0.50, therefore the majority of the markers were moderately to highly informative, according to the criteria proposed by Botstein *et al.* (1980). Among the five loci with PIC <0.50, three were located on chromosome four. Chromosomes four and six showed the lowest mean PIC value (0.43) among all chromosomes. Among the different sets, the highest PIC value was observed for wild barley (0.63) evidencing higher genetic diversity in these accessions. The second highest PIC value was found in the foreign genotypes (0.57), while PIC values varied from 0.27 (Companhia Antarctica Paulista) to 0.46 (CNPT-Embrapa) among the cultivars developed by the different breeding programs in Brazil. The PIC value considering all Brazilian genotypes (cultivars and breeding lines) was 0.51. It indicates a lower genetic diversity compared with foreign (0.57) and wild (0.63) accessions. Furthermore, the PIC value was lower among Brazilian genotypes developed in the last two decades in comparison to materials released in the 1980s (Figure 1A) suggesting a trend toward decreased diversity. The loss of genetic diversity is also supported by the statistically superior number of alleles detected in the genotypes developed in the 1980s (Figure 1B). Moreover, a change in the frequency of alleles was clearly observed between genotypes developed in the 1980s and 2000s where a reduced number of alleles per marker was observed in the genotypes developed in the 2000s (Figures 1C,D). The frequency of alleles per marker between genotypes from these two decades can be compared since a similar number of accessions developed in the 1980s and 2000s were analyzed (18 and 16, respectively).

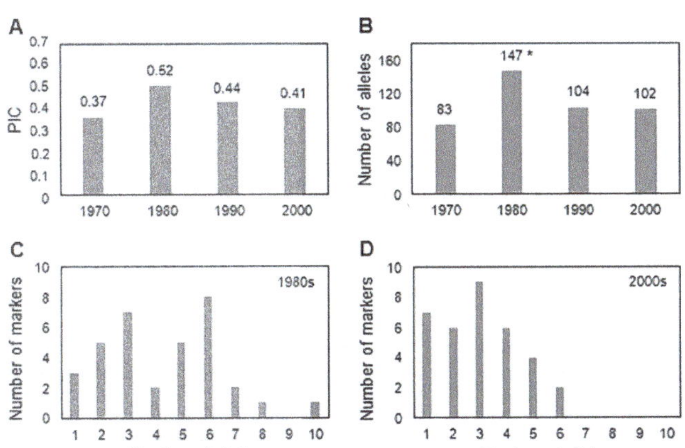

Figure 1: PIC values (A) and total number of alleles (B) of the genotypes developed in Brazil in the last four decades. Frequency of alleles per marker is shown for genotypes developed in the 1980s (C) and after 2000 (D). Number of genotypes per decade are: 5 (1970); ...

Genetic Similarity

In order to assess the clustering of barley accessions based on SSR polymorphism, we conducted a Principal Coordinates Analysis (PCoA). The scatter plots for the two first axes showed that accessions formed two principal groups (Figure 2A). The first group (small circle on the right-hand side of the PCA plot) was further subdivided into two. One subgroup (larger dotted circle in Figure 2A) contained all wild accessions, two CNPT-Embrapa cultivars (Paraí and BRS 180), and three foreign cultivars (Dayton, Quest and Murasakimochi), while the other subgroup (smaller dotted circle in Figure 2A) included three CNPT-Embrapa breeding lines (PFC 88209, PFC 88210, and PFC 88211), and one CNPT-Embrapa cultivar (Vacaria). The second group (large circle on the left-hand side of the PCA plot) was composed of all other accessions and showed higher variation. Within this second group, the CNPT-Embrapa cultivars were scattered over a large area, with most of them being located in a separate region.

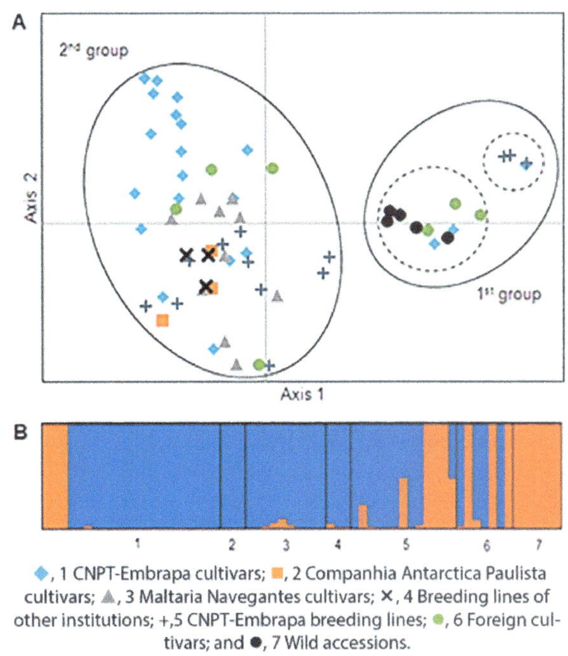

Figure 2: Analysis plots of the barley accessions based on the polymorphism of 34 microsatellite markers. (A) By Principal Coordinate Analysis (PCoA) two groups were identified, with the first one separated into two subgroups. Two symbols, in the first group, indicating ...

The AMOVA analysis revealed that 15% of the genetic variation (P <0.001) was distributed among the seven sets, representing the different breeding programs, countries and type of accession (wild/domesticated and breeding lines/cultivars), whereas 85% (P <0.001) was within the sets. When we consider only accessions developed in Brazil, the AMOVA showed 12% of the genetic variation among sets, and 88% (P <0.001) within sets (P <0.001). These results indicated that the genetic variation within the sets contributed to most of the genetic diversity detected.

The UPGMA cluster analysis produced three main groups (named as Groups A, B and C) (Figure 3). Group A consisted of all wild accessions except Hspo 584, which was clustered in the Group C. Besides Hspo 584, Group C also contained 10 other accessions represented by breeding lines and cultivars developed by CNPT-Embrapa and foreign genotypes. When comparing the accessions present in Groups A and C with the ones belonging to the first group detected by the PCoA analysis, only the line PFC 88212 was different between them revealing agreement patterns across analyses.

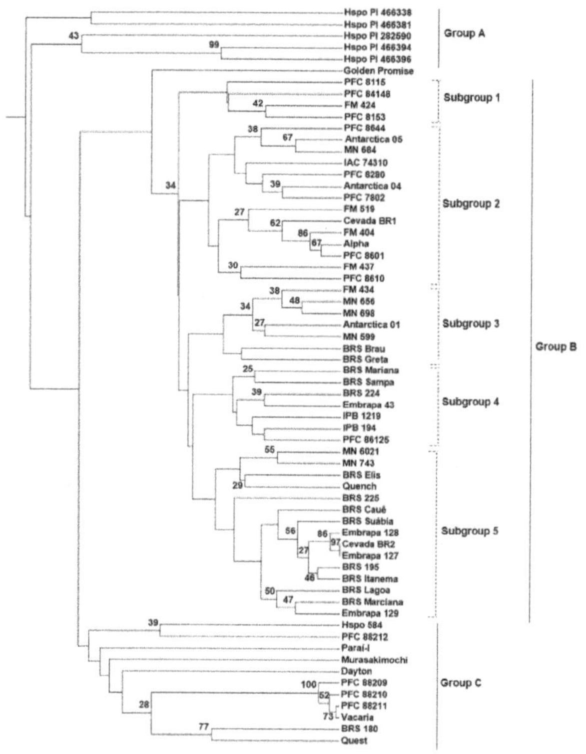

Figure 3: UPGMA tree of barley accessions constructed based on the matrix of shared microsatellite alleles. Bootstrap values are indicated at nodes. Only bootstrap values ≥ 25% are shown.

Group B in the dendrogram was the largest, consisting of 48 accessions representing four foreign materials and 44 Brazilian genotypes. This means that 86% of the Brazilian genotypes were clustered inside the same group. Group B was further substructured into five subgroups. Subgroup one consisted of three CNPT-Embrapa breeding lines and one cultivar developed by Maltaria Navegantes. Subgroup two contained breeding lines developed by CNPTEmbrapa and Instituto Agronômico de Campinas (IAC), cultivars released by CNPT-Embrapa, Companhia Antarctica Paulista and Maltaria Navegantes, and one foreign genotype (Alpha). Subgroup three consisted of cultivars developed by the three breeding programs in Brazil, while subgroup four was represented by cultivars and breeding lines developed by CNPT-Embrapa and by other institution. Finally, subgroup five included twelve CNPT-Embrapa cultivars, one cultivar developed by Maltaria Navegantes and two foreign genotypes (MN 6021 and Quench). Interestingly, Golden Promise appeared separated within this Group B. When considering the different breeding programs in Brazil, the genotypes developed by CNPT-Embrapa were the only ones distributed across more than one group and across all the subgroups within Group B.

The Bayesian clustering results, implemented in STRUCTURE software, were in agreement with the results of the PCoA and UPGMA analyses. In this analysis we observed that the best inferred number of clusters was $K = 2$, obtained by the Evanno's method (Figure 2B). The majority of accessions of each group were grouped preferentially in one or the other cluster. Remarkably, the great majority of genotypes clustered in Group B by UPGMA analysis were assembled to the same group by Bayesian clustering. In addition, those clustering in Groups A and C were assigned to the other group. Exceptions were PFC 88212 and PFC 8610 that presented dual ancestry.

DISCUSSION

Brazil ranks third among the countries with the largest beer production in the world. However, it produces only around 300 thousand tons of barley per year, which results in one of the largest barley import needs (de Mori and Minella, 2012). Therefore, increments in barley production are essential. The demand for higher yielding and betteradapted crop varieties has increased the necessity to exploit the genetic variation of genebank collections (Keilwagen *et al.*, 2014). Evaluation of the genetic diversity of barley accessions can be an important source of information in order to discover and exploit novel alleles to be used in breeding programs. Here, for the first time, the SSR variability of Brazilian barley genotypes is reported. Although other groups have previously published molecular analyses of Brazilian barley accessions (Echart-Almeida and CavalliMolina, 2000; Kroth *et al.*, 2005; Maris AF, 1992, Bachelor's thesis, Universidade Federal do Rio Grande do Sul, Porto Alegre, RS), the number of genotypes evaluated in this report is higher and also the use of SSR markers is advantageous when compared to RAPD or isoenzymes used previously. SSR marker analysis have proven to be a method of choice for marker-assisted selection in breeding and genetic diversity studies (Varshney*et al.*, 2007), largely because they are highly informative and abundant in genomes, codominantly inherited and multiallelic (Gupta *et al.*, 1996; Oliveira *et al.*, 2006).

We successfully amplified 34 barley SSR barley loci obtaining an average of eight alleles for each locus. This is similar to the results reported for Iranian barley landraces (Khodayari *et al.*, 2012), higher than the one reported for Indian barley (Jaiswal *et al.*, 2010) and lower than the ones published by others (Saghai Maroof *et al.*, 1994;Malysheva-Otto *et al.*, 2006; Yahiaoui *et al.*, 2008; Nandha and Singh, 2014). Here we obtained a mean PIC value of 0.57, however, when considering just the Brazilian genotypes the diversity index was lower than that of foreign and wild accessions (Table 2), illustrating the low diversity in the Brazilian accessions. One interesting aspect is that in the UPGMA results (Figure 3) clusters seen inside Group B correlate with cultivar release date. Most of the accessions (77%) in subgroups one and two are older genotypes developed between 1970 and 1988, while 79% of the genotypes in the other subgroups were developed between 1989 and 2013. Only one genotype developed after the year 2000 (MN 684) was clustered outside of subgroups three, four and five and, among the 13 Brazilian genotypes in subgroup five, only three were developed before the year 2000. This means that cultivars released in the last years are more similar, indicating a decrease in the genetic diversity. Other indicators of the tendency to decreased genetic diversity are the lower number of alleles, the lower frequency of alleles per marker, and lower PIC values obtained for Brazilian genotypes developed after the 1980s (Figure 1). The observation that Brazilian barleys are closely related is in agreement with other studies (Selbach and Cavalli-Molina, 2000;Kroth *et al.*, 2005).

In fact, crop breeding programs have drastically narrowed the genetic diversity in cultivated plants, reducing tolerance to environmental stresses (Nevo, 2007). This genetic diversity reduction may have occurred because of the selection of specific alleles that confer stability in the specific environment of Southern Brazil. An evidence of the use of common materials for obtaining the cultivars can be visualized by the analysis of Group B in the dendrogram (Figure 3). The genotypes Volla and Wpgm 626-46-25 are present in the genealogy of three accessions in subgroup one (Table 1). Alpha is an ancestor of Antarctica 04, FM 519, Cevada BR1, FM 404, and FM 437, which clustered in subgroup two (Table 1). In subgroup four, most of the accessions were developed by CNPT-Embrapa. The genotypes BRS Mariana and BRS Sampa, which were closely related, share PFC 9205 in their genealogy. The genotype IPB 194 was used in the cross that generated Embrapa 43, which was then used to obtain BRS 224 (Table 1). Most of the genotypes grouping in subgroup five have BRS 195 and Cevada BR2 in their genealogy (Table 1). Hence, barley breeding programs in Brazil should definitely consider the use of a more diverse set of materials in their crossings.

The UPGMA results (Figure 3) were in agreement with PCoA and STRUCTURE analysis (Figure 2). The three groups detected in the UPGMA plot showed a pattern correlated to the number of rows of grains on the ear and to barley type (wild or domesticated). For instance, Group A was exclusively formed by wild accessions (two-row genotypes), while all genotypes in Group C, except for the wild accession Hspo 584, were representatives of six-row barley. These results could suggest that Hspo 584 has a different geographical origin than the other wild accessions. Although it has been shown that SSR and SNP markers can clearly separate the two- and six-rowed barley groups (Hayes and Szucs, 2006; Khodayari *et al.*, 2012), exceptions have been reported where a few two- and six-rowed genotypes showed similarity (Marmiroli *et al.*, 1999; Brantestam *et al.*, 2007; Lamara *et al.*, 2013), as we also found here.

Wild barley (*H. vulgare* ssp. *spontaneum*), the progenitor of cultivated barley, is a selfing annual grass distributed over a wide ecological range that differs in water availability, temperature, soil type, altitude and vegetation, thus generating a high potential for adaptive genetic diversity against abiotic and biotic stresses. It has been reported that wild barley has developed unique mechanisms for surviving harsh environments, mainly through developing new genetic variations and alleles (Nevo and Chen, 2010). Here we detected a greater diversity in wild barley accessions when compared to domesticated ones. Higher numbers of SSR alleles and greater diversity levels of individual loci were also found for wild genotypes in comparison to cultivated barley (Saghai Maroof*et al.*, 1994; Nandha and Singh, 2014). Another evidence for greater genetic diversity in wild barley is the higher number of private alleles detected (Table S2), a result that has also been reported byNandha and Singh (2014). Interestingly, cultivars Vacaria, Paraí, and BRS 180 and breeding lines PFC 88209, PFC 88210, and PFC 88211 developed by CNPT-Embrapa, as well as three foreign cultivars (Dayton, Quest, and Murasakimochi) grouped together with one wild accession (Hspo 584) in all analyses performed here. The genetic diversity shared by these accessions and wild barley indicates the potential of these accessions as a source of alleles for breeding purposes. A close similarity between domesticated barley and one wild accession has also been reported by Nandha and Singh (2014).

In this study, the geographic origin was not clearly associated with the groups, despite previous observation that SSR markers can differentiate barleys according to their region of origin (Malysheva-Otto *et al.*, 2006). However, other associations can be done based on genotype similarity. For instance, Group C consisted of six accessions that are feed varieties (Paraí-I, PFC 88209, PFC 88210, PFC 88211, PFC 88212, and Vacaria), whereas all accessions within Group B are malting barleys.

The highest average PIC was found for chromosome three, the same one where the highest diversity parameters were detected by Malysheva-Otto *et al.* (2006). Chromosome four was the one with the highest number of markers evaluated, however it showed the lowest PIC values. Among these loci, one was monomorphic and the others had PIC values below 0.50. Interestingly, chromosome four contains the *HvAACT1* gene responsible for the aluminum tolerance in barley (Furukawa *et al.*, 2007; Wang *et al.*, 2007). Aluminum is an important constraint in Southern Brazil (Echart and Cavalli-Molina, 2001), where most of Brazilian barley is produced. It has been proposed that the low variability for aluminum tolerance in Brazilian barley indicates that significant increments through conventional breeding are unlikely (Minella, 2001). It is possible that the low variability found in chromosome four could explain the low level of aluminum tolerance diversity in Brazilian barley. Strategies must be considered to surpass the low variability for some traits in barley, allowing the increase in stress tolerance and, ultimately, yield.

In conclusion, we have detected a lower PIC value among barley genotypes bred in Brazil compared with foreign and wild genotypes. A tendency to a decrease in genetic diversity of Brazilian barley is occurring with modern materials released over the last two decades, showing lower number of alleles, lower frequency of alleles per marker, and lower PIC values in comparison to cultivars released in the 1980s. This could be partially explained by the use of common ancestors carrying important traits. The use of more diverse material could be of interest to exploit novel alleles in crop breeding programs without losing quality. Some of the genotypes used here, such as the ones present in Group C of the dendrogram, should be considered. Private alleles unique to Brazilian genotypes are important for future use in cultivar identification.

ACKNOWLEDGMENTS

This work was financially supported by Agrária and Ambev (Companhia de Bebidas das Américas). CAD and JRF received fellowships from the Brazilian National Council for Scientific and Technological Development (CNPq).

REFERENCES

1. Anderson JA, Churchill GA, Autrique JE, Tanksley SD, Sorrells ME. Optimizing parental selection for genetic linkage maps. Genome. 1993;36:181–186.
2. Baik BK, Ullrich SE. Barley for food: characteristics, improvement, and renewed interest. J Cereal Sci. 2008;48:233–242.
3. Bian M, Waters I, Broughton S, Zhang XQ, Zhou M,

Lance R, Sun D, Li C. Development of gene-specific markers for acid soil/aluminium tolerance in barley Hordeum vulgare L. Mol Breed. 2013;32:55–164.

4. Botstein D, White RL, Skolnick M, Davis RW. Construction of a genetic linkage map in man using restriction fragment length polymorphisms. Am J Hum Genet. 1980;32:314–331.
5. Brantestam AK, von Bothmer R, Dayteg C, Rashal I, Tuvesson S, Weibull J. Genetic diversity changes and relationships in spring barley (*Hordeum vulgare* L.) germplasm of Nordic and Baltic areas as shown by SSR markers. Genet Resour Crop Evol. 2007;54:749–758.
6. de Mori C, Minella E. Aspectos econômicos e conjunturais da cultura da cevada. Passo Fundo: Embrapa Trigo, Documentos online. 2012;139:28–28.
7. Doyle JJ, Doyle JL. A rapid DNA isolation procedure for small quantities of fresh leaf material.Phytochem Bull. 1987;19:11–15.
8. Earl EA, von Holdt BM. Structure Harvester: a website and program for visualizing Structure output and implementing the Evanno method. Conserv Genet Resour. 2012;4:359–361.
9. Echart CL, Cavalli-Molina S. Fitotoxicidade do alumínio: efeitos, mecanismo de tolerância e seu controle genético. Cienc Rural. 2001;31:531–541.
10. Echart-Almeida C, Cavalli-Molina S. Hordein variation in Brazilian barley varieties *Hordeum vulgare* L. and wild barley *H. euclaston* Steud. and *H. stenostachys* Godr. Genet Mol Biol.2000;23:425–433.
11. Evanno G, Regnaut S, Goudet J. Detecting the number of clusters of individuals using the software STRUCTURE: a simulation study. Mol Ecol. 2005;14:2611–2620.
12. Excoffier L, Lischer HEL. Arlequin suite ver. 3.5: A new series of programs to perform population genetics analyses under Linux and Windows. Mol Ecol Resour. 2010;10:564–567.
13. Excoffier L, Smouse PE, Quattro J. Analysis of molecular variance inferred from metric distances among DNA haplotypes: Application to human mitochondrial DNA restriction data.Genetics. 1992;131:479–491.
14. Furukawa J, Yamaji N, Wang H, Mitani N, Murata Y, Sato K, Katsuhara M, Takeda K, Ma JF. An aluminumactivated citrate transporter in barley. Plant Cell Physiol. 2007;48:1081–1091.
15. Falush D, Tephens MS, Pritchard JK. Inference of population structure using multilocus genotype data: Linked loci and correlated allele frequencies. Genetics. 2003;164:1567–1587.
16. Feuillet C, Langridge P, Waugh R. Cereal breeding takes a walk on the wild side. Trends Genet.2007;24:25–32.
17. Goyal A, Ahmed M. Barley: production, improvement, and uses. Crop Sci. 2012;52:2852–2854.
18. Gupta PK, Balyan HS, Sharma PC, Ramesh B. Microsatellites in plants: a new class of molecular markers. Curr Sci. 1996;45:45–54.
19. Hayes P, Szucs P. Disequilibrium and association in barley: thinking outside the glass. Proc Natl Acad Sci USA. 2006;103:18385–18386.
20. Jakobsson M, Rosenberg NA. CLUMPP: a cluster matching and permutation program for dealing with label switching and multimodality in analysis of population structure.Bioinformatics. 2007;23:1801–1806.
21. Karakousis A, Barr AR, Chalmers KJ, Ablett GA, Holton TA, Henry RJ, Lim P, Langridge P. Potential of SSR markers for plant breeding and variety identification in Australian barley germplasm. Crop Pasture Sci. 2003;54:1197–1210.
22. Keilwagen J, Kilian B, Ozkan H, Babben S, Perovic D, Mayer KF, Poskar CH, Ordon F, Eversole K, et al. Separating the wheat from the chaff - a strategy to utilize plant genetic resources from *ex-situ* genebanks. Sci Rep. 2014;4:5231–5231.
23. Khodayari H, Saeidi H, Roofigar AA, Rahiminejad MR, Pourkheirandish M, Komatsuda T. Genetic diversity of cultivated barley landraces in Iran measured using microsatellites. Int J Biosci Biochem Bioinform. 2012;2:278–290.
24. Kroth MA, Ramella MS, Tagliari C, Francisco AD, Arisi ACM. Genetic similarity of Brazilian hull-less and malting barley varieties evaluated by RAPD markers. Sci Agric. 2005;62:36–39.
25. Ladizinsky G. How many tough-rachis mutants gave rise to domesticated barley? Genet Resour Crop Evol. 1998;45:411–414.
26. Lamara M, Zhang LY, Marchand S, Tinker NA, Belzile F. Comparative analysis of genetic diversity in Canadian barley assessed by SSR, DarT, and pedigree data. Genome. 2013;56:351–358.
27. Ledovskoy Y, Abugalieva S, Turuspekov Y. Comparative assessment of the genetic variation in wild and cultivated barley based on SSR markers. Asian Australas J Plant Sci Biotechnol.2010;4:21–26.
28. Liu K, Muse SV. PowerMarker: integrated analysis environment for genetic marker data. Bioinformatics. 2005;21:2128–2129.
29. Liu ZW, Biyashev RM, Maroof MS. Development of simple sequence repeat DNA markers and their integration into a barley linkage map. Theor Appl Genet. 1996;93:869–876.
30. Malysheva-Otto LV, Ganal MW, Röder MS. Analysis of molecular diversity, population structure and linkage disequilibrium in a worldwide survey of cultivated barley germplasm*Hordeum vulgare* L. BMC Genet. 2006;7:e6–e6.
31. Marmiroli N, Maestri E, Liviero L, Massari A, Malcevschi A, Monciardini P. Application of genomics in assessing biodiversity in wild and cultivated barley. Mol Ecol. 1999;8:S95–S106.
32. Minella E. Reunião Anual de Pesquisa de Cevada, Guarapuava. Vol. 1. Passo Fundo: Embrapa Trigo; 2001. Desafios e potencialidades do melhoramento genético de cevada no Brasil; pp. 31–40.
33. Minella E. Melhoramento de cevada. In: Borém A, editor. Melhoramento de Espécies Cultivadas. 2nd edition. Editora UFV; Viçosa: 2005. pp. 275–299.
34. Nandha PS, Singh J. Comparative assessment of genetic

diversity between wild and cultivated barley using gSSR and EST-SSR markers. Plant Breed. 2014;133:28–35.

35. Nevo E. Evolution of wild wheat and barley and crop improvement: studies at the Institute of Evolution. Isr J Plant Sci. 2007;55:251–262.

36. Nevo E, Chen G. Drought and salt tolerances in wild relatives for wheat and barley improvement. Plant Cell Environ. 2010;33:670–685.

37. Nevo E, Fu YB, Pavlicek T, Khalifa S, Tavasi M, Beiles A. Evolution of wild cereals during 28 years of global warming in Israel. Proc Natl Acad Sci USA. 2012;109:3412–3415.

38. Oliveira EJ, Pádua JG, Zucchi MI, Vencovsky R, Vieira MLC. Origin, evolution and genome distribution of microsatellites. Genet Mol Biol. 2006;29:294–307.

39. Peakall R, Smouse PE. GENALEX 6: genetic analysis in Excel. Population genetic software for teaching and research. Mol Ecol Notes. 2006;6:288–295.

40. Peakall R, Smouse PE. GenAlEx 6.5: genetic analysis in Excel. Population genetic software for teaching and research - an update. Bioinformatics. 2012;28:2537–2539.

41. Pritchard JK, Stephens M, Donnelly P. Inference of population structure using multilocus genotype data. Genetics. 2000;155:945–959.

42. Ramsay L, Macaulay M, degli Ivanissevich S, MacLean K, Cardle L, Fuller J, Edwards KJ, Tuvesson S, Morgante M, Massari A, et al. Asimple sequence repeat-based linkage map of barley. Genetics. 2000;156:1997–2005.

43. Röder MS, Korzun V, Wendehake K, Plaschke J, Tixier MH, Leroy P, Ganal MW. A microsatellite map of wheat. Genetics. 1998;149:2007–2023.

44. Rosenberg NA. DISTRUCT: a program for the graphical display of population structure. Mol Ecol Notes. 2004;4:137–138.

45. Saghai Maroof M, Biyashev RM, Yang GP, Zhang Q, Allard RW. Extraordinarily polymorphic microsatellite DNA in barley: species diversity, chromosomal locations, and population dynamics. Proc Natl Acad Sci USA. 1994;91:5466–5470.

46. Schuelke M. An economic method for the fluorescent labelling of PCR fragments. Nat Biotechnol. 2000;18:233–234.

47. Selbach A, Cavalli-Molina S. RAPD characterization of Brazilian barley *Hordeum vulgare* ssp. *vulgare* varieties. Euphytica. 2000;111:127–135.

48. Soto-Cerda BJ, Peñaloza EH, Montenegro AB, Rupayan AR, Gallardo MH, Salvo-Garrido H. An efficient marker-assisted backcrossing strategy for enhancing barley *Hordeum vulgare* L. production under acidity and aluminium toxicity. Mol Breed. 2013;31:855–866.

49. Varshney RK, Marcel TC, Ramsay L, Russell J, Röder MS, Stein N, Waugh R, Langridge P, Niks RE, Graner A. A high density barley microsatellite consensus map with 775 SSR loci.Theor Appl Genet. 2007;114:1091–1103.

50. Wang J, Raman H, Zhou M, Ryan PR, Delhaize E, Hebb DM, Coombes N, Mendham N. High-resolution mapping of the *Alp* locus and identification of a candidate gene HvMATE controlling aluminium tolerance in barley *Hordeum vulgare* L. Theor Appl Genet. 2007;115:265–276.

51. Yahiaoui S, Igartua E, Moralejo M, Ramsay L, Molina-Cano JL, Lasa JM, Gracia MP, Casas AM. Patterns of genetic and eco-geographical diversity in Spanish barleys. Theor Appl Genet.2008;116:271–282.

CHAPTER 16

A NEW ALLELE FOR ALUMINIUM TOLERANCE GENE IN BARLEY (HORDEUM VULGARE L.)

Yanling Ma, Chengdao Li, Peter R. Ryan, Sergey Shabala, Jianfeng You, Jie Liu, Chunji Liu and Meixue Zhou

Keywords: Aluminium tolerance, Barley, HvAACT1

BACKGROUND

Aluminium (Al) is the most abundant metal in the earth crust. Al can be toxic to plants when the concentration of soluble Al increases in acidic soils (<pH 5.0) due to the formation of the phytotoxic Al^{3+} species. An early symptom of Al toxicity is the inhibition of root elongation which limits water and nutrient uptake [1, 2, 3]. The inhibition of root growth in wheat (*Triticum aestivum* L.) occurs within minutes or hours in simple hydroponic solutions [4] due to decreases in root cell division and elongation [5]. Longer treatments result in thickened roots, damaged root cap, and lesions in the epidermal and cortical tissues near the root apices [6, 7, 8]. Additional symptoms of stress in wheat and maize roots include the appearance of large swollen cortical cells near the root tip [9, 10].

An important mechanism of resistance in many species relies on the exclusion of Al from root tissues by the release of organic anions from the root apices. The organic anions such as malate and citrate chelate the harmful Al cations in the apoplast and prevent them damaging the root tissues [11, 12, 13]. Barley (*Hordeum vulgare* L.) is one of the most Al-sensitive cereal species yet it still shows genotypic variation [14]. A mechanism for Al tolerance described in barley relies on the Al-activated release of citrate from root apices. This is controlled by a single major locus called *Alp* on chromosome 4HL [15, 16, 17]. The gene underlying the *Alp*locus is *HvAACT1* which encodes a member of the multidrug and toxic compound extrusion (MATE) family [8,14, 17, 18]. This gene has been linked with tolerance in many genotypes of barley including Murasakimochi, Dayton, Honen, WB229, Svanhals, Br2 and Brindabella [16, 19, 20, 21, 22]. Tolerant genotypes of barley show a constitutively higher expression of *HvAACT1* in root apices than sensitive genotypes. Furthermore, constitutively over-expression of *HvAACT1* in transgenic barley and wheat plants significantly increases the Al-activated citrate efflux and their tolerance to Al in hydroponic solution and in acid soil [8].

Citation: Yanling Ma, Chengdao Li, Peter R. Ryan, Sergey Shabala, Jianfeng You, Jie Liu, Chunji Liu and Meixue Zhou, A new allele for aluminium tolerance gene in barley (Hordeum vulgare L.), DOI: 10.1186/s12864-016-2551-3.

The higher expression of *HvAACT1* in tolerant barley was recently linked to the presence of a 1023-bp transposable-element like insertion in the 5' untranslated region (UTR) of *HvAACT1* [23]. This insertion alters the usual distribution and level of *HvAACT1* expression such that it becomes constitutively high in the root apices [11, 23]. This mutation is only found in cultivated Al-tolerant barley genotypes from East Asia where acid soils are prevalent and likely represents an important mutation that has helped the expansion of barley from the Near East where the soils are rarely acidic [23].

A PCR-based marker can be used to detect the presence or absence of the insert in the 5' UTR of *HvAACT1* and this is a convenient method for screening barley for Al tolerance [19, 23]. The PCR product from tolerant barley genotypes is approximately 1 kb larger than the product from sensitive barley genotypes [23]. Another gene-specific marker called *HvMATE*-21 was designed to target polymorphism at the 3' UTR of *HvAACT1* and this marker has been used successfully to score more than 50 varieties differing in Al tolerance [18]. All tolerant varieties tested possessed the 21-bp deletion compared with the sensitive varieties. Using association analysis the *HvMATE*-21 marker could explain 66.9 % of phenotypic variation for Al tolerance [18]. Meanwhile, several simple sequence repeats (SSR) markers such as Bmac310, Bmag353 and HVM03 are closely linked with tolerance and commonly used for genetic analysis [16, 20, 24].

In this study, we identified a Chinese barley variety, CXHKSL, which was moderately tolerant to acid soil (Additional file 1: Table S1) but which gave a non-standard result for the 5' UTR *HvAACT1* marker. This indicated that Al tolerance in CXHKSL might be controlled or regulated in a different way. We investigated the Al tolerance mechanism in CXHKSL and mapped the trait using a double haploid (DH) population derived from CXHKSL and the Al-sensitive variety Gairdner.

METHODS

Genetic Materials

CXHKSL is a six-rowed Chinese variety that is tolerant to acid soils. The Al-sensitive variety, Gairdner, is an Australian malting barley. The Al tolerant variety, Dayton, was used as a control when investigating tolerance mechanisms of CXHKSL. One DH population consisting of 210 lines derived from a cross between CXHKSL and Gairdner was used for QTL mapping study.

Al Tolerance and Root Growth

The relative Al tolerance of the different varieties and selected double haploid lines (DHLs) were evaluated with hydroponic culture methods. Sterilized seeds were germinated in the dark for 2 days at 4 °C and then 2 days at 28 °C. Root length of the seedlings was measured and they were placed in an aerated nutrient solution containing 500 μM KNO_3, 500 μM $CaCl_2$, 500 μM NH_4NO_3, 150 μM $MgSO_4$, 10 μM KH_2PO4, 2 μM Fe:EDTA, 11 μM H_3BO_3, 2 μM $MnCl_2$, 0.35 μM $ZnCl_2$ and 0.2 μM $CuCl_2$. For barley, Al tolerance was estimated by measuring net root length after 4 days in 0, 1, and 4 μM $AlCl_3$ (pH=4.3), respectively. Relative root length (RRL) was estimated as: (net root growth in Al treatment/net root growth in control solution) × 100 % [8]. Meanwhile, 4-day old control and 4 μM $AlCl_3$ treated seedlings roots were stained with haematoxylin for 15 min and rinsed for 10 min to compare the density of Al accumulation at root apices. Haematoxylin could form a purple-red complex with Al and provides an indirect of non-complexed Al in root apices, with the intensity of staining correlated with sensitivity of Al toxicity [25].

Al tolerance was also scored using acid soil collected from the Northern Tasmania (pH=4.3). Three seeds of each DHL and parent varieties were sown in the acid soil in each replicate. Two independent experiments including six replicates were conducted in April and June 2013, respectively. Four replications were applied in each experiment. Both root length and root morphology were used to assess Al tolerance. Root length (mm) of each seedling was measured seven days after sowing. Meanwhile, root tips were screened for the absence or presence of thickening caused by Al toxicity.

Assaying Citrate Efflux and Malate Efflux from Root Apices

Seedlings were grown for 4 days in the nutrient solution described above (without added $AlCl_3$). To study if the expression of *HvAACT1* need longer Al treatment duration, half of the plants of each genotype were subjected to 0.2 mM $CaCl_2$ solution containing 10 μM $AlCl_3$ (pH=4.3) for overnight pre-treatment. Ten root apices (3-5 mm) with 4 replicates were excised from the same line and washed in 1 ml 0.2 mM $CaCl_2$ solution (pH=4.3) for on a platform shaker (60 rpm). After 30 min washing, 1 ml 0.2 mM $CaCl_2$ solution (pH=4.3) with 30 μM $AlCl_3$was added and shaken for 2 h at 60 rpm. The solutions were centrifuged to dryness on a rotary vacuum drier for citrate efflux detection. The enzyme assay used to determine citrate concentration is described by Wang et al. [16]. The initial citrate content in each sample was calculated from a standard curve. Malate concentration was measured with an enzyme assay as described previously [26].

Molecular Marker Analysis

Three primer pairs were used to investigate allelic variation in the 5'UTR of *HvAACT1*. These were to detect the presence or absence of a ~1 kb transposon-like insertion previously described in Al-tolerant genotypes of barley. The first pair of primers was from Fujii et al. [23] with forward sequence 5'-GGTCCAACACTCTACCCTCCTT and reverse 5'-GGTGCGAG -TTGCCCCTAGCTATTA. The second pair of primers described by Bian et al. [18] was forward 5'-CTTCATTTCAACCAAGCACTCC and reverse 5'-GCTTTTGGTCGAACAAA- GTATCG. The third pair of primers was designed to amplify a slightly larger fragment that included the above two pairs of primers comprised, forward 5'-TGTCGATATGGTGCTCTT -CG and reverse 5'-AGCTCCATGACAATTCTGGG. PCR reactions were performed at 20 μl-volume including 10 μl HotstarTaq™master mix (Qiagen), 2 μl primer mix (1:1 mix of forward and reverse primers at 10 nM), 3 μl DNA template, and 5 μl H_2O. Cycling conditions were as follows: 1 cycle of 1 min at 95 °C, 35 cycles of 1 min at 95 °C, 30 s at 60 °C, 40 s at 72 °C, and finally with an extension step of 1 min at 72 °C. All PCR reactions were run at C1000™Thermal cycler (BIO-RAD). PCR products were separated at 1 % agarose and visualized by staining with 1 % Red safe under Gel Doc™ XR^+ imagining system (BIO-RAD).

Another *HvAACT1*-specific marker, *HvMATE*-21, was used to genotype the population as well as three closely*HvAACT1*-linked SSR markers: Bmac310, Bmag353 and HVM03 [14, 16, 19]. *HvMATE*-21 was a PCR marker that detected the presence or absence of a 21-bp fragment in the 3' UTR of *HvAACT1*. PCR reactions were carried out in a total volume of 15 μl containing 25~30 ng genomic DNA, 0.5 M of forward and reverse primers, 7.5 μl GoTaq® Hot Start Colorless Master Mix, 2X (Promega). The amplification of SSRs were performed by: 1 cycle of 3 min at 94 °C, 35 cycles of 1 min at 94 °C, 1 min at the annealing temperature 55 °C and 1 min at 72 °C, with a final extension step of 5 min at 72 °C. The PCR profiles for *HvMATE*-21 were almost the same as that for SSR markers except the annealing temperature was 60 °C. All PCR reactions were run on Mastercycler Gradient 5331 (Eppendorf AG, Germany). The PCR products were separated on 5 % denatured polyacrylamide gels and visualized by a rapid silver staining method [27].

Isolation and Sequence Analysis of Coding Region of HvACCT1 Gene in CXHKSL and Dayton

The published complete coding DNA sequence (CDS) of *HvAACT1* gene (Genebank: AB302223.1) was retrieved from the National Center for Biotechnology Information (NCBI, http://www.ncbi.nlm.nih.gov/gene) and aligned with barley reference genome data using IPK blast server (http://webblast.ipk-gatersleben.de/barley/viroblast.php). Based on the best hit sequence, a total of 4 pairs of primers (Additional file 2: Table S2) were used to amplify the whole *HvACCT1* open reading frames (ORFs). The amplified PCR products was purified and cloned with pGEM®-T Vector System (Promega). The final CDS were constructed using sequencing results from 12 independent clones (3 clones for each pair of primers). Sequence analysis was completed with software DNAMAN (version 7.0; Lynnon Biosoft, USA). The sequence data of CDS have been deposited to NCBI Genbank Database with accession number of KU725980 for variety CXHKSL and KU725981 for Dayton.

HvAACT1 Expression

RNA was isolated from root apices (also from plants used for citrate efflux measurement) by RNeasy™ plant kit (Qiagen) and purified by inclusion of RNase-free DNase (Qiagen). One microgram total RNA was used to synthesize cDNA by reverse transcriptase system (Invitrogen). 1.0 µl oligo primer was added into 11.5 µl reaction mixture including 1 µg RNA. The mixture was incubated at 70 °C for 10 min, and transferred to ice immediately. Each aliquot included 4 µl buffer, 2 µl 0.1 M DTT, 1 µl dNTP mix, and 0.5 µl superscript III Reverse Transcriptase was added into the mixture, and incubated at 42 °C for 1 h. RNA degradation step was performed by addition of 0.25 µl RNase H (Thermo Scientific™) and incubated at 37 °C f or 30 min.

Quantification real time polymerase chain reaction(RT-PCR) was run in a C1000TM Thermal cycler (BIO-RAD) with 10 µl reaction mixture containing 4.0 µl of cDNA diluted to 1:40, 5 µl of SYBR Green Jumpstart Taq Readymix (Sigma) and 1 µl primer mix (1:1 mix of forward and reverse primers at 10 nM). Three pairs of primers used to measure expression of *HvAACT1* (*HvAACT1*-forward 5'-AGCAGCCAAGACCTTGAGAA and reverse 5'-AGCAG GAATCCACAACCAAG; New-*HvAACT1*-1-forward ACGGGGCTCTACCTCTTT -GT and reverse 5'-GGCAATAGAAACACCAACAGC; New-*HvAACT1*-2-forward CTGTGTCACTC TGGCATCGT, and reverse 5'-AAGCTGCAGAACACGAGAGGT). The constitutively expressed barley glyceraldehyde-3-phosphate dehydrogenase (HvGAPDH) gene and barley homologous to eukaryotic translation elongation factor 1A (HveEF-1A) gene was used as reference genes. The sequences of primers are as follows: *HvGAPDH*-forward: 5'-GTGAGG CTGGTGCTGATTACG and reverse 5'-TGGTGCAGCTAGCATTTGAGAC, *HveEF-1A*-forward 5'-TTTCACTCTTGGTGTGAAGCAGAT and reverse 5'-GACTTCCTTCACGAT-TTCATCGTAA. Cycling conditions were 3 min at 95 °C, followed by 40 cycles at 95 °C for 10 s, 60 °C for 20 s, 68 °C for 10 s. At the end, a melting curve of the amplified fragments was produced by increasing the temperature every 0.5 °C from 60 °C to 95 °C.

Data Analysis and QTL Mapping

All phenotypic data was analysed by SPSS software package (Version 20.0, IBM), including all basic statistics calculation and Chi-Square Goodness of Fit Test. For genetic linkage map analysis, the genetic distances between molecular markers were calculated using software JoinMap 4.0 [28]. The mean values of root lengths of DHLs were used to detect QTL affecting root length under Al toxicity with software MapQTL6 [29]. Interval mapping (IM) method was first used to identify the major QTL. By selecting significantly linked markers as cofactors, multiple QTL mapping (MQM) mapping method based on the multiple-QTL model was used. A set of 1000 permutations was performed to identify the LOD threshold corresponding to a genome-wide false discovery rate of 5 % ($P<0.05$) [29].

RESULTS

A PCR-based marker is available to score for the presence or absence of the ~1 kb insert in the 5'UTR of the*HvAACT1* associated with higher expression in the root apices of Al-tolerant barley [23]. The marker generates a larger product in Al-tolerant genotypes than in sensitive genotypes. A Chinese variety named CXHKSL was identified which was tolerant to acid soil but did not show the typical marker result associated with tolerance. In fact the standard PCR reaction failed to produce any PCR band in CXHKSL while the results obtained from control varieties including the highly tolerant variety, Dayton, and sensitive variety, Gairdner, were as expected (Fig. 1a). The failure to generate a band was not related to the quality or quantity of DNA extracted from CXHKSL since other PCR reactions were successful including the *HvMATE*-21 marker which targets the 3' UTR of *HvAACT1* (data not shown). Polymorphisms in CXHKSL may have reduced annealing temperature of one or both primers and so two additional primer pairs were designed to target this polymorphic region in the 5' UTR of *HvAACT1* (Fig. 1b, c). One set had primers farther upstream and downstream from the first set and therefore would amplify a slightly larger product (Fig. 1c). The second set of primers (Fig. 1b) was previously described by Bian et al. [18]. Using these additional primers, PCR products of the expected size were reliably generated from Dayton and Gairdner. By contrast CXHKSL generated no products at all (Fig. 1b, c). By contrast the *HvMATE*-21 marker for CXHKSL, Dayton and Gairdner were as described by Bian et al. [18]: the tolerant genotype possessed a 21-bp deletion but sensitive genotypes did not (data not shown). The coding region of*HvAACT1* in both CXHKSL and Dayton consisted of 1668 bp which is the same as previously published CDS [15]. CXHKSL was similar to Dayton with only one nucleotide being different for *HvAACT1* coding region (Additional file 3: Figure S1) and the difference didn't cause any changes in amino acids. Two and three SNPs were detected between the published CDS and CXHKSL and between the CDS and Dayton, respectively.

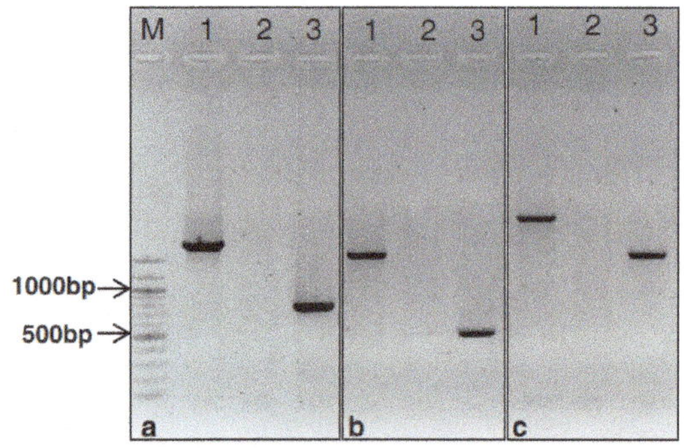

Figure 1: Testing presence of the 1-kb transposon-like insertion in the 5'UTR of *HvAACT1* associated with Al tolerance. Primers sets used in the PCR reactions are listed: (a) from Fujii et al. [23] (b) from Bian et al. [18] and (c) a new set of primers farther upstream and downstream from those used in (a) (see Methods). Varieties are Dayton (1), CXHKSL (2) and Gairdner (3).

The Al tolerance of CXHKSL was further investigated in hydroponic experiments and compared with Dayton and Gairdner. Relative root length (RRL) of CXHKSL was ~75 % after four days in 4 μM $AlCl_3$ which was significantly greater than Gairdner (~30 % RRL) but less than Dayton (~110 % RRL) (Fig. 2). This ranking was similar to the Al-related tissue damage on roots in the three varieties (Fig. 3). Root apices of Gairdner became significantly thicker than the others after four days in 4 μM $AlCl_3$ which is a typical symptom of Al toxicity. Haematoxylin staining was also more intense in the root apices of Gairdner than in Dayton or CXHKSL, indicating greater Al accumulation in the roots of Gairdner. The Al-dependent efflux of citrate was then measured from these barley lines. Citrate efflux from CXHKSL was less than Dayton (~40 pmol·apex^{-1}·h^{-1}) but greater than Gairdner which correlated well with their relative tolerance to Al (Fig. 4a). Pre-treatment with Al prior to these measurements did not increase citrate efflux further compared with roots without pre-treatment (Fig. 4a). These results indicate that the mechanism of Al tolerance in CXHKSL is associated with Al-activated citrate efflux from roots which is consistent with previous reports for barley [14, 16]. We also investigated whether Al-activated malate release from the roots of CXHKSL was apparent but we found no indication of efflux from this in any barley lines tested (Fig. 4b). An Al-tolerant wheat line included as a positive control in this experiment showed malate efflux of 0.9 nmol apex^{-1}h^{-1} which is similar to published values [26].

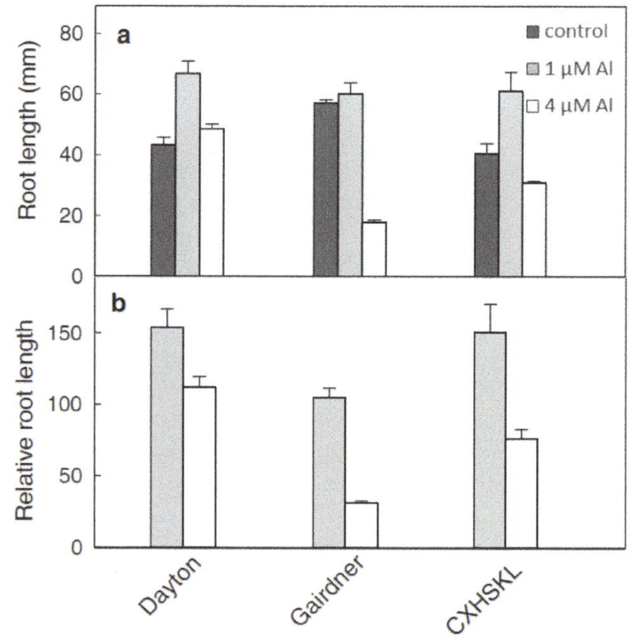

Figure 2: Comparison of Al tolerance in hydroponic culture. a Net root growth of seedlings after four days in nutrient solution containing 0, 1 or 4 μM $AlCl_3$ (pH =4.3). b Relative root length at each Al concentration in contrast with controls. Data showed means and standard error (SE) (n=4-7).

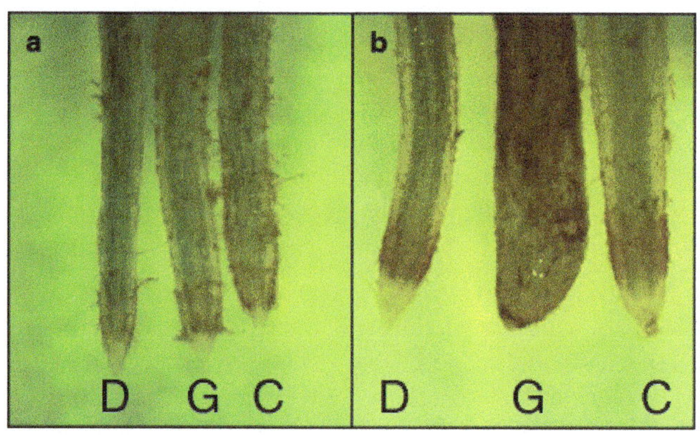

Figure 3: Comparison of haematoxylin staining of barley roots in hydroponic experiments following different treatments. Shown are root apices after 4 d in (a) control solution and (b) 4 μM $AlCl_3$. Genotypes shown are Dayton (D), Gairdner (G) and CXHKSL (C). Darker staining indicated greater Al accumulation at the root apices.

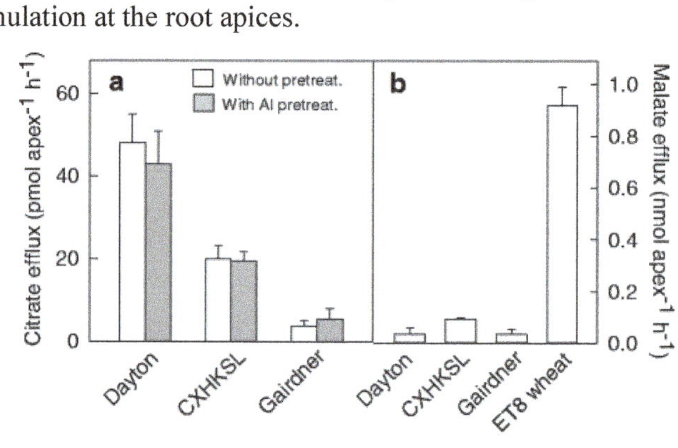

Figure 4: Citrate and malate efflux from the root apices of the barley varieties. a Citrate efflux was measured in presence

of 30 μM $AlCl_3$ with and without an overnight pre-treatment in 10 μM $AlCl_3$. b Malate efflux measured in the presence of 30 μM $AlCl_3$ without pre-treatment. ET8 is an Al-tolerant wheat line used as a positive control for malate efflux. Data show means and SE (n=4).

The chromosomal location of the tolerance locus in CXHKSL was investigated using a doubled haploid population generated by crossing CXHKSL and Gairdner. A total of 210 DHLs were grown in the acid soil and the roots were scored based on both root length and root tip damage (thickness). Root length of the CXHKSL parent was 75±8 mm and showed no damage of root tips (Fig. 5; Fig. 6a). Root length of the Gairdner was 40±5 mm and the root apices showed swelling, thickening and clear signs of damage (Fig. 5; Fig. 6b). A total of 65 DHLs showed no thickened roots or visible tissue damage (Fig. 6c) while 145 DHLs showed those strong phenotypes (Fig. 6d). When both criteria of growth and damage were used to score the population, we found that 128 DHLs had root lengths ≤50 mm and showed thickened root apices and 52 DHLs had root lengths ≥70 mm without visible thickening of the root apices (Fig. 5; Fig. 6). The remaining 30 DHLs had root lengths from 55 to 65 mm of which 17 DHLs showed thickening of the root apices (Fig. 6d).

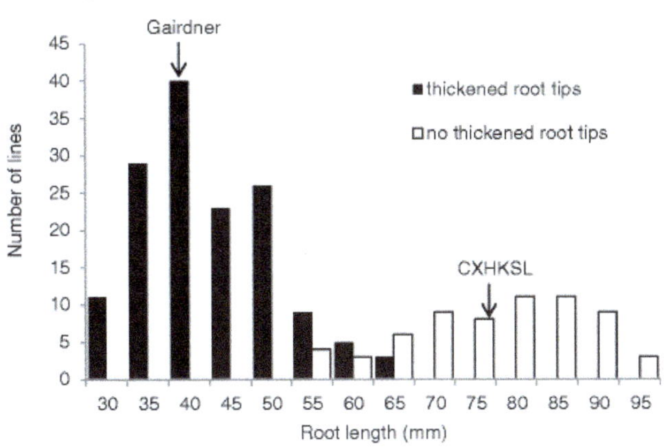

Figure 5: Distribution frequency of root length of 210 DHLs including parents after growth in an acid soil (pH=4.3). Solid bars indicated genotypes with thickened root tips and white bars indicate genotypes without thickened root tips.

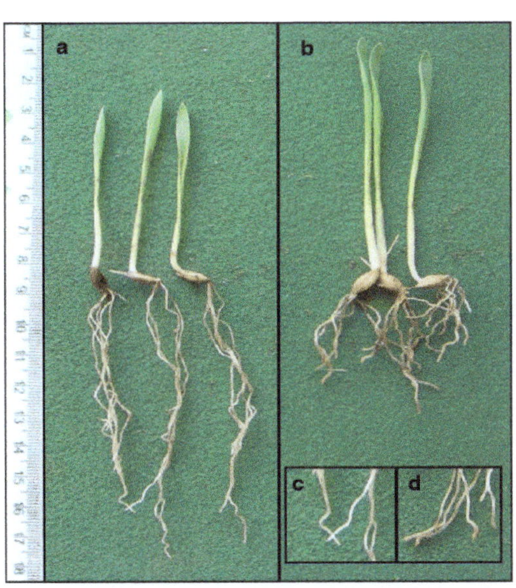

Figure 6: Comparison of Al tolerance of two parental varieties grown in acid soil (pH=4.3) by root length and damages to root tips. Root length was compared between (a) CXHKSL and (b) Gairdner. Typical root apices from an Al-tolerant genotype is shown in (c) and typical root apices from a sensitive genotype with obvious thickening and damage is show in (d).

Preliminary genetic analysis localised the Al tolerance phenotype in CXHKSL to a single major locus on chromosome 4H (data not shown). Therefore a more detailed genetic linkage map on chromosome 4H was generated using three SSR markers linked to Al tolerance (HVM03, Bmag353, Bmac310) as well as the *HvMATE*-21 marker which targets an 21-bp deletion in tolerant genotypes in the 3'UTR of *HvAACT1*. The map spanned a total length of 17.8 cM and the order of markers (Fig. 7) was similar to Bian et al. [18]. Analysis of root length under Al toxicity using this linkage map identified a significant QTL with a LOD score of 56.44 (Fig. 8). The closest marker, *HvMATE*-21, accounted for 71.0 % of the phenotypic variation, while Bmag353 and Bmac310 explained 61.0 % and 50.5 % of the variation respectively (Fig. 8). We conclude that Al tolerance in CXHKSL maps to the *HvAACT1* gene as reported for other tolerant barley lines. This was further tested with the*HvAACT1* 5' UTR marker on a selection of tolerant and sensitive DHLs. All sensitive DHLs tested amplified a band similar to the Gairdner parent which is consistent with expectations. Similarly, all tolerant DHLs tested failed to produce a band which is the same as the CXHKSL parent (data not shown).

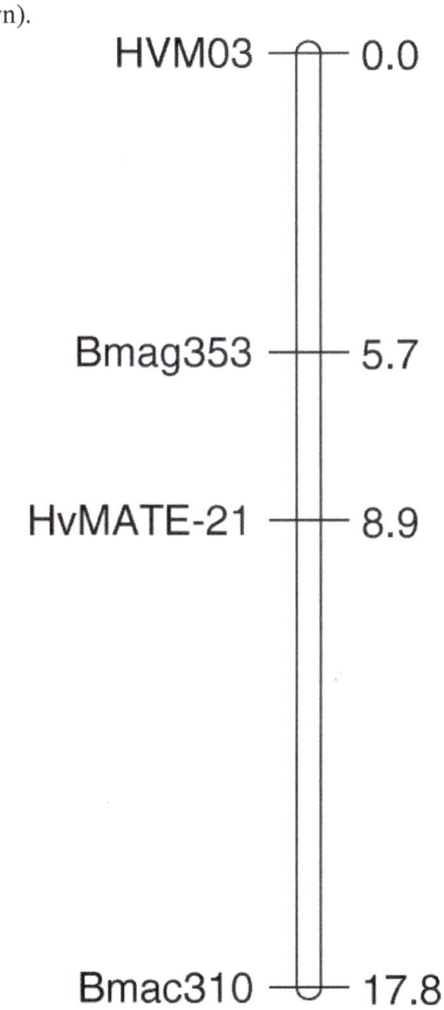

Figure 7: The linkage map on chromosome 4H with four molecular markers used to score 210 DHLs. Numbers on the right side represented genetic distances in unit of centiMorgan (cM).

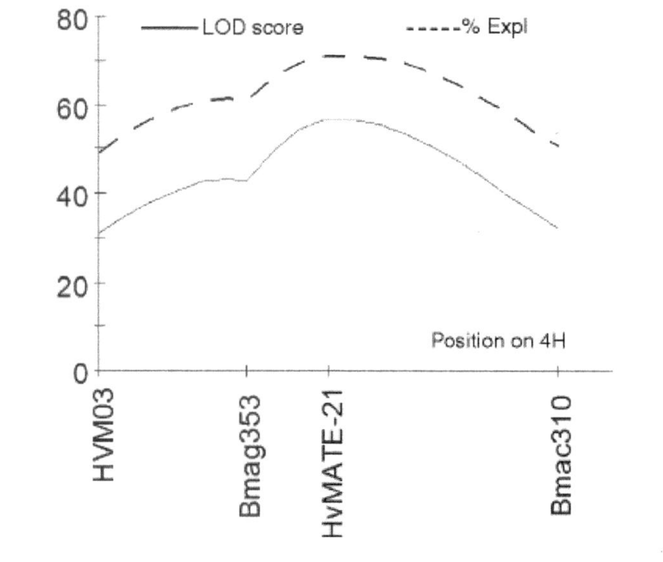

Figure 8: QTL detected for Al tolerance on chromosome 4H using root length variation in acid soil. The continuous line represented for the LOD score and the dashed line for phenotype variation (%) explained by each marker.

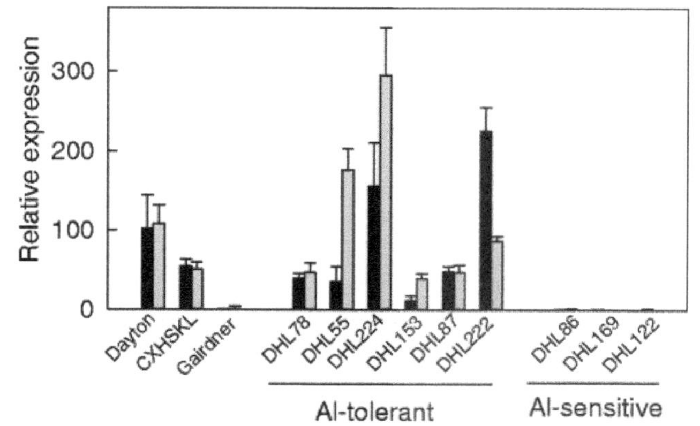

Figure 10: Relative expression of *HvAACT1* in Al-sensitive and Al-tolerant DH lines generated from Gairdner and CXHSKL measured with quantitative RT-PCR. The Al-tolerant variety Dayton was included. Expression of Gairdner was designated as 1.0. Data showed means of relative expression level using *HvGAPDH* (*black bars*) and *HveEF-1A*(*shaded bars*) as reference genes, and SE from three biological replicates with two technical replicates.

Citrate efflux was also measured from the selection of tolerant and sensitive DHLs. Six tolerant DHLs were examined and efflux was measured from all of them. Five of these had efflux of 18 to 28 pmol·apex^{-1}·h^{-1}which was similar to the CXHKSL parent (Fig. 9). Efflux from the remaining tolerant line was lower than the other five but greater than the sensitive lines measured which were <5 pmol·apex^{-1}·h^{-1} (Fig. 9). The level of*HvAACT1* expression was also determined in selected DHLs to determine whether this was linked with the other phenotypes of tolerance and citrate efflux. These measurements used two different reference genes*HvGAPDH* and *HveEF-1A*. Expression of *HvAACT1* was detected in CXHKSL, Dayton and the tolerant DHLs tested but no expression was detected in Gairdner or the three sensitive DHLs tested (Fig. 10). These results suggest that Al-tolerance in CXHKSL is controlled by a novel allele of the *HvAACT1* gene.

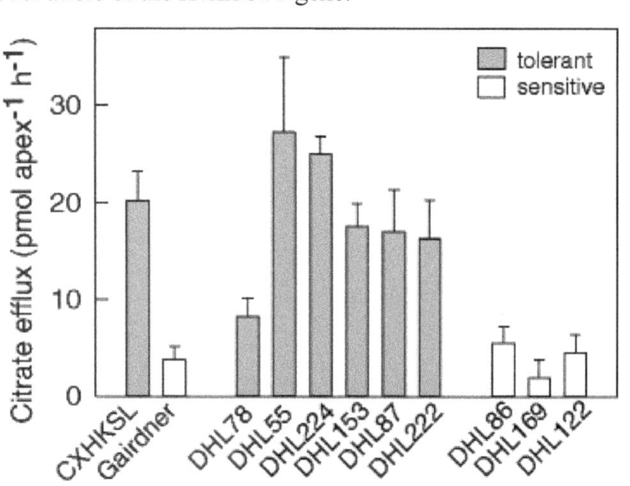

Figure 9: Citrate efflux from the root apices of the barley lines. Citrate efflux from CXHKSL and Gairdner as well as Al tolerant and sensitive DHLs in the presence of 30 μM $AlCl_3$ without Al pre-treatment. Data showed means and SE (n=4).

DISCUSSION

The only mechanism of Al tolerance in barley described to date relies on the release of citrate from the root apices via the HvAACT1 transporter [14, 15, 16]. In our study, we characterised a Chinese barley variety CXHKSL which is moderately tolerant to Al stress in hydroponics and acid soil but did not generate the expected result for a standard marker that targets the 1 kb insert in the 5'UTR of *HvAACT1* correlated with Al tolerance. Dayton and Gairdner amplified fragments of expected sizes from the 5'UTR of *HvAACT1* while no PCR products were detected in CXHKSL. The determinant role of promoter variation in Al tolerance was further confirmed by the fact that only one SNP was detected for CDS of *HvAACT1* gene between CXHKSL and Dayton without causing changes in protein sequence. We showed that the Al tolerance mechanism in CXHKSL likely relies on the Al-activated efflux of citrate reported for other tolerant barley. Using a DH population generated by crossing CXHKSL and the Al-sensitive variety Gairdner and a set of markers linked with Al tolerance, we found that the tolerance locus in CXHKSL mapped to *HvAACT1*. Collectively these results suggest that CXHKSL possesses a novel allele of Al-tolerance gene *HvAACT1*.

Fuji et al. [23] reported that ~1 kb insertion in the upstream 5'UTR of *HvAACT1* alters the distribution and level of gene expression in Al-tolerant cultivars. We tried to detect the presence of this insert using the published pair of primers but no fragments were amplified in CXHKSL or the tolerant DH lines tested. However, the expression level of *HvAACT1* in CXHKSL, Dayton and Gairdner was positively correlated with the relative Al tolerance of these varieties with Dayton>CXHKSL>Gairdner. We also found that *HvAACT1* expression was higher in the tolerant DHLs than sensitive DHLs and that the marker targeting the 5' UTR of *HvAACT1*segregated with tolerance (absence of a band in CXHKSL). This is consistent with the central role of *HvAACT1* in Al tolerance in these barley lines [23]. The absence of a PCR product with the 5'UTR *HvAACT1* marker in CXHKSL could be due to polymorphisms which prevented primer binding to the DNA. We tested this possibility by designing the additional primers from the same region but these also failed to generate a product in CXHKSL (Fig. 1c). The absence of a PCR product for this marker could also be due to a deletion in CXHKSL or the presence of a much larger insert. It is clear that polymorphisms exist in the 5'UTR of *HvAACT1* in CXHKSL compared to the published sequences for other tolerant barley [23].

The gene-specific marker *HvMATE*-21 was more efficient in predicting the phenotypic variation (71.0 % in this study, Fig. 8) under Al toxicity than other commonly-used SSR markers, Bmac310 and Bamag353 [16, 19, 20]. Meanwhile, segregation distortion occurred in the DH population which skewed the distribution of root lengths toward the sensitive parent Gairdner and away from Mendelian expectations. No significant monogenic segregation ratio 1:1(^{2}=36.82> $^{2}_{0.05}$=3.84) was observed. This distortion can occur when the generation of fertile hybrids are prevented as a result of the methods used to produce the lines [18, 30].

Al tolerance of wheat primarily relies on the Al-dependent malate efflux from root apices which is controlled by the Al-activated anion transporter encoded by the *TaALMT1* gene [31]. The closest homologue of this gene in barley is *HvALMT1* which is located on chromosome 2H and does not contribute to the Al resistance [32]. However, over-expression of *HvALMT1* gene in barley with a constitutive promoter can increase the efflux of malate and Al tolerance in barley and wheat [33]. The contribution of malate efflux to Al tolerance was investigated in CXHKSL and other tolerant and sensitive genotypes but no significant malate release was detected.

CONCLUSIONS

In the present study, we demonstrated that CXHKSL possesses a novel allele for the major Al tolerance gene*HvAACT1* but the mechanism of tolerance is similar to other tolerant barley lines.

ACKNOWLEDGEMENTS

This work is supported by the Grains Research and Development Corporation of Australia.

AUTHORS' CONTRIBUTIONS

YM conducted the bulk of experiments and wrote the paper draft. MZ was responsible for experimental design, data interpretation and writing. JY and JL conducted the experiments. CDL, PR, SS and CJL contributed to data interpretation and writing. All authors have read and approved the manuscript.

REFERENCES

1. Kochian LV. Cellular mechanisms of aluminum toxicity and resistance in plants. Annu Rev Plant Physiol Plant Mol Biol. 1995;46:237–60.
2. Melakeberhan H, Bird GW, Jones AL. Soil pH affects nutrient balance in cherry rootstock leaves. HortScience. 2001;36:916–7.
3. Parker DR. Root growth analysis: an under-utilised approach to understanding aluminium rhizotoxicity. Plant Soil. 1995;171:151–7.
4. Ryan PR, Shaff JE, Kochian LV. Aluminium toxicity in roots: correlation between ionic currents, ion fluxes and root elongation in Al-tolerant and Al-sensitive wheat varieties. Plant Physiol. 1992;99:1193–200.**View**
5. Amenos M, Corrales I, Poschenrieder C, Illes P, Baluska F, Barcelo J. Different effects of aluminum on the actin cytoskeleton and brefeldin A-sensitive vesicle recycling in root apex cells of two maize varieties differing in root elongation rate and aluminum tolerance. Plant Cell Physiol. 2009;50:528–40.**View**
6. Foy CD. Physiological effects of hydrogen, aluminium, and manganese toxicities in acid soil. In: Pearson RW, Adams F, editors. Soil acidity and liming. Wisconsin: Amer Soc. Agronomy; 1984. p. 57–97.
7. Ryan PR, DiTomaso JM, Kochian LV. Aluminium toxicity in roots: an investigation of spatial sensitivity and the role of the root cap. J of Exp Bot. 1993;44:437–46.
8. Zhou G, Delhaize E, Zhou M, Ryan PR. The barley MATE gene, *HvAACT1*, increases citrate efflux and Al^{3+} tolerance when expressed in wheat and barley. Ann Bot. 2013;112:603–12.
9. Ciamporova M. Morphological and structural responses of plant roots to aluminium at organ, tissue, and cellular levels. Biol Plantarum. 2002;45:161–71.
10. Zelinova V, Haluskova L, Huttova J, Illes P, Mistrik I, Valentovicova K, et al. Short-term aluminium-induced changes in barley root tips. Protoplasma. 2011;248:523–30.
11. Delhaize E, Ma J, Ryan PR. Transcriptional regulation of aluminium tolerance genes. Trends Plant Sci. 2012;17:341–8.
12. Maron LG, Guimaraes CT, Kirst M, Albert PS, Birchler JA, Bradbury PJ, et al. Aluminum tolerance in maize is associated with higher MATE1 gene copy number. Proc Natl Acad Sci U S A. 2013;110:5241–6.
13. Ryan PR, Tyerman SD, Sasaki T, Yamamoto Y, Zhang WH, Delhaize E. Identification of aluminium-resistance genes in plants provides an opportunity for enhancing the acid-soil tolerance of crop species. J Exp Bot. 2011;62:9–20.
14. Ma J, Nagao S, Sato K, Ito H, Furukawa J, Takeda K. Molecular mapping of a gene responsible for Al-activated secretion of citrate in barley. J Exp Bot. 2004;55:1335–41.
15. Furukawa J, Yamaji N, Wang H, Mitani N, Murata Y, Sato K, et al. An aluminum-activated citrate transporter in barley. Plant Cell Physiol. 2007;48:1081–91.
16. Wang J, Raman H, Zhou M, Ryan PR, Delhaize E, Hebb DM, et al. High-resolution mapping of the *Alp* locus and identification of a candidate gene *HvMATE* controlling aluminium tolerance in barley (*Hordeum vulgare* L.). Theor Appl Genet. 2007;115:265–76.
17. Ma J, Ryan PR, Delhaize E. Aluminium tolerance in plants and the complexing role of organic acids. Trends Plant Sci. 2001;6:273–8.
18. Ryan PR, Delhaize E, Jones DL. Function and mechanism of organic anion exudation from plant roots. Annu Rev

Plant Physiol Plant Mol Biol. 2001;52:527–60.

19. Bian M, Waters I, Broughton S, Zhang X, Zhou M, Lance R, et al. Development of gene-specific markers for acid soil/aluminium tolerance in barley (*Hordeum vulgare* L.). Mol Breed. 2013;32:155–64.
20. Raman H, Moroni JS, Sato K, Read BJ, Scott BJ. Identification of AFLP and microsatellite markers linked with an aluminium tolerance gene in barley (*Hordeum vulgare* L.). Theor Appl Genet. 2002;105:458–64.
21. Tang Y, Sorrells ME, Kochian LV, Garvin DF. Identification of RFLP markers linked to the barley aluminum tolerance gene*Alp*. Crop Sci. 2000;40:778–82.
22. Wang J, Harsh R, Zhang G, Mendham N, Zhou M. Aluminium tolerance in barley (*Hordeum vulgare* L.): physiological mechanisms, genetics and screening methods. J Zhejiang Univ Sci B. 2006;7:769–87.
23. Fujii M, Yokosho K, Yamaji N, Saisho D, Yamane M, Takahashi H, et al. Acquisition of aluminium tolerance by modification of a single gene in barley. Nat Commun. 2012;3:713.
24. Raman H, Karakousis A, Moroni JS, Raman R, Read BJ, Garvin DF, et al. Development and allele diversity of microsatellite markers linked to the aluminium tolerance gene *Alp* in barley. Aust J Agric Res. 2003;54:1315–21.
25. Delhaize E, Ryan PR, Hebb DM, Yamamoto Y, Sasaki T, Matsumoto H. Engineering high-level aluminum tolerance in barley with the *ALMT1* gene. Proc Natl Acad Sci U S A. 2004;101:15249–54.
26. Ryan PR, Delhaize E, Randall PJ. Characterisation of Al-stimulated efflux of malate from the apices of Al-tolerant wheat roots. Planta. 1995;196:103–10.
27. Liang HW, Wang CZ, Li Z, Luo XZ, Zou GW. Improvement of the silver-stained technique of polyacrylamide gel electrophoresis (in Chinese). Hereditas (Beijing). 2008;30:1379–82.
28. Van Ooijen JW. JoinMap 4.0, Software for the calculation of genetic linkage maps. Wageningen: Plant Research International; 2006.
29. Van Ooijen JW. MapQTL 6, Software for the mapping of quantitative trait loci in experimental populations of diploid species. Wageningen, The Netherlands: Kyazma BV; 2009.
30. Zhu CS, Wang CM, Zhang YM. Modeling segregation distortion for viability selection I. Reconstruction of linkagemaps with distorted markers. Theor Appl Genet. 2007;114(2):295–305.
31. Sasaki T, Ryan PR, Delhaize E, Hebb DM, Ogihara Y, Kawaura K, et al. Sequence upstream of the wheat (*Triticum aestivum* L.) ALMT1 gene and its relationship to aluminum resistance. Plant Cell Physiol. 2006;47:1343–54.
32. Gruber BD, Ryan PR, Richardson AE, Tyerman SD, Sunita R, Hebb DM, et al. HvALMT1 from barley is involved in the transport of organic anions. J Exp Bot. 2010;61:1455–67.
33. Gruber BD, Delhaize E, Richardson AE, Roessner U, James RA, Howitt SM, et al. Characterisation of *HvALMT1* function in transgenic barley plants. Funct Plant Biol. 2011;38:163–75.

CHAPTER 17

TRANSGENIC PRODUCTION OF AN ANTI HIV ANTIBODY IN THE BARLEY ENDOSPERM

Goetz Hensel, Doreen M. Floss, Elsa Arcalis, Markus Sack, Stanislav Melnik, Friedrich Altmann, Twan Rutten, Jochen Kumlehn, Eva Stoger, Udo Conrad

INTRODUCTION

The concept of molecular farming was first aired over 20 years ago, but despite its much trumpeted potential (see review by [1]), the first plant-derived products have only recently reached the market. Examples include recombinant human glucocerebrosidase synthesized in carrot suspension cells [2], and human growth hormone and cytokines synthesized in the barley grain [3]. Genes encoding various monoclonal antibodies have been successfully expressed in planta [4–7], and have been subjected to clinical trials [8]. These achievements clearly demonstrate that plant-based production systems have been recognized as alternative platform to microbial and mammalian cells with specific advantages including the lack of human pathogens [9]. Although transient transgene expression systems have attracted greater investment than stable ones, largely because the development effort is less demanding, currently marketed products have all been based on stably transformed plants or plant cells.

The cereal grain is a particularly attractive vehicle for producing recombinant protein, because its evolution as a protein storage organ assures an environment in which proteins remain stable for long periods without any need for refrigeration or aseptic conditions. The specific advantages of barley are its tolerance of a wide range of environmental conditions, its largely autogamous mode of reproduction, its G.R.A.S. (generally recognized as safe) status, its well established agronomy, and its ease of genetic transformation. Many conventional barley breeding programs use doubled haploid technology to drive homozygosity in elite germplasm [10], and the combination of this methodology with genetic transformation provides a highly convenient means to fix transgenes, as shown by [11] in tobacco.

The major barley grain storage proteins are the hordeins, which are transported first into the lumen of the endoplasmic reticulum (ER), and are ultimately deposited in the protein storage vacuole (PSV) [12, 13]. In barley 50% of the grain's protein are hordeins, while 75% of oat grain proteins are globulins[14]. The genes encoding such storage proteins have therefore become the prime targets for characterization of endosperm-specific promoters. Vickers *et al.* [15], in a comparison of the effectiveness of a number of endosperm-specific promoters to transiently express transgenes in both barley and wheat, were able to identify a cis-acting element, which made a substantial contribution to the level of transgene expression in the endosperm. Here the synthesis of an anti-HIV monoclonal antibody in the barley grain is described. The approach involved the use of the oat *GLO1* (*GLOBULIN1*) promoter to drive the transgene, and the addition of signal peptides to direct and stabilize the recombinant protein.

MATERIALS AND METHODS

Plasmids, Strains and Generation of Transgenic Barley Plants

The oat *GLO1* promoter sequence was amplified from the genomic DNA of oat cv. 'Solidor' (IPK Genebank accession AVE 1500) using the primer pair GH-AsGlo1-EcoRV-F1

Citation: Hensel G, Floss DM, Arcalis E, Sack M, Melnik S, Altmann F, et al. (2015) Transgenic Production of an Anti HIV Antibody in the Barley Endosperm. PLoS ONE 10(10): e0140476. doi:10.1371/journal.pone.0140476.

and GH-AsGlo1-PstI-R1 (sequences given in S3 Table). The resulting 977 bp *Eco*RV/*Pst*I fragment was cloned into pCK21/4 (C. Kaydamov, unpublished) in front of the Legumin type B 5'UTR and signal peptide from *Vicia faba* fused to GFP and the endoplasmic reticulum retention signal KDEL (*LeB4* 5'UTR::*LeB4*SP:*gfp*:*KDEL*) cassette to produce the construct pGH220. An *Sfi*I fragment harboring the entire expression cassette was then subcloned into p6d35S (DNA Cloning Service, Hamburg, Germany) to generate the construct pGH223 (*d35S*P::*HPT-AsGLO1*P::*LeB4*SP:*gfp*:*KDEL*). Plasmids harboring the coding sequences for the 2G12 heavy chain (HC) and light chain (LC) were based on pGH221 which is the same as pGH220 but with bacterial chloramphenicol resistance. The primer pair LC-Nco/LC-Sal (S3 Table) was used to amplify the 2G12 LC sequence from a template of pRTRA 35S 2G12LC [10] and the resulting*Nco*I/*Sal*I fragment was substituted for *GFP* in pGH221 to form the construct pGH221-LC. Similarly, 2G12 HC was amplified using the primer pair HC-Nco/HC-Sal (S3 Table) from a template of pRTRA 35S 2G12HC [11], and the resulting *Nco*I/*Sal*I fragment introduced into pGH221 to generate pGH221-HC. Finally, the two *Sfi*I fragments were cloned into p6d35S to generate the binary vectors pGH248 (HC) and pGH249 (LC). A schematic representation of pGH223, pGH248 and pGH249 is given in S1 Fig. The binary plasmids were all introduced into*Agrobacterium tumefaciens* strain AGL1 [16], which was co-cultured with immature embryos of barley cv. 'Golden Promise', as described elsewhere [17].

Embryo Rescue and Segregation for Hygromycin Resistance

Aseptically dissected immature embryos from 21–28 day old T_1 or T_2 caryopses were cultured for 24 h on B5 medium [18] to induce their germination. The germinating embryos were transferred onto K4N medium [19] supplemented with 100 mg/L hygromycin, and held for up to ten days at 24°C under a 16 h photoperiod.

Generation of Doubled Haploid Lines

Microspores at the highly vacuolated, pre-mitotic stage were isolated from immature spikes following [20]. The ploidy level of regenerants was determined using a Ploidy Analyser 1 (PARTEC, Muenster, Germany), following the manufacturer's protocol. To induce whole genome duplication where it was not spontaneous, haploid plantlets were exposed to 0.1% w/v colchicine, following the [21] procedure.

Grain Protein Extraction and the Quantification of GFP

Soluble proteins were extracted from mature grains in 2% (w/v) SDS, 5% (v/v) β-mercaptoethanol, 0.001% (w/v) pyronin, 10% (v/v) glycerol, 0.063 M Tris-HCl [22]. Following their separation by SDS-PAGE, the proteins were transferred onto a membrane, which was challenged with a monoclonal anti-GFP mouse IgG_1 antibody (Roche Diagnostics, Mannheim, Germany) and subsequently by an anti-mouse IgG (H+L) alkaline phosphatase conjugate antibody (Promega, Madison, WI, USA). Conjugated proteins were visualized by supplying Western Blue® Stabilized Substrate for Alkaline Phosphatase (Promega, Mannheim, Germany). GFP was quantified densitometrically using TotalLab™ software (TL120, Nonlinear Dynamics Ltd., Newcastle upon Tyne, UK).

For calculating the antibody yield in %TSP, 0.1 g grain flour was extracted with 400 µl extraction buffer and total protein content of the sample was estimated using standard methods.

PCR and DNA Gel Blot Analysis

The template used in PCRs amplifying from plant genomic DNA was extracted from ~100 mg snap-frozen leaf, following [23]. Each PCR was seeded with 100 ng template, using primers detailed in S3 Table. Amplicons were separated by agarose gel electrophoresis and visualized by ethidium bromide staining. DNA gel blot analysis were carried out following the [24] procedure, and were based on 30 µg aliquots of electrophoresed *Hind*III-digested genomic DNA transferred onto a positively charged nylon membrane (Roche Diagnostics, Mannheim, Germany). The membranes were hybridized with a DIG-dUTP-labeled *HPT* probe (Roche Diagnostics, Mannheim, Germany).

Light and Electron Microscopy

Transverse median sections of mature barley grains were cut with a razor blade, and placed in a droplet of 50 mM phosphate buffer (pH 7.0) under a cover slip. Fluorescence microscopy was carried out using a LSM510 META confocal laser scanning microscope (Carl Zeiss, Jena, Germany). GFP fluorescence was analyzed with a 488 nm laser line in combination with a 505–530 nm band pass filter. Genuity of the GFP signal was confirmed by lambda stack analysis. Cell wall autofluorescence was captured with a 364 nm laser line in combination with a 385 nm long pass filter. For the localization of recombinant 2G12, thin slices were cut from transgenic and wild type endosperm using a razor blade, floated in 0.1 M phosphate buffer (pH 7.4), fixed in 4% (w/v) paraformaldehyde, 0.5% (v/v) glutaraldehyde in 0.1 M phosphate buffer (pH 7.4) and processed as described by [25]. Both 1 µm thick sections mounted on glass slides for fluorescence microscopy and sections showing silver interference colors mounted on gold grids for electron microscopy were pre-incubated in 5% (w/v) fraction V bovine serum albumin in 0.1 M phosphate buffer (pH 7.4) and then challenged with an recommended dilution of either polyclonal goat anti-HC or anti-LC (Sigma, Carlsbad, CA, USA), polyclonal rabbit anti-wheat gliadin (Sigma, Carlsbad, CA, USA) or polyclonal anti-PDI. The sections were then exposed to a solution of secondary antibody in 0.1 M phosphate buffer (pH 7.4). For fluorescence microscopy experiments, the secondary antibody was either donkey anti-goat IgG Alexa Fluor® 488 (Life Technologies, Carlsbad, USA) or donkey anti-rabbit IgG Alexa Fluor® 526 (Life Technologies, Carlsbad, USA), while for the electron microscopy experiments, the antibody was either rabbit anti-goat IgG 10 nm gold (British BioCell International, Cardiff, UK) or goat anti-rabbit IgG 10 nm gold (British BioCell International).

Immunoblotting

Hordeins were extracted from 50 mg of flour milled from cv.

Golden Promise grain in 1 mL 55% (v/v) isopropanol, 2% (v/v) β-mercaptoethanol for 1 h at 65°C. After centrifugation, a 50 µL aliquot was combined with 750 µL acetone, and the mixture was re-centrifuged. The resulting pellet was dissolved in 60 µL LDS buffer and loaded onto a 10% (w/v) SDS polyacrylamide gel. Bands corresponding to the B and C hordeins were identified according to [13]. Polyclonal rabbit anti-wheat gliadins serum (Sigma, Carlsbad, CA, USA) was used for detection. Total soluble proteins were extracted from individual grains in phosphate buffered saline (PBS; 1:3 w:v) and then centrifuged (17,500 x *g*, 4°C). For sequential extractions, the pellets were washed six times with PBS (NaCl 137 mM, KCl 2.7 mM, Na_2HPO_4 10 mM, KH_2PO_4 1.8 mM) as described above and the sixth supernatant was kept as a negative control. The pellet was then re-extracted in an equal volume 1x PBS supplemented with, 5% (v/v) β-mercaptoethanol, and 8 M urea. The resulting supernatant was boiled for 10 min, and then subjected to electrophoresis through a 12% (w/v) SDS-PAGE under reducing conditions. Immunoblotting analyses were carried out according to standard protocols.

Biacore Analysis

Binding ratios to bacterial surface proteins protein-A, protein-L and HIV glycoprotein with a molecular weight of 120 kDa (gp120) were determined on a BiacoreT200 instrument (GE Healthcare, Braunschweig, Germany) using HBS-EP as running buffer and assay conditions reported previously [5]. The antigen binding kinetics was measured using a protein-A capture assay [11, 26]. HIV-1 gp120 was obtained from the Center of AIDS Reagents (CFAR/NIBSC, Potters Bar, Hertfordshire, UK) and CHO-derived 2G12 was provided by Polymun Scientific GmbH (Klosterneuburg, Austria).

Glycosylation

The N-glycan analysis was applied to tryptic glycopeptides, as described by [27]. Briefly, the heavy chain of the SDS-PAGE separated IgG was excised from an SDS-PAGE gel, S-alkylated, digested with sequencing grade modified trypsin (Promega, Mannheim, Germany) and subsequently separated through a reverse phase capillary column (Biobasic C18, ThermoScientific, Vienna, Austria) and a Q-TOF Ultima mass spectrometer (Waters, Manchester, UK).

RESULTS

The oat *GLOBULIN1* Promoter Produces Endosperm-Specific Gene Expression

A set of 50 independent primary transgenics harboring pGH223 (*d35S*P::*HPT-AsGLO1*P::*LeB4*SP:*gfp*:*KDEL*) was generated (S1 Fig). Inspection of T_1 mature grains revealed a variable level of GFP accumulation in the endosperm, but no case of any detectable signal in the embryo (Fig 1A). In the subsequent generation, GFP was accumulated in the grain to a level of up to 1.2 mg per g dry weight (Fig 1B). GFP could also be detected by Coomassie Brilliant Blue staining of SDS-PAGE separations of total protein extracted from the grain (Fig 1C), a result which enabled a simple means to confirm the stable inheritance and expression of the transgene. The number of transgene copies introduced in each transgenic event was determined by DNA gel blot hybridization, which was also informative with respect to the zygosity (hemi- vs homo-) of the transgene(s) (S1 Table). The accumulation of GFP did not vary significantly between the single and multicopy transgene homozygotes, lying in the range 598–808 µg per g dry weight.

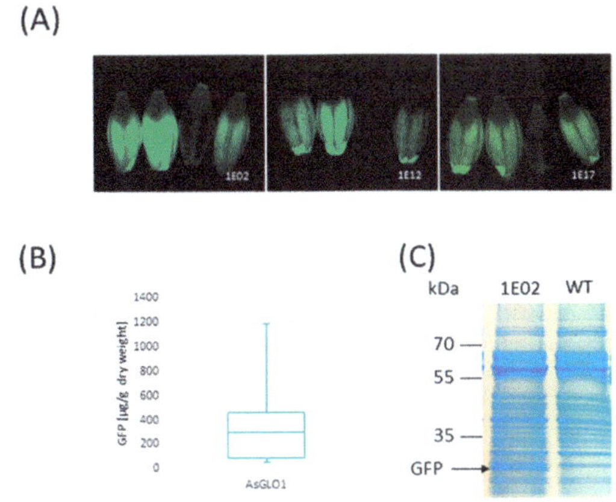

Figure 1: Establishment of an endosperm transgene expression system.

(A) Fluorescence microscopy of four grains expressing *GFP* driven by the oat *GLO1*promoter. (B) Quantification of GFP based on Western blotting with an anti GFP antibody. (C) Total soluble protein profile obtained by Coomassie Brilliant Blue staining of 12% SDS-PAGE separations of extracts of a T_3 selection of line BG136/1E02-14-4 expressing *GFP*.

http://dx.doi.org/10.1371/journal.pone.0140476.g001

The Site of GFP Deposition in the Grain of Transgenic Plants

Confocal laser scanning microscopy (CLSM) of mature grains set by the transgenic plant 1E17 (harboring the construct *d35S*P::*HPT-AsGLO1*P::*LeB4*SP:*gfp*:*KDEL*) confirmed an accumulation of GFP in both aleurone layer and starchy endosperm (Fig 2A). In the grains set by a homozygous selection of plant B13/10E28 (which expressed *GFP* under the control of the maize *UBIQUITIN1* promoter), GFP was abundant in the aleurone layer but hardly detectable in the starchy endosperm (Fig 2B and 2C). At the cellular level the GFP signal was uniformly distributed throughout the cytoplasm of endosperm cells, consistent with its deposition to small, probably vesicle-like compartments (Fig 2D). Wild type grains lacked any GFP activity (Fig 2E).

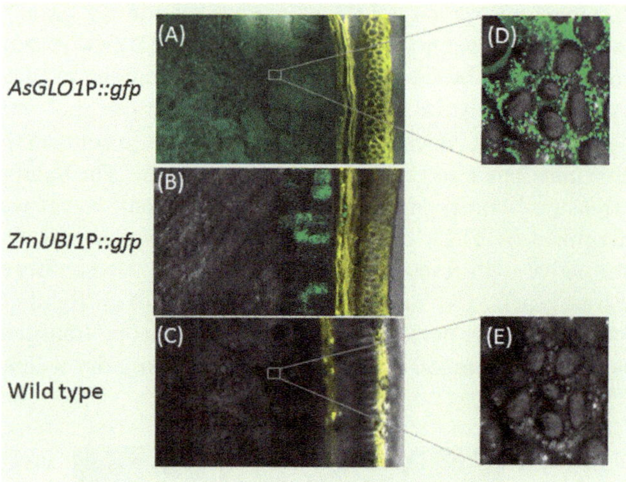

Figure 2: Transverse grain sections demonstrating the site of GFP accumulation (green) and cell wall autofluorescence (yellow).

(A) GFP accumulation throughout the aleurone and endosperm of grains set by plants harboring the *AsGLO1*P::*GFP* transgene. (B) In a transgenic barley line constitutively expressing *GFP* (*ZmUBI1*P::*GFP*), GFP fluorescence was restricted to the aleurone layer. (C) In a wild type grain, no GFP signal was detectable. (D) Higher magnification of an endosperm cell reveals discrete deposits of GFP within the cytoplasm of the transgenic line shown in (a). (E) The cytoplasm of a wild type endosperm cell lacks any GFP signal. Bar: 50 µm in (A-C), 10 µm in (D-E).

http://dx.doi.org/10.1371/journal.pone.0140476.g002

Stable Barley Transgenic Lines Expressing 2G12 in the Endosperm

Transgenic barley plants harboring both the coding sequence for the heavy (HC) and light (LC) chains of the anti-HIV monoclonal antibody 2G12, driven by the oat *GLO1* promoter and including both the LeB4 signal peptide and SEKDEL retention signal (S1 Fig) were generated using the co-transformation strategy described by [28]. The inoculation of 180 immature embryos with an *Agrobacterium tumefaciens* strain harboring both pGH248 (LC) and pGH249 (HC) produced 29 regenerants. A multiplex PCR designed to reveal the presence of both transgenes identified six of these 29 plants as co-transformants, with ten carrying only the HC gene and 13 only the LC gene. A Western blot analysis suggested some variation in the amount of recombinant protein accumulated (data not shown). Transgene homozygotes were generated from three of the primary HC/LC carriers (BG208/1E05, BG208/1E06 and BG208/2E07) via pollen embryogenesis. Confirmed spontaneously doubled haploid regenerants carrying both transgenes were obtained from both BG208/1E05 (six lines) and BG208/1E06 (11 lines); the six regenerants from BG208/2E07 did not spontaneously undergo whole genome duplication, so required a colchicine treatment to restore self fertility. A sample of progeny of each doubled haploid line was tested for the expression of the two 2G12 antibody genes and the quantity of recombinant protein present in the grain was assessed. No free light or heavy chains were detected in Western blots (Fig 3 and data not shown). The proportion of total soluble grain protein represented by 2G12 lay between 0.1 to 0.4% (S2 Table). A DNA gel blot hybridization-based check of the transgene copy number in each line showed that both BG208/1E05 and BG208/1E06 harbored a single copy of both the HC and the LC genes (Fig 4). The transgenes strictly co-segregated, indicating a single integration event. The number of hybridizing fragments revealed in the BG208/2E07 material suggested the integration of six distinct transgenes in the T_0 plant.

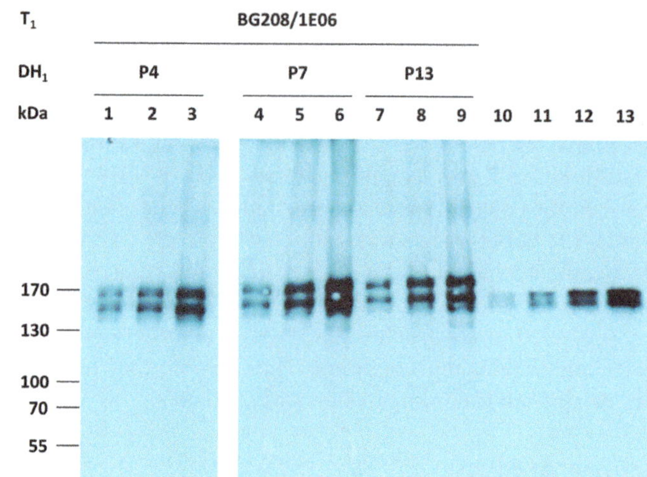

Figure 3: Detection of 2G12 antibodies in grain extracts.

Grain extracts and CHO cell-derived 2G12 standards were separated by gradient SDS-PAGE under non-reducing conditions. 2G12 antibodies were detected by their binding to anti-human IgG Fcg-peroxidase. Lanes 1–3: 0.5, 1.0 and 2.0 µg of protein extracted from BG208/1E06-P4. Lanes 4–6: 0.5, 1.0 and 2.0 µg of protein extracted from BG208/1E06-P7. Lanes 7–9: 0.5, 1.0 and 2.0 µg of protein extracted from BG208/1E06-P13. Lanes 10–13: 0.5, 1.0, 2.0 and 4.0 ng 2G12 obtained from CHO cells.

http://dx.doi.org/10.1371/journal.pone.0140476.g003

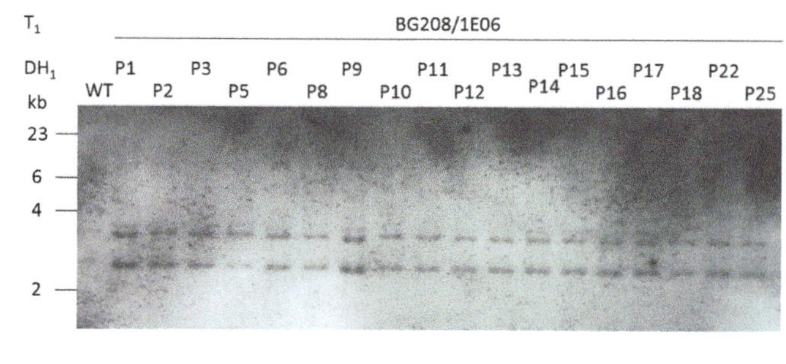

Figure 4: DNA gel blot analysis profiles of doubled haploid lines expressing the 2G12 monoclonal antibody.

A 30 µg aliquot of genomic DNA was digested with *Hin*dIII, separated through a 0.8% agarose gel, blotted onto a Nylon membrane and hybridized with *HPT*. WT: wild type.

http://dx.doi.org/10.1371/journal.pone.0140476.g004

Recombinant 2G12 Antibody was Deposited in Endosperm Protein Bodies

The pattern of endosperm maturation allows for cells at a variety of developmental stages to be visualized within a given section. Cells in the sub-aleurone layer (sal) form later than those in the inner part of the endosperm and have a higher relative protein content (Fig 5A and 5B). At the mid-developmental stage, the barley protein bodies harbored low electron density spherical protein inclusions embedded within an electron dense protein matrix (Fig 5C). The former contains B, C, γ1 and γ2 hordeins [29], and can be stained with toluidine blue (Fig 5B). The subcellular distribution of the recombinant protein in the sal of BG208/1E05 grain was compared with that of the hordeins; the latter were identified by an antibody which recognizes B and C hordeins [30] and thereby identifies protein bodies in the barley endosperm (S2 Fig). The analysis revealed that, although associated with the protein bodies, the recombinant 2G12 did not co-localize with the hordeins, but rather accumulated at the periphery of the hordein aggregates (Fig 5D–5F). This pattern of deposition was confirmed by immunoelectron microscopy, which detected label exclusively along the periphery of the developing protein bodies (Fig 5G and 5H). Given the recombinant 2G12 protein was engineered to carry a KDEL tag to encourage its retention in the ER, a comparison was made with the subcellular localization of protein disulphide isomerase (PDI), an endogenous HDEL-containing, ER-resident endosperm protein [31]. PDI was detectable in the ER and was evenly distributed within the protein bodies (Fig 5I and 5K), and there was no discernible effect on this distribution induced by the presence of the 2G12 transgenes (data not shown). The 2G12 and PDI deposition sites were non-identical: they occupied two distinct phases inside and along the margin of the protein bodies (Fig 5J and 5L).

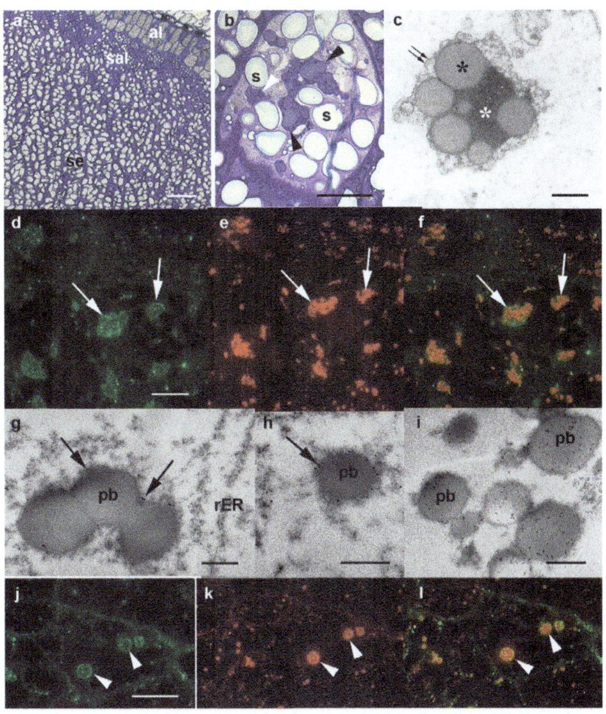

Figure 5: Deposition of 2G12 antibody in the barley endosperm.

(A-C) Wild type grain, (D-L) transgenic grains. (A) Light microscopy shows three distinct layers, namely the aleurone layer (al), the sub-aleurone layer (sal) and the endosperm (se). (B) Toluidine blue staining of a mid-development sal cell. Note the oblong starch grains (s) and the large protein bodies (black arrowheads). The protein bodies are multiphasic and take up the stain differentially (white arrowheads). (C) Transmission electron microscopy, general, non-specific contrasting. The protein bodies lie within a membrane (double arrow) displaying their multiphasic nature: moderately electron-dense, spherical protein bodies (black asterisks) embedded in a highly electron-dense, amorphous matrix (white asterisk). (D-F) Fluorescence microscopy demonstrates the co-localization of 2G12 (D, arrows) and hordeins (E, arrows). The antibody chains accumulate in the periphery of the protein bodies alongside the hordeins (F, merged, arrows). (G-I) immuno-electron microscopy-based localization of 2G12. Label (arrowed) associated with the periphery of the protein bodies (pb). (I) Localization of PDI within the protein bodies (pb) that appear evenly labelled. (J-L) Co-localization of 2G12 antibody (J, arrowheads) and PDI (K, arrowheads). 2G12 was detected exclusively in the periphery of the protein bodies, labelled with PDI (L, merged, arrowheads). Bars: 20 μm in (A), 5 μm in (B, D-F, J-L), 1 μm in (C, G-I).

http://dx.doi.org/10.1371/journal.pone.0140476.g005

N-Glycosylation of the Recombinant 2G12 Antibody

An analysis of BG208/1E05 grains revealed the presence of a mixture of N-glycan structures (Fig 6). The main peak corresponded to a trimmed glycan lacking terminal N-acetyl glucosamine residues and carrying a β(1,2)-linked xylose core with an α(1,3)-linked fucose residue. This structure is commonplace in vacuolar glycoproteins and is indicative of the passage of the recombinant protein through the Golgi apparatus *en route* to the storage vacuole. A smaller quantity of complex N-glycans possessing terminal N-acetylglucosamines was also detected, diagnostic of glycoproteins which have been exported from the Golgi apparatus or secreted into the apoplast. The presence of significant amounts of this N-glycan structure is quite unusual in starchy cereal endosperm. The proportion of mannosidic N-glycans was very low, whereas structures consisting of just a single GlcNAc residue were abundant. This pattern is consistent with the trimming of oligomannosidic glycans by endoglycosidase activity, as has been observed in both the maize [5, 32] and the rice [33] endosperm. A significant quantity of non-glycosylated HC protein was also detected.

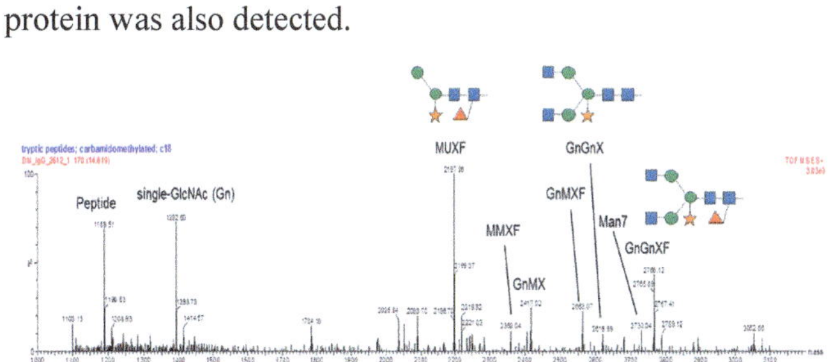

Figure 6: N-glycan analysis of grain extracts.

The deconvoluted spectrum of the elution region of tryptic peptides and glycopeptides of the Fc peptide EEQYNSTYR shows the presence of a non-glycosylated peptide, carrying only a single GlcNAc residue. Peak heights roughly reflect the relative molar ratios. So smaller species may be somewhat over-represented. Abbreviations follow the proglycan system. GlcNAc, Man, Xyl and Fuc are represented, respectively, by blue squares, green circles, orange stars and red triangles.

http://dx.doi.org/10.1371/journal.pone.0140476.g006

The Assembled Recombinant 2G12 Antibody is Functional in a BIACORE Antigen Binding Assay

A comparison between barley and Chinese hamster ovary (CHO) recombinant 2G12 was based on a binding ratio assay under mass-transport limiting conditions [5]. Various dilutions of the antibody were passed over high-capacity protein-A (Fc2), protein-L (Fc3) and gp120 (Fc4) surfaces, the blank subtracted binding signals were plotted against one another and the binding ratios were then determined by linear regression. A relative binding ratio was derived by dividing the barley 2G12 value by the CHO 2G12 one. Whereas the relative protein-L-to-protein-A binding ratio was 100.6±0.8% (i.e., the proteins were essentially identical), the relative HIV-1 ^{BaL}gp120-to-protein-A and HIV-1 ^{BaL}gp120-to-protein-L binding ratios were, respectively, 87.5±1.1% and 87.0±0.6%, indicating that the barley recombinant antibody included an inactive fraction which did not bind the antigen as effectively. The binding kinetics were explored in more detail using a protein-A capture assay based on monomeric soluble HIV-1 ^{BaL}gp120 (EVA607, CFAR/NIBSC, Potters Bar, Hertfordshire, UK) as the analyte. The analysis compared two different batches of both CHO- and barley-derived 2G12, along with aliquots sampled from a single barley 2G12 preparation stored at either -20°C or +4°C. The samples were diluted to ensure that they bound immobilized protein-A equivalently (166–174 RU), with the remaining minor differences corrected by normalizing the binding curves to 170 RU. The resulting binding curves were fitted using a simple 1:1 binding model (Table 1). Only minor differences were observed in the kinetic rate constants (k_{ass} and k_{diss}), the equilibrium dissociation constant K_D and the fitted R_{max} values, thus the binding behaviour of the antibodies from plants and CHO cells is principally comparable.

Table 1: Summary of kinetic analysis

http://dx.doi.org/10.1371/journal.pone.0140476.t001

Sample	k_{ass} [1/Ms]	k_{diss} [1/s]	K_D [nM]	R_{max} [RU]	Chi^2	U
^{barley}2G12 (4°C)	$8.57 \cdot 10^4$	$1.33 \cdot 10^{-4}$	1.55	63.1	1.68	7
^{barley}2G12 (-20°C)	$9.15 \cdot 10^4$	$1.28 \cdot 10^{-4}$	1.40	65.4	1.50	7
^{CHO}2G12 I-8 (older batch)	$10.4 \cdot 10^4$	$1.19 \cdot 10^{-4}$	1.14	65.5	2.08	9
^{CHO}2G12 A1 (newer batch)	$10.2 \cdot 10^4$	$1.44 \cdot 10^{-4}$	1.41	61.9	1.87	7

doi:10.1371/journal.pone.0140476.t001

DISCUSSION

To date, a number of endosperm-specific expression systems have been developed, exploiting the promoters of genes encoding a variety of endosperm storage proteins from barley, wheat and rice (for review see [34]). Few attempts have been made to compare their effectiveness based on the quantification of reporter protein abundance. Patel *et al.* [35] showed that the transgenic expression in barley of the bacterial gene *XynA* was fivefold greater when driven by the *GluB-1* (from rice) promoter than it was by the *Hor2-4* (from barley) promoter, and were able to recover a maximum yield of recombinant protein equivalent to 0.037% of the total grain protein. Meanwhile, Vickers *et al.* [15] expressed *GFP* in barley driven by the oat *GLO1*promoter, and achieved a yield equivalent to 10% of the total grain protein. The *GLO1* promoter was also employed in the present study, with the inclusion of signal sequences derived from LeB4 to encourage uptake of the transgene product into the ER and KDEL for its retention there. The 2G12 HC and LC coding sequences used here have previously been introduced as transgenes into tobacco, maize and rice [4–6, 33], although not with the same promoters as used here. The proportion of the recombinant antibody found in protein extracts of the tobacco leaf and seed varied from 0.3–0.4%, raised to about 1% by incorporating elastin-like peptide (ELP) fusion technology [11]. In maize, the rice *GLUTELIN1* promoter achieved the accumulation of up to 75 μg/g 2G12 in the grain [6]. The range of 2G12 accumulation observed here in the barley grain was 0.1–0.4% of TSP corresponding to 33–160 μg/g grains (Table 2), in line with the amounts of an anti-glycophorin single chain antibody fused to an HIV epitope and recovered from barley grains by [36]. To compare the barley grain production system with other production platforms based on stable nuclear transgenic plants, we have reproduced published data referring to antibody 2G12 in Table 2. For stable tobacco-based systems, we refer to the report of Sack et al. [37]. Higher yields have been achieved with transient expression systems based on Agro-infiltration of N. benthamiana leaves [25, 38]. However, the specific amount of the transgenic protein produced per biomass unit is only one parameter that can be used to describe the productivity of an expression system. More meaningful assessments will also include the space, time and labor input required for completing a production cycle, as well as the flexibility in terms of upscaling and batch processing. Greenhouse-based production of 2G12 in barley grains will result in 1 g antibody per 10 square meters. With three possible harvests per year a 120 square meter greenhouse would be sufficient to produce 36 g of antibody 2G12. This is similar to the figures calculated for stable transgenic tobacco [36]. A main benefit of the barley grain production system is the stability of antibodies in seeds allowing conventional harvesting and storage at ambient conditions. The potential of an in planta recombinant antibody production system using cereal crops has yet to be fully exploited. Considerations also have to comprise regulatory issues including the definition of master seed banks, maintenance of pure genotypes and the quality control procedures for the recombinant antibody [38, 39]. Biosafety issues are closely linked to the level of containment and have been discussed elsewhere [40].

Table 2: Comparison of different stable transgenic expression systems for the production of 2G12 mAB in plants

http://dx.doi.org/10.1371/journal.pone.0140476.t002

Species	Tag	Organ	Promoter	Yield		Reference
Tobacco	-	Leaf	CaMV *35S*	17–18 μg/mL	(45 μg/g)*	[37]
	KDEL KDELELP, KDEL	leafseedLeaf	CaMV *35S* Faba bean *USP*CaMV *35S*	0.4% of TSP0,3% of TSP1% of TSP	(2.8 μg/g)**(1.7 μg/g) **(6.9 μg/g)**	[11][11] [11]
	ELP, KDEL	Seed	Faba bean *USP*	1% of TSP	(5.8 μg/g)**	[11]
Maize	KDEL-	Kernels/ endospermKernels/ endosperm	Rice *Glut1*Rice *Glut1*	30 μg/g75 μg/g		[5][6]
Rice	-	Kernels/ endosperm	Rice *Glut1*	46 μg/g		[33]
Barley	KDEL	Grain/ endosperm	Oat *Glo1*	0.4% of TSP	(160 μg/g)	this study

* line selected over several generations and optimized conditions for growth and harvest

**units were converted based on the average TSP values obtained with the respective extraction method [10]

doi:10.1371/journal.pone.0140476.t002

Barley doubled haploids can be produced either via the pollen embryogenesis or the gynogenesis (following wide crosses with *H. bulbosum*) pathway [10]. The major feature of doubled haploidy is that genotypes become fixed in just a single recombination step. In the doubled haploid lines BG208/1E05 and BG208/1E06, the two HC and LC 2G12 genes were co-inherited, as shown by an identical integration footprint for the two genes. This was not the case for the BG208/2E07 material, where the number of transgene copies present varied, indicating a lack of linkage. As might be expected from highly homozygous material, there was little plant to plant variation in the yield of 2G12 antibody, fulfilling the need to breed material which delivers a stable and predictable level of transgene product as prerequisite for molecular farming. 2G12 is planned to be administered not by injection but by vaginal application to prevent reinfections. This fact minimizes the risk for allergic reactions caused by crop-based allergens. Future scale up and downstream processing development will be required to meet the standards for therapeutic proteins and should include the removal of hordeins by ("polishing" with) size exclusion chromatography and ion exchange chromatography. Despite the inclusion of a KDEL tag on both 2G12 antibody chains, the recombinant protein accumulated alongside the hordeins within the PSV. The deposition of KDEL-tagged proteins in this site has been noted previously by [25]. Cereal prolamins are thought to be routed directly from the ER to the vacuole, bypassing the Golgi apparatus [41]. Thus it seems possible that recombinant proteins engineered to accumulate in the ER may be passively drawn into the PSV by the bulk flow of prolamins. However, the deposition of 2G12 was largely within the periphery of the protein bodies which harbor the hordeins, unlike that of the KDEL-tagged recombinant human serum albumin expressed in wheat, which was rather evenly dispersed within the prolamins [25]. The major 2G12 deposition site coincided with that of hordein $\gamma 1$ and $\gamma 2$, both of which have the capacity to form intermolecular disulphide bonds [29]. A portion of the antibody chains of recombinant 2G12 generated in the maize kernel remained insoluble, ostensibly due to their covalent interaction with zein [42]. It may therefore be that the antibody chains can form hetero-oligomers with hordein $\gamma 1$ and $\gamma 2$, since this would explain the observed localization of 2G12. However, unlike in the maize situation, there was no evidence of any insoluble/covalently linked antibody (data not shown). The localization of 2G12-KDEL within the barley PSV also contrasts with its predominant localization within the maize ER-derived zein bodies [5]. Note, however, that PSVs are less abundant in the maize endosperm than in the endosperm of other cereals, especially at later developmental stages [32].

The N-glycan profile of 2G12-KDEL included a major proportion of structures consisting of just a single GlcNAc residue. The phenomenon likely reflects the presence of endo-N-acetyl-beta-D-glucosaminidase activity in the barley grain [43]. Maize grain and tobacco seed-derived KDEL-tagged 2G12 antibody harbored mainly oligomannosidic N-glycans and single GlcNAc residues, and only a small proportion of complex glycan structures [5, 11]. The most abundant N-glycan structure in the barley 2G12 product was a paucimannosidic glycan with Golgi apparatus-specific modifications (MUXF), typical of vacuolar localized proteins. While still consistent with its deposition in the PSV, this glycan pattern is somewhat unusual for a KDEL-tagged protein, as it suggests that a major part of it must have passed through the Golgi apparatus *en route* to the PSV. Nevertheless, the KDEL sequence was not removed from the protein (data not shown). The transgene construct N-terminal signal peptide sequence used for barley transformation replaced the native IgG sequence used for maize with the LeB4 signal sequence. Furthermore, the KDEL sequence was attached in a different manner, since an extra four residues were added upstream of the SEKDEL sequence in the barley transgene. It is possible that one or both of these differences may have interfered with trafficking or with the exposure of the KDEL signal.

A higher proportion of the 2G12-KDEL product remained non-glycosylated in the barley product than in the maize one (11%, see [6]). The equivalent proportion in rice was even higher [33], as it was also in *Arabidopsis thaliana* seeds producing high amounts of a scFvFc antibody; the potential explanation in the latter case was an overloading of the glycosylation machinery [44]. The equilibrium and kinetic rate constants of the barley 2G12-KDEL and the CHO 2G12 antibodies were highly comparable, with the differences lying well within the range observed between different antibody production batches. This is in agreement with earlier analyses of 2G12 which have suggested that its functionality is independent of both the glycoform and source of production [5, 6, 11, 33, 37].

In summary, the endosperm-specific expression system based on the oat *GLO1* promoter in combination with the LeB4 signal peptide has been shown to be effective in barley, and is appropriate as a technology for producing recombinant protein in the barley grain. The test protein, an anti-HIV antibody, was produced in a functional and soluble form and accumulated within the periphery of the PSV, alongside the hordeins. Its level of accumulation exceeded those achieved in transgenic tobacco, maize and rice.

ACKNOWLEDGMENTS

We thank Dr. Catrin Kaydamov (IPK, Germany) for providing plasmid pCK21/4, Dr. Stefanie Goedeke (IPK, Germany) for providing a sample of recombinant GFP protein produced in *E. coli* and R. Strasser (BOKU, Vienna) for the gift of polyclonal anti-PDI. The excellent technical assistance of Cornelia Marthe, Andrea Müller, Ingrid Otto and Isolde Tillack is gratefully acknowledged.

AUTHOR CONTRIBUTIONS

Conceived and designed the experiments: GH DF JK ES UC. Performed the experiments: GH EA MS SM TR. Analyzed the data: GH DF FA JK ES UC. Wrote the paper: GH MS ES.

REFERENCES

1. Fischer R, Buyel JF, Schillberg S, Twyman RM. Molecular Farming in Plants: The Long Road to the Market. In: Howard JA and Hood EE, editors. Commercial Plant-Produced Recombinant Protein Products. Biotechnology in Agriculture and Forestry 68. Berlin Heidelberg: Springer; 2014. pp. 27–41.
2. Aviezer D, Brill-Almon E, Shaaltiel Y, Hashmueli S, Bartfeld D, Mizrachi S, et al. A plant-derived recombinant human glucocerebrosidase enzyme—a preclinical and phase I investigation. PloS ONE 2009; 4: e4792. doi: 10.1371/journal.pone.0004792. pmid:19277123
3. Magnusdottir A, Vidarsson H, Björnsson JM, Örvar

BL. Barley grains for the production of endotoxin-free growth factors. Trends in Biotechnol. 2013; 31: 572–580. doi: 10.1016/j.tibtech.2013.06.002

4. Floss DM, Sack M, Stadlmann J, Rademacher T, Scheller J, Stöger E, et al. Biochemical and functional characterization of anti-HIV antibody-ELP fusion proteins from transgenic plants. Plant Biotechnol J. 2008; 6: 379–391. doi: 10.1111/j.1467-7652.2008.00326.x. pmid:18312505
5. Rademacher T, Sack M, Arcalis E, Stadlmann J, Balzer S, Altmann F, et al. Recombinant antibody 2G12 produced in maize endosperm efficiently neutralizes HIV-1 and contains predominantly single-GlcNAc N-glycans. Plant Biotechnol J. 2008; 6: 189–201. pmid:17979949 doi: 10.1111/j.1467-7652.2007.00306.x
6. Ramessar K, Capell T, Christou P. Molecular pharming in cereal crops. Phytochem Rev. 2008; 7: 579–592. doi: 10.1007/s11101-008-9087-3
7. Strasser R, Stadlmann J, Schähs M, Stiegler G, Quendler H, Mach L, et al. Generation of glyco-engineered *Nicotiana benthamiana* for the production of monoclonal antibodies with a homogeneous human-like N-glycan structure. Plant Biotechnol J. 2008; 6: 392–402. doi: 10.1111/j.1467-7652.2008.00330.x. pmid:18346095
8. Ma J, Drossard J, Lewis D, Altmann F, Boyle J, Christou P, et al. Regulatory approval and a first-in-human phase I clinical trial of a monoclonal antibody produced in transgenic tobacco plants. Plant Biotechnol J. 2015 doi: 10.1111/pbi.12416
9. De Muynck B, Navarre C, Boutry M. Production of antibodies in plants: status after twenty years. Plant Biotechnol. J. 2010; 8: 529–563. doi: 10.1111/j.1467-7652.2009.00494.x. pmid:20132515
10. Kumlehn J. Haploid technology. In: Kumlehn J, Stein N, editors. Biotechnological approaches to barley improvement. Biotechnology in agriculture and forestry 69. Berlin: Springer; 2014. pp. 379–392.
11. Floss DM, Sack M, Arcalis E, Stadlmann J, Quendler H, Rademacher T, et al. Influence of elastin-like peptide fusions on the quantity and quality of a tobacco-derived human immunodeficiency virus-neutralizing antibody. Plant Biotechnol J. 2009; 7: 899–913. doi: 10.1111/j.1467-7652.2009.00452.x. pmid:19843249
12. Cameron-Mills V. The structure and composition of protein bodies purified from barley endosperm by silica sol density gradients. Carlsberg Res Commun. 1980; 45: 557–576. doi: 10.1007/bf02932923
13. Cameron-Mills V, von Wettstein D. Protein body formation in the developing barley endosperm. Carlsberg Res Commun. 1980; 45: 577–594. doi: 10.1007/bf02932924
14. Shewry PR. Improving the protein content and composition of cereal grain. J Cereal Sci. 2007; 46: 239–250. doi: 10.1016/j.jcs.2007.06.006
15. Vickers CE, Xue G, Gresshoff PM. A novel cis-acting element, ESP, contributes to high level endosperm-specific expression in an oat globulin promoter. Plant Mol Biol. 2006; 62: 195–214. pmid:16915522 doi: 10.1007/s11103-006-9014-1
16. Lazo GR, Stein PA, Ludwig RA. A DNA transformation-competent Arabidopsis genomic library in *Agrobacterium*. Biotechnol. (N. Y.) 1991; 9: 963–967. doi: 10.1038/nbt1091-963
17. Hensel G, Valkov V, Middlefell-Williams J, Kumlehn J. Efficient generation of transgenic barley: the way forward to modulate plant-microbe interactions. J Plant Physiol. 2008; 165: 71–82. pmid:17905476 doi: 10.1016/j.jplph.2007.06.015
18. Gamborg O, Miller R, Ojima K. Nutrient requirement of suspension cultures of soybean root cells. Exp Cell Res. 1968; 50: 151–158. pmid:5650857 doi: 10.1016/0014-4827(68)90403-5
19. Kumlehn J, Serazetdinova L, Hensel G, Becker D, Loerz H. Genetic transformation of barley (*Hordeum vulgare* L.) via infection of androgenetic pollen cultures with*Agrobacterium* tumefaciens. Plant Biotechnol J. 2006; 4: 251–261. pmid:17177801 doi: 10.1111/j.1467-7652.2005.00178.x
20. Coronado MJ, Hensel G, Broeders S, Otto I, Kumlehn J. Immature pollen-derived doubled haploid formation in barley cv. Golden Promise as a tool for transgene recombination. Acta Physiol Plant. 2005; 27: 591–599. doi: 10.1007/s11738-005-0063-x
21. Thiebaut J, Kasha KJ. Modification of the colchicine technique for chromosome doubling of barley haploids. Can J Genet Cytol. 1978; 20: 513–521. doi: 10.1139/g78-060
22. Popelka JC. Development of a genetic transformation protocol for rye (*Secale cereale*L.) and characterisation of transgene expression after biolistic or *Agrobacterium*-mediated gene transfer. Dissertation, University of Halle, 2002;http://sundoc.bibliothek.uni-halle.de/diss-online/02/02H059/index.htm
23. Pallotta MA, Graham RD, Langridge P, Sparrow DHB, Barker SJ. RFLP mapping of manganese efficiency in barley. Theor Appl Genet. 2000; 101: 1100–1108. doi: 10.1007/s001220051585
24. Sambrook J, Fritsch EF, Maniatis T. Molecular cloning—a laboratory manual. New York: Cold Spring Laboratory Press, 1989.
25. Arcalis E, Marcel S, Altmann F, Kolarich D, Drakakaki G, Fischer R, et al. Unexpected deposition patterns of recombinant proteins in post-endospermic reticulum compartments of wheat endosperm. Plant Physiol. 2004; 136: 3457–3466. pmid:15489278 doi: 10.1104/pp.104.050153
26. Rosenberg Y, Sack M, Montefiori D, Forthal D, Mao L, Hernandez-Abanto S, et al. Rapid high-level production of functional HIV broadly neutralizing monoclonal antibodies in transient plant expression systems. PloS ONE, 2013; 8(3): e58724. doi: 10.1371/journal.pone.0058724. pmid:23533588
27. Stadlmann J, Pabst M, Kolarich D, Kunert R, Altmann F. Analysis of immunoglobulin glycosylation by LC-ESI-MS of glycopeptides and oligosaccharides. Proteomics 2008; 8: 2858–2871. doi: 10.1002/pmic.200700968.

pmid:18655055

28. Kapusi E, Hensel G, Coronado MJ, Broeders S, Marthe C, Otto I, et al. The elimination of a selectable marker gene in the doubled haploid progeny of co-transformed barley plants. Plant Mol Biol. 2013; 81: 149–160. doi: 10.1007/s11103-012-9988-9. pmid:23180016
29. Rechinger KB, Simpson DJ, Svendsen I, Cameron-Mills V. A role for gamma 3 hordein in the transport and targeting of prolamin polypeptides to the vacuole of developing barley endosperm. Plant J. 1993; 4: 841–853. pmid:7506098 doi: 10.1046/j.1365-313x.1993.04050841.x
30. Møgelsvang S, Simpson DJ. Protein folding and transport from the endoplasmic reticulum to the Golgi apparatus in plants. J Plant Physiol. 1998; 153: 1–15. doi: 10.1016/s0176-1617(98)80038-2
31. Wilkinson B, Gilbert HF. Protein disulfide isomerase. Biochim Biophys Acta Protein Proteomics 2004; 1699: 35–44. doi: 10.1016/j.bbapap.2004.02.017
32. Arcalis E, Stadlmann J, Marcel S, Drakakaki G, Winter V, Rodriguez J, et al. The changing fate of a secretory glycoprotein in developing maize endosperm. Plant Physiol. 2010; 153: 693–702. doi: 10.1104/pp.109.152363. pmid:20388665
33. Vamvaka E, Twyman RM, Murad AM, Melnik S, Teh AY, Arcalis E, et al. Rice endosperm produces an underglycosylated and potent form of the HIV-neutralizing monoclonal antibody 2G12. Plant Biotechnol J. 2015; doi: 10.1111/pbi.12360.
34. Hensel G, Himmelbach A, Chen W, Douchkov DK, Kumlehn J. Transgene expression systems in the Triticeae cereals. J Plant Physiol. 2011; 168: 30–44. doi: 10.1016/j.jplph.2010.07.007. pmid:20739094
35. Patel M, Johnson JS, Brettell RIS, Jacobsen J, Xue GP. Transgenic barley expressing a fungal *xylanase* gene in the endosperm of the developing grains. Mol Breed. 2000; 6: 113–123.
36. Schuenmann PHD, Coia G, Waterhouse PM. Biopharming the SimpliRED™ HIV diagnostic reagent in barley, potato and tobacco. Mol Breed. 2002; 9: 113–121.
37. Sack M, Rademacher T, Spiegel H, Boes A, Hellwig S, Drossard J, et al. From Gene to Harvest: Insights into upstream process development for the GMP production of a monoclonal antibody in transgenic tobacco plants. Plant Biotechnol J. 2015; accepted doi: 10.1111/pbi.12438
38. Sainsbury F, Sack M, Stadlmann J, Quendler H, Fischer R, et al. Rapid transient production in plants by replicating and non-replicating vectors yields high quality functional anti-HIV antibody. PLoS One. 2010;5(11):e13976. doi: 10.1371/journal.pone.0013976. pmid:21103044
39. Naqvi S, Ramessar K, Farré G, Sabalza M, Miralpeix B, Twyman RM, et al. High-value products from transgenic maize. Biotechnol Adv. 2011;29(1):40–53. doi: 10.1016/j.biotechadv.2010.08.009. pmid:20816943
40. Spök A, Twyman RM, Fischer R, Ma JK, Sparrow PA. Evolution of a regulatory framework for pharmaceuticals derived from genetically modified plants. Trends Biotechnol. 2008 26(9):506–17. doi: 10.1016/j.tibtech.2008.05.007. pmid:18676047
41. Levanony H, Rubin R, Altschuler Y, Galili G. Evidence for a novel route of wheat storage proteins to vacuoles. J Cell Biol. 1992; 119: 1117–1128. pmid:1447291 doi: 10.1083/jcb.119.5.1117
42. Arcalis E, Stadlmann J, Rademacher T, Marcel S, Sack M, Altmann F, et al. Plant species and organ influence the structure and subcellular localization of recombinant glycoproteins. Plant Mol Biol. 2013; 83: 105–117. doi: 10.1007/s11103-013-0049-9. pmid:23553222
43. Vuylsteker C, Cuvellier G, Berger S, Faugeron C, Karamanos Y. Evidence of two enzymes performing the de-N-glycosylation of proteins in barley: expression during germination, localization within the grain and set-up during grain formation. J Exp Bot. 2000; 51: 839–845. pmid:10948209 doi: 10.1093/jexbot/51.346.839
44. Van Droogenbroeck B, Cao J, Stadlmann J, Altmann F, Colanesi S, Hillmer S, et al. Aberrant localization and underglycosylation of highly accumulating single-chain Fv-Fc antibodies in transgenic Arabidopsis seeds. Proc Natl Acad Sci USA 2007; 104: 1430–1435. pmid:17227846 doi: 10.1073/pnas.0609997104

INDEX

S

T

U

W